The Springer Series in Solid-State Sciences features fundamental scientific books prepared by leading and up-and-coming researchers in the field. They strive to communicate, in a systematic and comprehensive way, the basic principles as well as new developments in theoretical and experimental solid-state science.

We welcome submissions for monographs or edited volumes from scholars across this broad domain. Topics of current interest include, but are not limited to:

- Semiconductors and superconductors
- Quantum phenomena
- Spin physics
- Topological insulators
- Multiferroics
- Nano-optics and nanophotonics
- Correlated electron systems and strongly correlated materials
- Vibrational and electronic properties of solids
- Spectroscopy and magnetic resonance

More information about this series at http://www.springer.com/series/682

Springer Series in Solid-State Sciences

Volume 196

Arash Rahimi-Iman

Semiconductor Photonics of Nanomaterials and Quantum Structures

Applications in Optoelectronics and Quantum Technologies

Springer

Arash Rahimi-Iman
Department of Physics
Philipps-Universität Marburg
Marburg, Hessen, Germany

ISSN 0171-1873 ISSN 2197-4179 (electronic)
Springer Series in Solid-State Sciences
ISBN 978-3-030-69351-0 ISBN 978-3-030-69352-7 (eBook)
https://doi.org/10.1007/978-3-030-69352-7

This Springer imprint is published by the registered company Springer Nature Switzerland AG
The registered company address is: Gewerbestrasse 11, 6330 Cham, Switzerland

Foreword

Semiconductor photonics have for long delivered key platforms to enhance technologies we use in our daily lives. Nowadays, more and more functionalities and improvements rely on nanomaterials and quantum structures, which experts all around the world study and develop for numerous applications.

When I first met the author of this book, Arash, in the year of 2014 during his visit to the Zhejiang University, I was impressed by his work on the demonstration of electrically-driven polariton laser, which became a prominent topic of our scientific encounter. Soon after that, our friendship was not only established because we shared a similar background on cavity quantum electrodynamics, but also due to his broad scholarly interests as well as benign and sincere character.

Our academic partnership was established during his next visit in one of the following years, starting with a joint undergraduate course, *Semiconductor quantum structures for photonic devices*, at the Zhejiang University. This course turned out to be a very successful one, highly rated by all the students in the class; and closely related to the subject of this book, it turns out years later. It was hence repeated and exchange students from both sides were a further step, which resulted in fruitful accomplishments.

One major topic, which this book addresses right from the beginning, is the content of a Sino-German Symposium which took place in 2018 in Hangzhou. During this memorable event, I fortunately could assist Arash and co-host Huizhen Wu in the superb organization of the first Functional Nano-Materials Sciences Symposium at the Zhejiang University, and then even was invited to partner in a Sino-German Cooperation Group on functional nano-materials sciences research which they successfully established shortly after.

Though the outbreak of COVID-19 has impeded international collaborations and travel since last year, our distance actually is shortened by the internet while I, together with several graduate students from Zhejiang University, are enjoying Marburg's web-based course *Quantum technology* taught by Arash at the Philipps-Universität Marburg this winter; a small remedy for not having the joint course in Hangzhou this year. I see that some of the relevant topics are also summarized in this book.

Apart from his enthusiasm and creativity, what impressed me most is his ability to explain the complicate quantum technologies in a simple way and make them understandable to the group of students without the background of quantum mechanics. As a researcher dedicated in this field for many years, I feel that I can still learn something while sitting in such introductory and topical course. This makes me confident that this exciting new book will benefit a wide readership ranging from university students to researchers who are interested in the fundamentals and applications of semiconductor nanomaterials and quantum structures, particularly the attractive class of 2D materials.

Hangzhou, China
February 2021

Wei Fang

Preface

This book delivers a topical overview of the vast field of semiconductor photonics of nanomaterials and quantum structures, for which a number of applications in optoelectronics and quantum technologies are highlighted together with a selection of fundamental research aspects. In this context, the reader will get acquainted with functional nanomaterials, monolayer semiconductors, and a hand-full of common as well as useful optical techniques for the characterisation of semiconductor quantum structures. Beginning with an introduction into functional nanomaterials sciences (Chap. 1), insights into the world of two-dimensional (2D) materials (Chap. 2), light–matter interactions (Chap. 3) and quantum technologies (Chap. 4) are provided based on a variety of examples and studies. Numerous useful experimental methods are summarised consecutively (Chap. 5), whereas the role of quantisation effects (Chap. 6) and nanostructuring possibilities (Chap. 7) are also briefly highlighted for pedagogical purposes. At the end, a conclusion-and-outlook section related to my own research activities pursued within and beyond a Habilitation project is found (Chap. 8).

Throughout the book, a plethora of references is provided. Given the choice of examples and the scope of explanations, the content of this work may serve as an introduction to the covered research topics, and the interested reader is encouraged to extend the view on this subject with the help of numerous books and review articles available in the precious pool of literature, many of which are referred to within the chapters of this book. Inevitably, the link to my own works is often established, as it lies in the nature of a Habilitation to summarise own works in the discussed fields in the context of ongoing developments. Nonetheless, keeping the fraction at about a tenth of the overall citations, a fair, meaningful, and diverse coverage shall be offered which puts great emphasis on the remarkable achievements and discoveries out there by the global science community related to the topic of this book. It is my wish to make the presented topic tangible to a broad readership, ranging from research fellows, lecturers, and students to interested science and engineering professionals in the interdisciplinary domains of photonics, materials sciences, nanotechnology, and quantum physics.

The exciting world of quantum physics and light–matter interactions has been the motivation behind a vast amount of research and has recently brought up this work embedded in this frame with the aim to achieve a Habilitation in physics. I could start praising the wonderful and useful achievements that have been enabled by the evolution of various technologies, which have utilised nanostructuring, electronics and optics together with principles from quantum mechanics, towards what is nowadays being widely labelled as quantum technologies of the first and second generation. But what was more important in the first place well before the beginning of this Habilitation project, was the motivation to dig deep into the principles of nature and to unravel concepts which may play an important role in the evolution of our physical realm, referred to as the cosmos in the large scale and the nanocosmos in the small scale.

Although the tools used throughout one's life may often not bring closer to us the furthest distant objects or events, as well as the smallest possible known length scales or actions attributed to Max Planck's historic studies, one can indeed shed light on important aspects of the else hidden world through knowledge-driven actions fed by enthusiasm and fascination. Solid-state research and particularly the work with (semiconductor) quantum structures can add considerable knowledge in this regard, whereas concepts and principles are tested on different, and most importantly, accessible length scales (also time scales) and environments that not seldom mimic with unimaginable precision those realms we are fundamentally more interested in. Often, who studies the small, understands better the large, who explores the large, reveals the small, and who examines both the small and large, may even unravel details about the very small. In a physical world full of resonances (eigenstates), excitations (particles), and energy transfer (interactions), this might hold true even more. Some reasoning from ancient philosophies and cultures might deliver an interesting access to the building blocks and the fabric of our natural habitat, e.g. through comparison and imagination. And with an intuition for the meaning of the Sanskrit expression 'Aom' ('Om/Aum'), perceived to be characterising the sound of our universe, and the far-eastern 'Tao',[1] describing the 'Nothingness' and 'Existence' of the same origin and the endless cycles of all things in a vital flow, quantum mechanics being governed by wave mechanics and the uncertainty principle also lets one remember to adapt to the flow of nature and to tolerate the intangible in the orchestra play of the universe, while striving for a deeper understanding.

Time is short to touch upon the concept of causality, its philosophical Sanskrit counterpart 'Karma', and the meaningful Kramers–Kronig relations well known from solid-state physics. By the way, what would the vivid world be like without time? 'Time' manifests itself in the energy exchanges within a system governed by interacting modes. And upon excitation, it is 'time' that describes the relaxation of an entity towards its ground state, it is 'time' that characterises the dephasing of a prepared oscillator interacting with its environment (a fully isolated and, thus, unattenuated oscillator would preserve its state indefinitely). A world free of energy exchange would seemingly lack temporal features, as time expresses evolution of systems which are not in a steady state and which are rich in dynamics, interactions

[1] See *The Book of Tao and Teh* by Lao Zi, special thanks to Wei Fang (ZJU, China) from my side.

and, thus, energy exchange between states. Thus, a natural motivation for time-resolving investigations and the consideration of temporal characteristics is given.

It seems that who wants to understand nature needs to understand resonances, and excitations of the respective system related to those resonances, and the dynamically resulting conglomerates which give birth to a complex world full of interactions, energy exchange, and transformations. They all resemble different states of one and the same overall entity, the different facets of which can unfold a plethora of impressions and wonders, described by the approximative laws of physics, which define rules based on observations and a sense for logic, which originates from the aspiration of order and predictability.

Besides the daily routine of modern-day scientists driven by paper publishing, project acquisition, and device/method optimisation, the exploratory and philosophical side of research deserves some attention as well. I allow myself to tap a bit the philosophy part related to my activities as a Ph.D./Dr. rer. nat. in the consideration of the subjects of nature in the lines above, as one hardly devotes the academic work time to a detached view on the own research. A tiny segment of this work may be philosophical before our attention is diverted towards elaborate summaries of scientific work within the scope of a Habilitation project. I believe that this may be a good approach (in addition to promoting the fascinating technologies typically obtained through scientific work) to substantially motivate the next generation of young scientists to commit themselves to often very narrow-focused research, and to trigger the sustainable interest of, in particular, the youth in knowledge-based topics and science. All the other times, enough words about scientific achievements and a bit of science fiction are spent in review articles, papers, and press releases.

Let us now briefly think about excitations, such as excitons and exciton complexes in materials. In quasi-perfect crystals with suppressed decay channels, their lifetime can be even noticeable for us, and formation of multi-particle electronically bound species is facilitated, analog to molecule formation with atoms. Or, in crystalline quantum materials, their binding energies can be high enough to enable the study of their properties in the presence of a big bath of lattice vibrations, with which they can couple to form even other entities such as polarons. Or, at low temperatures with reduced decoherence rates, reversible energy exchange with other oscillators can shape their behaviour drastically, as observed in the strong-coupling regime of a matter polarisation wave (coherent excitons) with the propagating electromagnetic mode (light in matter) or with photonic modes of high-quality optical microcavities, giving rise to hybrid light–matter quasi-particles (exciton–polaritons).

Individually considered, matter excitations such as in 2D crystals with their zoo of excitonic complexes across opposing valleys and spin–orbit split bands can act as a distant analog for the elementary particles of our world, from which compound particles are formed, e.g. protons and neutrons composed of the quarks—excitations of their quantum field described in the frame of quantum chromodynamics. In simpler terms, the well-known Coulomb-bound electron–hole pair referred to as exciton is the popular analog of a hydrogen-like (or more precisely, positronium-like) system within solids. All this focuses on the particle nature, whereas we do not want to forget that every excitation resembles a quant occupying a quantum-mechanical mode of the system, similar to photons occupying an electromagnetic mode of the vacuum field. Occupation of a mode means energy is stored in that mode in integer numbers of the resonance energy based on the occupation number, i.e. an excitation is present. That excitation propagates like a wave but interacts with its environment like a discrete entity, all according to quantum physics, discretely swapping from one mode to another in what we refer to as energy transfer,[2] or partially being reduced in energy in what can be seen in scattering processes, provided that the proper final states are there. Amazingly, under certain conditions even virtual particles can be involved, and the role of virtual states becomes prominent in processes such as Raman scattering, two-photon absorption, spontaneous parametric down conversion, and so forth (here using examples from optics). We often think about stimulated emission as an effect of quantum physics, where optical feedback (the presence of an occupied mode) literally causes a cloning of existing photons (bosons obey Bose–Einstein statistics and therefore the so-called bunching effect occurs) when stimulated emission is more likely than absorption for a given transition. But, we seldom regard spontaneous emission as stimulated emission driven by fluctuations of the electromagnetic vacuum field (virtual photons). In fact, the mere presence of 3D photonic modes governs the radiation behaviour of electric dipoles in free space. And since quantum structures allow one to tailor the (e.g. photonic) density of states, one can achieve a modification of excitation lifetimes (e.g. radiative recombination times).

So, one can easily consider oneself playing with energy exchanges between different modes of different types in a world full of resonances. The wave nature can easily be used to tailor confinement conditions for particles and to obtain quantised energy levels from previously seemingly continuous (i.e. quasi-continuous)[3] states, based on constructive and destructive interference effects towards standing waves[4] and the fulfilment of boundary conditions for them. Quantum potential wells and low-dimensional structures provide us incredible means to alter the properties of all kind of systems—most commonly optical/photonic or electronic ones, but can also be vibrational/phononic or what have you. On the one hand, optical quantum boxes can even be used to modify the photonic density of states to that extent that radiative decay of electronic excitations can be suppressed or enhanced (see Purcell effect in the literature), or that the empty cavity field can couple to a single exciton (see vacuum Rabi splitting in the literature). On the other hand, superlattices, be it vertical or in-plane, are nowadays frequently used to induce minibands for electronic or optical states. Naturally, neither the concept of quantum boxes nor superlattices is confined to one particle type, as 1D/2D/3D artificial photonic crystals, electron or polariton traps/wires, or phonon tunnelling barriers and so forth demonstrate. Remarkably, often the standard 2D quantum-well double-heterostructure is behind

---

[2]Note that probability distributions describing a measurement outcome change continuously, as far as a continuous flow of time is concerned, whereas a performed measurement causes a discrete result as a consequence of the interactions with the applied probe system.

[3]Quasi-continuity refers to the fact that very densely packed discrete states mimic a band of seemingly continuous states.

[4]Standing waves are nothing other than time-independent solutions of a confinement condition that are else known as the (steady-state) modes of a resonator.

the achievement of many useful device platforms (sometimes in the form of quantum-well molecules and crystals), whereas quantum wires and quantum dots have their own share in the success story of quantum devices.

When quantum entities act as an ensemble (coherently), Bose–Einstein-like condensates (quantum fluids) of quasi-particles in matter, such as excitons and polaritons, open up a "magical realm [...] filled with fascinating phenomena and manifestations [...] [including] polariton supernovas, black and white holes as well as other cosmological parallels such as polariton Hawking radiation" (my colleague Lorenzo Dominici, expert on polariton fluids, phrased it once nicely for me[5]), which to fully comprehend I needed to be well trained in astrophysics and particle physics. In fact, fundamental studies on exciton and exciton–polariton condensates in solids become highly attractive when taking into account that the vacuum is considered in the literature to contain a condensate of Higgs bosons and quark condensates (leading to a symmetry break, as the condensed system gives preference to one phase over others). They also become highly attractive owing to mesonic condensates—possibly important for characteristics of neutron stars—being considered nuclear analogs of exciton condensates. Given the significance of condensates, and being fascinated by related effects such as superfluidity (and superconductivity), I am thrilled to learn more about possibilities of how to manipulate and control the behaviour of polariton condensates—given their usefulness as optically accessible testbed—e.g. with transient electromagnetic fields.

All this shows that a view on often intangible things from different angles can be complementary, inspirational, and quite fulfilling, though not every single work on a subject may excite the whole research community equally. Nevertheless, with quantum structures and semiconductor (2D) materials one has at least various playgrounds to test physics on microscopic scales, and the study of excitations in matter definitely provides a rich pathway to the exploration of quantum mechanics and many-body physics. In addition, research on these systems can lead to considerable improvements in existing technologies, promising, for instance, more energy-efficient, more miniaturised, and more cost-effective optoelectronic devices, ranging from photo-transistors/solar cells to light-emitting diodes, or enabling, for example, novel quantum-optical devices, such as single-photon sources or polariton lasers.

In the following, a selection of scientific work in this vast domain is summarised and explained in the context of current research activities. Naturally, the discussion of phenomena, theories, methods, and recent studies cannot be complete, and it will be hardly possible to go into more detail within the scope of this work about these topics. However, wherever possible and appropriate, selected representative references to the widely available literature are provided which can deliver more insights and explanations about the relevant matter. Thus, I am hopeful that the interested reader can follow up on those topics through the use of these citations.

Marburg, Germany
November 2020

Arash Rahimi-Iman

[5] In *Polariton Physics*, Springer Nature Switzerland AG, Cham, 2020.

Acknowledgments

Regarding this book project, I am grateful to Zachary Evenson and his editorial team of Springer Nature for their support and endeavours, which has enabled submission of this academic work to the renowned Solid-State Sciences Series. (November 2020).

Regarding the academic work, I would like to summarise the acknowledgments the way it was provided with the submission of this work to the Physics Department in Marburg, as follows. (July 2020).

The privilege to work with young and senior scientists in different domains of semiconductor physics, laser technology, and materials sciences in the role of a 'Habilitand' (a hybrid of lecturer, young group leader, principal investigator) at an academic institution shall be acknowledged. The opportunities on a mission to promote science and education, as well as to teach and train the next generation of researchers and engineers, are highly appreciated by me. I very much enjoyed to support the overall academic education of young researchers and students either as a direct supervisor and as their team leader or via various lectures and seminars, internal collaborations, as well as interdisciplinary projects.

Building-up and leading the local VECSEL team since 2013 to its culmination as an internationally recognised research force with continuous scientific output and highlights in its VECSEL achievements would have been truly out of reach without the invaluable support, as well as strong commitment of the numerous team members, who worked in this field. Establishing a new local research branch on 2D materials in 2015 and leading it to the forefront of nanophotonics and quantum-structures research with growing visibility, based on an effective involvement of available resources and personnel, would have been similarly impossible without those many helping hands and minds who joined my research efforts. Ranging from internship students to postdoctoral staff, by pulling the same rope as one grown-together entity, by promoting an atmosphere of team work, mutual friendship, and scientific curiosity, many obstacles have become insignificant, various challenges have been overcome and a plethora of achievements have been demonstrated.

While not all ideas have materialised, not all experiments have borne fruits and not all goals have been reached yet due to the limited time and resources, the remarkable outcome of the joint endeavours has become a source of huge reward for the efforts of all team members and an inspiration for future explorations. Many thanks are devoted to all contributors who have had a visible and/or invisible hand in this success story. The knowledge gained and the accomplishments made throughout this academic journey have established the link between the previous, the outgoing, and the current team members. The connectivity of all participants, as well as the essence of the research, is not only preserved by a vast amount of publications. It is also preserved by nice memories and the gratitude towards each other.

The following is an attempt to thank everyone who had a positive role in the completion of this Habilitation project, and even if the following list likely is not complete, I also would like to thank those supportive and helpful people related to this academic work which may have not been recalled here. Please note that many of the (former) students mentioned in the following have received their Master or Ph.D. degrees by now and entered the next stage of their professional careers.

- Many thanks are devoted to the head of the Semiconductor Photonics group, Prof. M. Koch, for inviting me into his group to engage in a Habilitation project, as well as for his kind and enduring support over the last years. This provided an outstanding research opportunity as well as a fruitful platform for the preparation of this work and for independent explorations that have led to important findings and key publications of my Habilitation project. Without his help and trust, many of the works on the path to a complete Habilitation work presented would have not taken form.
- I would like to thank my 'nanomaterials and quantum-structures' team's former and current members, among them S. C. Lippert, L. M. Schneider, R. Li, O. M. Abdulmunem, M. Hünecke, D. Renaud, X. Lin, Dr. K. Hassoon, Dr. S. Edalati-Boostan, M.-U. Halbich, G. Q. Ngo, F. Wall, O. Mey, L. Gomell, M. J. Shah, M. A. Nouh, Dr. A. Usman, Dr. H. Masenda, C. Palekar, S. Machchhar, and many more, for their continuous help in different fields, and L. M. Schneider, M. J. Shah, M. A. Nouh and C. Palekar for assistance with the bibliographic work. Similarly, I thank co-supervised students J. Kuhnert and S. Schmitt from the Semiconductor Spectroscopy group of Prof. W. Heimbrodt, who worked with me on 2D materials.
- I would also like to thank my 'semiconductor disk laser' team's former and current members, among them M. Wichmann, M. A. Gaafar, F. Zhang, Dr. K. A. Fedorova, D. Al Nakdali, J. Quante, M. Vaupel, S. Kefer, S. Kress, C. D. Kriso, M. M. Alvi, M. T. Munshi, S. Wang, H. Guoyu, O. Mohiuddin, A. Barua, M. Gao and many more, for their continuous help in different fields. Similarly, I thank co-supervised student C. Möller from the group of Prof. W. Stolz, who contributed to the team's VECSEL work.
- Furthermore, help and commitment by our group's other supportive spectroscopy team members, among others R. Scott, M. Drexler, R. Woscholski, A. Velauthapillai, K. Shakfa, M. Gerhard, M. Bilal, is acknowledged. The eager participation in lab activities by my student interns is also appreciated, among others by M. Khalaf, S. Firoozabadi, R. Guo, O. Farooqui, M. Naramulu, G. Muhkerjee.
- Many thanks go to my (former) project employees Dr. R. Gente, Dr. S. Edalati-Boostan, S. C. Lippert, L. M. Schneider, C. D. Kriso, M.-U. Halbich and M. J. Shah, who supported my research over different periods of time.

- I am grateful to Ms. A. Ehlers, who accompanied me many years from the first steps in Marburg as an invaluable, precise, and warmhearted assistant in the Experimental Semiconductor Physics group. Her professional experience, which she brought from Bosch into the research group previously led by Prof. W. Rühle, made her retirement create a noticeable gap in our group's administration. Also, I would like to thank Ms. M. Strobel, assistant of the group, who had to balance her time carefully between two locations in Marburg, for her administrative support in processing numerous project related issues, such as staff employment and other financial matters.
- Furthermore, I would like to thank Dr. G. Urbasch, Dr. B. Fischer and Dr. J. Balzer for the helpful discussions and the collaborative work in the Experimental Semiconductor Physics group, S. Sommer and N. Born from the Terahertz-part of the group for the helpful discussions, and technician of the group R. Rink for his constant help that enabled the smooth expansion and operation of various pieces of lab equipment. Many thanks are devoted to the mechanical and electrical workshops at the Department of Physics and their staff for their frequent help with technical constructions and problem solving.
- I am grateful to faculty members Prof. em. G. Weiser and Prof. W. Heimbrodt for enriching discussions on semiconductor optics and support in the field of semiconductor spectroscopy, respectively. Prof. Weiser's insightful talks, which provided important views and pedagogical summaries on light–mater interactions in organic and inorganic semiconductor systems, given in my team's internal seminars have always been valued by my students and me. Moreover, I very much value the occasional meetings we have had and his kind nature. With Prof. Weiser's encouragement and affirmations, I could easily follow my scientific instincts and push forward my investigations, knowing that they are of interest to him as a semiconductor expert. Also, I am thankful for his helpful improvement recommendations regarding my work's summary (Sect. 8.1 in this book). Furthermore, the fruitful cooperation with Prof. Heimbrodt, his benevolence, his lasting support with regard to using his equipment, and the kind help of various of his team members, particularly J. Kuhnert and M. Wilhelm, is acknowledged.
- I would also like to thank guest researchers Prof. G. Town, Dr. W. Lai, and many others for the fruitful collaborations and scientific discussions, visiting Ph.D. students R. Li, W. Kong, X. Lin, H. Zhu, H. Guoyu, H. Alaboz and visiting scholars H. Keskin, D. Abdrabou for their commitment in the laboratories of the group and the positive role in my teams. Similarly, I would like to thank visiting DAAD Rise students D. Renaud, N. Wilson, C. McGinn for their help in the lab and the Marburg group's (former) Ph.D. students M. Drexler and C. Lammers, who supported the DAAD Rise internship announcement as well as supervision, and C. Lammers, D. Renaud, D. Berkhahn and G. Otto for their assistance during my public 'laser' talks.
- Special thanks are devoted to Prof. H. Z. Wu from the Department of Physics and the State Key Laboratory of Silicon Materials of the Zhejiang University (ZJU) in Hangzhou, China, for hosting me in his group multiple times, for inviting me to give numerous guest lectures, for the fruitful collaboration on nanomaterials

and quantum structures, for the valuable discussions and for enabling the joint grant application for a Sino–German symposium, as well as cooperation group. I am also grateful to his team members for their help and commitment, such as R. Li, W. Kong, B. Zhang, P. Lu, and many more, some of whom even joined my team as visiting Ph.D. students. Without them, joint research work on graphene, perovskites, nanoparticles, or 2D GaTe would be impossible.

- Similarly, special thanks are devoted to Prof. W. Fang and Prof. L. Tong from the State Key Laboratory of Modern Optical Instrumentation and College of Optical Science and Engineering of the ZJU in Hangzhou, China, for hosting me in their group multiple times, for inviting me to set up a joint course, for the fruitful collaboration on nanophotonics, and for the valuable discussions. I am also grateful to their team members for their help and commitment, such as Dr. W. D. Shen, X. Lin, P. Qing, J. Gong, N. Yao.
- I would like to thank Prof. W. Stolz and his co-workers from the Materials Sciences Center in Marburg and NAsP III/V GmbH, among others Dr. P. Hens, S. Reinhard, P. Ludewig, A. Ruiz Perez, and many more, for the supply with high-performance quantum-well semiconductor disk laser chips or other MOCVD-grown samples. These enabled outstanding results and the investigation of the novel self-mode-locking effect in VECSELs. In this context, semiconductor disk laser modelling support, which is behind a long-term optimisation of laser chip structures, by Prof. S. W. Koch and his co-workers, also in collaboration with his international partners from Tucson, AZ, USA, is acknowledged. Also, helpful semiconductor crystal characterisation by Prof. K. Volz, Dr. K. Gries, and co-workers from the Materials Sciences Center is acknowledged.
- Special thanks are devoted to Dr. M. Jetter, Prof. P. Michler, R. Bek, H. Kahle, M. Grossmann, and co-workers from the Center for Integrated Quantum Science and Technology (IQST) and SCoPE, Institut für Halbleiteroptik und Funktionelle Grenzflächen, for the fruitful cooperation on semiconductor disk lasers and also for the supply with VECSEL chips. This enabled the study of self-mode-locking in the visible spectral range and the investigation of nonlinear lensing in VECSELs.
- I would also like to thank Prof. E. U. Rafailov, Dr. K. A. Fedorova, and co-workers from the Aston University, Birmingham, UK, for the fruitful collaboration on quantum-dot semiconductor disk lasers and also for the supply with VECSEL chips and SESAMs. This made possible the demonstration of record-high output powers from quantum-dot VECSELs and even self-mode-locking with a quantum-dot VECSEL delivering sub-picosecond light pulses by my team members, mainly D. Al Nakdali and M. A. Gaafar, respectively.
- The fruitful collaboration on organic semiconductors for optoelectronics with Prof. U. Lemmer from the Karlsruhe Institute of Technology, and his team members, such as Dr. I. A. Howard, A. P. Arndt, and co-workers, is acknowledged. On the Marburg side, I am thankful to team members M. Gerhard, M.-U. Halbich, and M. Bilal, who were successfully involved and who carried out the experiments.
- Also, the fruitful collaboration on novel materials with Prof. S. Dehnen and Dr. J. Heine from the Department of Chemistry, and their team members, such as G.

Thiele, B. Wagner, and co-workers, is acknowledged. On the Physics side, I am thankful to team members S. C. Lippert, M.-U. Halbich, who were successfully involved and who carried out the experiments.

- Special thanks are devoted to Prof. T. F. Heinz, who generously hosted me twice in his group for multiple weeks both at the Columbia University and after his group's transfer also at the Stanford University, for the excellent opportunity to evidence 2D-materials research at the forefront, for the helpful discussions, and for his support. This truly seeded my activities on 2D materials. The latter visit was particularly refreshing, because I could evidence all the changes and modernisation on the campus and in the Physics Department. Moreover, I would like to thank Y. Peterman and R. Sasaki, assistants of the Heinz group and formerly assistants of the YY ('Yoshi Yamamoto') group, who had always time for my administrative inquiries as a visiting scholar to both groups. I should once again acknowledge the important support by (former cooperation partner) Prof. Y. Yamamoto, who enabled previous stays in his group that became important milestones on my academic journey.
- I am grateful to Dr. Y. D. Kim, Dr. A. Chernikov, Ö. B. Aslan, and others for invaluable discussions on 2D materials and their properties that took place during my stay in the Heinz group at the Columbia University, and Dres. Z. Ye, D. Sun, Y. Li, and co-workers for the same during my stay in the Heinz group at the Stanford University.
- Special thanks are devoted to Prof. E.-H. Yang, Dr. K. N. Kang, and co-workers from the Stevens Institute of Technology, NJ, USA, for the fruitful collaboration, valuable support, and indispensable sample supply. His group's CVD-grown vdW structures enabled numerous works in my team within joint research.
- Special thanks are devoted to Prof. J. C. Hone, Prof. K. Barmak, Dr. D. A. Rhodes, O. Ajayi, S. S. Esdaille, and co-workers from the Columbia University, NY, USA, for the fruitful collaboration, valuable support, and indispensable sample supply. Their groups' high-quality 2D materials, encapsulated stacks, and monolayer heterostructures opened up new possibilities in my team's research and led us to new frontiers.
- I would like to thank Dr. T. Stroucken, Prof. S. W. Koch, Dr. U. Huttner, L. Meckbach, and co-workers in Marburg for the helpful cooperation concerning theoretical predictions of band structures of vdW monolayers, bilayers, and heterostructures, as well as monolayer dielectric functions, and Dr. O. Vänskä from the group of Prof. M. Kira for the calculations for III/V-semiconductor heterostructures and for many helpful discussions.
- Many thanks are devoted to Prof. C. Draxl, Humboldt Universität Berlin, who enabled a cooperation project on 2D heterostructures (DFG RA2841/5-1), which Dr. S. Edalati-Boostan and I targeted for joint research, by hosting the theory part of the project in her group. Also, I would like to thank project partner Dr. Edalati-Boostan and co-workers in the ongoing project for their efforts to shed light on the theoretical aspects of 2D-semiconductor heterobilayers. In this context, research support by Prof. W. Heimbrodt (cooperation partner in the ongoing SFB1083,

part B3) is acknowledged, for instance, concerning Raman spectroscopy of 2D materials, joint supervision of project students, and helpful discussions.

- Similarly, I am grateful to Prof. H. G. Roskos from the Goethe-Universität Frankfurt, and his group members, such as Dr. A. Soltani, F. Walla, M. Wiecha, and co-workers, for their help and support in the study of 2D materials and nanophotonic structures with their s-SNOM technique, which allowed raster-probe optical nanoscopy of various vdW monolayers and stacks prepared in my team, for instance, on top of circular nanogratings. This collaboration also enabled uncomplicated AFM topography analysis and fruitful discussions.
- Furthermore, I am grateful to Prof. G. Witte and his group members, such as D. Günder, who generously helped to acquire AFM topography maps of 2D materials and to use facilities such as his group's glovebox.
- Also, I would like to thank Prof. V. Evtikhiev and his group from the Ioffe Institute in St. Petersburg, Russia, for the kind support with nanostructuring efforts regarding III/V semiconductor structures.
- Furthermore, I would like to thank Prof. P. J. Klar and his Micro-/Nanostructure Physics Group at the Justus-Liebig-Universität Giessen for the helpful and uncomplicated access to the MiNa clean-room facilities via the Semiconductor Photonics group Marburg subscriptions which enabled the production, microstructuring, and contacting of various types of samples. Particularly, frequent technical support by Dr. T. Henning and co-workers is acknowledged.
- Special thanks are devoted to Dr. L. Dominici and Dr. D. Sanvitto, from the Institute of Nanotechnology of the National Research Council (NANOTEC-CNR) in Lecce, Italy, for enabling a valuable collaboration on the coherent manipulation and control of exciton–polaritons in optical microcavities, for their helpful discussions and support of our joint investigations in an ongoing project (DFG RA2841/9-1)—promising novel insights. Particularly, I want to thank both Italian partners for their trustful, friendly and positive attitude, which encouraged me to follow my project aims and promoted the opportunity of joint research targeted together years ago.
- I am also grateful to Prof. S. Höfling, Dr. C. Schneider, and co-workers from the University of Würzburg for the successful long-term cooperative research on microcavity polaritons and their supply of excellent MBE-grown samples from the Gottfried-Landwehr-Labor für Nanotechnologie (formerly known as Mikrostrukturlabor der Technischen Physik). In this context, many thanks are devoted to former supervisors Professors A. Forchel, S. Reitzenstein, L. Worschech, M. Kamp for their guidance and support in various quantum optics and nanophotonics works at the Technische Physik, Würzburg, to cooperation partners, such as Dr. A. V. Chernenko (Chernogolovka, Russia), Dr. N. Y. Kim (Stanford University, USA), and many others, to all my former students (be it directly supervised or co-supervised under the guidance of my 'Doktorvater' Prof. S. Reitzenstein) and co-workers, many of whom carried on the inherited work on polariton lasers and condensates, such as J. Fischer, M. Amthor, A. Schade, S. Brodbeck, and K. Winkler.

- A pleasant return to the subject of my Master thesis in combination with my works on VECSELs with the result of a record-high single-photon flux became feasible thanks to the effective and straightforward cooperation with Prof. S. Reitzenstein, Dr. T. Heindel, A. Schlehahn, and co-workers, and the enthusiasm of team members M. A. Gaafar and M. Vaupel from Marburg.
- I am grateful to Dr. S. L. Tait and Prof. K. Kern from the Max Planck Institute for Solid State Research, Stuttgart, Germany, for having laid the foundations of my nanomaterials research with their supervision of my Bachelor project on scanning-tunnelling microscopy and organic molecules on copper crystal surfaces long time ago, and Prof. J. Geurtz in Würzburg for having promoted semiconductor physics at an early stage.
- Financial support by the Federal Ministry of Education and Research (BMBF) in the frame of the German Academic Exchange Service's (DAAD) program Strategic Partnerships and Thematic Networks is acknowledged, which provided me the means to frequently visit the Zhejiang University in Hangzhou, China, for course teachings of Master students at the Department of Physics, for the establishment of research collaborations with different groups as well as joint lecture courses for senior Bachelor students at the Department of Optical Science and Engineering. I would also like to thank the International Office in Marburg, particularly Ms. S. Halliday, for administrative support on this matter.
- Multiple fellowships by the German Academic Exchange Service (Deutscher Akademischer Austausch Dienst, DAAD) and the German National Academic Foundation (Studienstiftung des Deutschen Volkes) during all phases of my personal academic development helped to obtain the necessary flexibility, mobility, and freedoms for continuous progress in the domain of science.
- A scholarship from the Adolf-Häuser Stiftung for a short-term visit to the Aston University is acknowledged which provided an important link to long-term cooperation partners and facilitated the establishment of collaborative work with the group of Prof. E. U. Rafailov.
- Financial support by the Deutsche Forschungsgemeinschaft (DFG) for my completed, prolonged and current projects via DFG grants RA2841/1-1, RA2841/1-3 ("Investigating Self-Mode-Locking in Semiconductor Disk Lasers"), RA2841/5-1 ("Theoretical and experimental study of optical properties of 2D material heterostructures", collaborative Marburg–Berlin 2D heterostructures research), and RA2841/9-1 ("Manipulation and Control of Coherent States in Exciton–Polariton Systems (MacExP)", collaborative Marburg–Lecce exciton–polariton research) is acknowledged. I am thankful to the DFG for having enabled me to perform independent and self-determined scientific work.
- Financial support by the Philipps-Universität Marburg via the Forschungsförderfond is acknowledged, which provided seed funding for my young-group-leader project "Heterostructures and Interfaces in Two-Dimensional Material Systems" with the aim of an expansion of 2D-materials research.
- Financial support by the DFG SFB1083 "Structure and Dynamics of Internal Interfaces" is acknowledged, which enabled the expansion of a micro-photoluminescence setup for 2D-materials research.

- Financial support by the DFG GRK1782 "Functionalization of Semiconductors" via the Experimental Semiconductor Physics groups is acknowledged, which enabled many of the internal research collaborations on III/V-semiconductor quantum-well structures and devices.
- Financial support by the Sino-German Center for Research Promotion is acknowledged, which enabled me to carry out a unique and unforgettable 1st Sino-German Symposium on Functional Nano-Materials Sciences in Hangzhou (2018) together with Prof. H. Z. Wu. Special thanks are devoted to him (and his group) for the co-organisation of the FNMS2018 symposium and for his (and his group's) endeavours related to hosting the event at the ZJU in Hangzhou, China. Moreover, many thanks go to all FNMS2018 participants, who enabled this memorable experience with excellent discussions and with their symposium contributions to the 16 scientific sessions, among them dozens of students, who made the student poster session a unique platform for discussions, explorations, and exchange of knowledge.
- Financial support by the Sino-German Center for Research Promotion is also acknowledged with respect to the formation of a Sino-German Cooperation Group, the Functional Nano-Materials Sciences Cooperation Group (FNMS-COOP), which is coordinated by Prof. H. Z. Wu and me jointly, providing us the means and framework to carry out collaborative research work and multiple workshops. Many thanks are devoted to the participating teams of this cooperation group from China and Germany with in total more than 32 scholars, represented by their group leaders Prof. J. He (National Center for Nanoscience and Technology, Beijing), Prof. W. Kong (College of Physics Science and Technology, Hebei University, Baoding, with cooperation-group members from the Hefei University of Technology), Prof. W. Fang (College of Optical Science and Engineering, ZJU, Hangzhou), Prof. H. G. Roskos (Goethe-University Frankfurt) and Prof. P. J. Klar (Justus-Liebig-Universität Giessen).
- Last but not least, I am very grateful for the continuous support from my family, which enabled my successful progress and the pursuit of the Habilitation.

In fact, many of the people listed above may bear today additional titles and responsibilities, since many have received, for instance, their Ph.D. or Master degrees or other titles by now and entered the next stage of their professional careers, while I refer to the situation during the cooperation and interaction time (in many cases related to thesis supervision time).

Throughout all the years in the role of a team leader and lecturer at the university, I have encountered many supporters among the students, colleagues, and senior faculty members. Moreover, I have learned to value fruitful collaborations and evidenced the importance of mutual help, team play, and the big impact of the work climate on the success of the individual person and of the whole team. I find it crucial to preserve this understanding and experience, and it has been a major aim of my actions to promote a good environment for my team members. Thus, every success of my students feels like my own success, and no achievement is independent from the constructive input of various contributors. If my scientific work was in any sense successful, then only owing to the invaluable help of my project teams and my project partners.

Marburg, Germany
July 2020

Arash Rahimi-Iman

Contents

Abbreviations and Symbols

Abbreviations

0D	Zero-dimensional
1D	One-dimensional
2D	Two-dimensional
2DEG	2D electron gas
2DM	Two-dimensional material
3D	Three-dimensional
II/VI	Group II–VI semiconductors
III/V	Group III–V semiconductors
arb.u.	Arbitrary unit
AFM	Atomic-force microscopy
Al	Aluminium
As	Arsenic
ASE	Amplified spontaneous emission
APD	Avalanche photodiode
BBO	Beta-barium borate
BEC	Bose–Einstein condensation/condensate
BEKE	Bound-electron Kerr effect
BL	Bilayer
BN	Boron nitride, existing in three crystal modifications α (hexagonal), β (cubic) and γ (wurtzite)
cav	Cavity
CB	Conduction band
Cd	Cadmium
cQD	Colloidal quantum dots
cQED	Cavity quantum electrodynamics
CCD	Charge-coupled device (e.g. CCD camera)
CMOS	Complementary metal-oxide-semiconductor (e.g. CMOS camera)
CVD	Chemical-vapour-deposition
CW	Continuous wave

DBR	Distributed Bragg reflector
DFG	Difference-frequency generation (physical meaning), also, Deutsche Forschungsgemeinschaft (funding agency)
DOS	Density of states
e^-	Electron
EID	Excitation-induced dephasing
EL	Electroluminescence
EP	Exceptional point
FCN	Free-carrier nonlinearities
FET	Field-effect transistor
FF	Far-field
FP	Fabry–Pérot (resonator/etalon)
FRET	Förster resonant energy transfer
FSR	Free spectral range
FWHM	Full width at half maximum
Ga	Gallium
h^+	Hole, defect electron
hBN	Hexagonal boron nitride, see BN
HBT	Hanbury–Brown-and-Twiss (experimental configuration)
high-Q	High-quality
HS	Heterostructure, also, heterosystem
iCCD	intensified CCD
In	Indium
IR	Infrared
LA	Longitudinal acoustic
LASER	Light amplification by stimulated emission of radiation
LCF	Longitudinal confinement factor
LED	Light-emitting diode
LO	Longitudinal optical
LP	Lower polariton, referring to the lower energy mode/branch of a polaritonic system
μEL	Micro-electroluminescence
μPL	Micro-photoluminescence
MBE	Molecular beam epitaxy
ML	Monolayer
MM	Multi-mode
MO	Microscope objective
MoS_2	Molybdenum disulfide
$MoSe_2$	Molybdenum diselenide
$MoTe_2$	Molybdenum ditelluride
MOVPE	Metal-organic vapour-phase epitaxy
NA	Numerical aperture
NF	Near-field
NIR	Near-infrared
OC	Output coupler

PL	Photoluminescence
PNP	Plasmonic nanoparticle
PMMA	Poly(methyl methacrylate), a polymer also known as plexiglass or acrylic
pn	Positive-negative (...junction/diode)
PSB	Phonon sidebands
PVD	Physical-vapour-deposition
Qubit	Quantum bit
Q-factor	Quality factor
QCSE	Quantum-confined Stark effect
QCL	Quantum-cascade laser
QD	Quantum dot
QLED	Quantum-dot light-emitting diode
QW	Quantum well
QY	Quantum yield
s-SNOM	Scattering-type SNOM
SC	Semiconductor
SE	Spontaneous emission
SEM	Scanning electron microscopy
SESAM	Semiconductor saturable-absorber mirror
SFG	Sum-frequency generation
SHG	Second-harmonic generation
SNOM	Scanning near-field optical microscopy
SPDC	Spontaneous parametric down conversion
SPS	Single-photon source
STEM	Scanning transmission electron microscopy
STM	Scanning tunnelling microscopy
SU-8	A negative photoresist, transparent, epoxy-based
TA	Transverse acoustic
TCSPC	Time-correlated single-photon counting
TDS	Time-domain spectroscopy
TECSEL	Terahertz external-cavity surface-emitting laser
TMDC	Transition-metal dichalcogenide
Te	Tellurium
THz	Terahertz
TO	Transverse optical
TR	Time-resolved (e.g. photoluminescence)
UP	Upper polariton (...mode/branch)
UMR	Philipps-Universität Marburg
UV	Ultra violet
vac	Vacuum
VACNTs	Vertically aligned carbon nanotubes
VB	Valence band
VCSEL	Vertical-cavity surface-emitting laser
VECSEL	Vertical-external-cavity surface-emitting laser

VIS	Visible (spectral range)
vdW	van der Waals
WS_2	Tungsten disulfide
WSe_2	Tungsten diselenide
WTe_2	Tungsten ditelluride
X	Exciton, neutral exciton
X*	Charged exciton, also referred to as trion
XP	Exciton–polariton
X_d	Dark exciton
XX	Biexciton, neutral biexciton
ZJU	Zhejiang University (Hangzhou)

Symbols

h	Planck's quantum of action
$\hbar$	$= \frac{h}{2\pi} \approx 1.054572 \cdot 10^{-34} J\,s$
k_B	$\approx 8.617333 \cdot 10^{-5}$ eV/K, Boltzmann constant
c	$= 2.99792458 \cdot 10^8 \frac{m}{s}$, speed of light
π	≈ 3.141593
$\mathbf{k}$	Wave-vector
$k_{\|\|}$	Transversal component of the wave-vector, corresponding to an in-plane momentum, for instance, in the plane of a quantum well; $[k_{\|}] = \mu\text{m}^{-1}$
$k_\perp$	Longitudinal component of the wave-vector, corresponding to an out-of-plane momentum, for instance, vertical to the plane of a quantum well
e	$\approx 1.602177 \cdot 10^{-19}$ C, elementary charge
m_e	$\approx 9.109384 \cdot 10^{-31}$ kg, free electron mass
ε_0	$\approx 8.854188 \cdot 10^{-12} \frac{\text{As}}{\text{Vm}}$, dielectric constant, vacuum permittivity
ε_r	Relative dielectric constant (vacuum: 1 per definition); high frequency constant: $\varepsilon_r(\infty)$, background constant ε_b; in general, frequency-dependent complex dielectric function $\varepsilon_r(\omega)$
ε	$= \varepsilon_0 \cdot \varepsilon_r$, permittivity of a medium
$\tilde{n}(\omega)$	Frequency-dependent complex refractive index
n	Refractive index; due to its relationship to $\tilde{n}$, in general frequency dependent (also used as symbol representing the particle density $n = N/V$, number N divided by volume V)
κ	Extinction coefficient; due to its relationship to $\tilde{n}$, in general frequency dependent
α	Absorption coefficient; due to its relationship to the extinction coefficient and thereby $\tilde{n}$, in general frequency dependent (also used as symbol representing the linearity factor in input–output measurements)

E	$= \hbar\omega = h\nu = \hbar c\|\mathbf{k}\|$, energy
ω	Angular frequency
ν	Frequency
λ	$= hc/E = 2\pi/\|\mathbf{k}\|$, wavelength
E_0	Energy of a mode's (or system's) ground state
E_g	Band gap energy
γ	Linewidth, FWHM
m	Mass (of a particle)
μ	Reduced mass (also used as symbol representing the chemical potential in distribution functions)
E	Electric field (vector)
B	Magnetic flux (vector)
r	Position (vector)
T	Temperature
P	Pump power/rate
j	Current density
θ	Angle
$\|\varphi\rangle$	Wave-function (quantum-mechanical state)
H	Hamiltonian of a system
g	Coupling constant (coupling strength)
$2\hbar\Omega_0$	$= 2g = E_{\mathrm{Rabi}}$, (vacuum) Rabi splitting, with $2\Omega_0 = \Omega_{\mathrm{Rabi}}$ the Rabi oscillation frequency
M	Transition-matrix element
f	Oscillator strength (in optics descriptions f is typically used as focal length)
τ	Lifetime (also used as symbol representing the delay time as a parameter in some experiments)
ΔE	$= E_0 - E_{\mathrm{ab}}$, spectral detuning of two oscillators (here of the light field and an electronic transition from state a to b, respectively)
Q	$= E/\gamma$, quality factor
$\|X\|^2$	Excitonic fraction of a polariton (one of the Hopfield coefficients)
$\|C\|^2$	Photonic fraction of a polariton (one of the Hopfield coefficients)
$g^{(2)}(\tau)$	Second-order temporal autocorrelation function

Chapter 1
Introduction

Abstract This overarching introductory chapter provides a brief topical overview of semiconductor photonics with a focus on nanomaterials and quantum structures, as well as a general motivation for the works pursued within the author's Habilitation project in this field which are discussed throughout the course of this book. In this context, also the structure of this work is summarised. Following that, scientific advances regarding functional nanomaterials with an emphasis on optoelectronics and quantum technological applications are highlighted. A number of selected examples from experimental and theoretical works in this field shed light on novel material systems, material engineering and physics, as well as optoelectronic devices.

1.1 A Topical Overview

Quantum structures and functional nanomaterials have had a unique impact on modern technologies which evolved in recent decades owing to the remarkable achievements in semiconductor physics, nanotechnology and materials sciences. Particularly, they enabled considerable advances in the domains of photonics [1–3], nanoelectronics [4–6], and quantum technologies [7–10]. The availability of sophisticated and high-quality optical structures has given rise to numerous studies of light–matter interactions [11–16]. Similarly, the abilities in terms of deterministic epitaxial growth [17–19] and post-processing capabilities [20] have enabled the design and realisation of a plethora of coherent-light emitting chip-scale devices [21–35] and novel quantum-optical concepts [36–45] with eyes towards faster, energy-efficient and reliable hardware for secure telecommunications and powerful computations.

Beyond Moore's law, both new technologies and material systems are required to provide the constant increase in device sophistication, operation effectiveness and data processing capabilities. These ambitious aims drive various scientific fields, one of which can be regarded the research domain of functional nanomaterials—itself a vast field intersecting with other fields in physics, materials sciences and

A. Rahimi-Iman, *Semiconductor Photonics of Nanomaterials and Quantum Structures*,
Springer Series in Solid-State Sciences 196,
https://doi.org/10.1007/978-3-030-69352-7_1

chemistry. Naturally, its progress has a huge impact on other natural sciences and modern technologies, such as information technologies, biology, or medicine, and has implications for various industries. Obviously, the well explored domain of quantum structures with its utilisation of low-dimensional systems for quantum effects (tailoring the eigen-states and therewith the resonances/energies $\omega(k) = E(k)/\hbar$)[1] and potential landscapes of the order of the respective particles' wavelengths (de-Broglie wavelength $\lambda = h/p$ for matter waves)[2] for confinement, superpositions and waveguiding has had a strong influence on photonic, electronic and optoelectronic device engineering in the past decades.

Material engineering and physics have become the subject of intense research towards the achievement of flexible, cheap and miniature optoelectronic devices, both on the fundamental and the applied level. Within the overall research domain concerning low-dimensional materials, which was previously dominated by conventional semiconductor heterostructure technology and by quantum dots (QDs), two-dimensional materials (2DMs) have gained a large share, recently. Despite the popularity of novel and emerging materials, these well-established fields have matured over the last decades as part of excessive integrated-circuit, light-source and photodetector development. Correspondingly, they have brought up various novel devices ranging from single-photon sources (SPSs) over quantum-cascade lasers (QCLs) to quantum-dot light-emitting diodes (QLEDs), while the utilisation of novel materials promises further improvements of such devices and even the discovery of more exotic device concepts. All in all, the pool of attractive optoelectronic materials has nowadays become extended from classical inorganic III/V semiconductors to organic semiconductors, hybrid material systems, van-der-Waals (vdW) materials and the perovskite material class.

Inherently, the systematic characterisation of functional nanomaterials and quantum structures remains an important constituent of many explorations concerning novel or nanostructured material systems. Typically, the study of relevant platforms is preceded or accompanied by theoretical predictions. Moreover, in the field of optoelectronics, optical spectroscopy has proven itself indispensable when it comes to material characterisation, be it in the ultraviolet (UV), visible (VIS), infrared (IR) or terahertz (THz) spectral region. Furthermore, in order to enable new and improved device concepts, sensitivity and responsivity to photons of various energy are important aspects and probed with common optoelectronic techniques.

It is clear that different optoelectronic devices can benefit in the short or long term from many of the discussed materials and nanostructured systems. These can, on the one hand, range from particularly sensitive or broadband photodetectors to wavelength-flexible coherent or broadband light sources. On the other hand, as part of ongoing efforts to develop quantum technologies of the second generation, efficient

[1] The availability of states of a system as a function of energy is expressed by the dimensionality-dependent density of states $D(E)$, which reflects the excitation spectrum based on the energy–phase-space relationship $E(k)$ and plays a major role in the interactions and dynamics of particles in that system.

[2] Derived from the quantum-mechanical description of the classical particle momentum $p = mv = \hbar k = \hbar 2\pi/\lambda$, with particle mass m, and momentum p, velocity $v = \sqrt{2E/m}$ and wave-number k in one-dimensional (vectorless) representation.

quantum emitters and deterministic nonclassical light sources are expected to utilise novel material platforms and tailored nanostructures. In this context, SPSs, quantum computers or sensors and so forth are strongly expected to gain from technological leaps and advances in functional nanomaterials sciences. These forecast benefits are indeed not only limited to newer quantum technologies, since even quantum technologies of the first generation still remain candidates for significant improvement based on the employment of novel or functionalised nanomaterials, such as for particularly efficient or compact light-emitting diodes (LEDs), laser diodes (LDs), or transistors. Also, QCLs are believed to benefit from advanced material systems, which can be regarded as located between these two generations.

This wider field of functional nanomaterials sciences is reflected in this work by different and complementary studies, by both fundamental and applied research. A summary of related disciplines shall support the discussion of the highlighted examples. Beginning with three separate introductory chapters following this brief general introduction (this chapter), work on this field is motivated and a brief overview of the topics of relevance provided. These chapters summarise activities in the fields of 2D materials (Chap. 2), light–matter interactions for photonic applications (Chap. 3), and work in the wider field of quantum technologies (Chap. 4). With regard to the characterisation of nanomaterials and quantum structures, optical measurement techniques, which have been employed in the context of the here presented work, are summarised (Chap. 5). In the following, the effects of quantisation on important properties of optoelectronic systems are highlighted (Chap. 6) and various relevant structuring possibilities explained (Chap. 7). A brief conclusion and outlook completes the presentation of numerous optical studies addressed within this work (Chap. 8).

The significance of this research direction can be easily anticipated when taking a look at the successful First Sino–German Symposium on Functional Nano-Materials Sciences (FNMS2018), which was carried out 2018 in Hangzhou, China. A brief overview of the prominent subjects in this field discussed within the frame of the FNMS2018 symposium (cf. Fig. 1.1) shall be given in the following (after the author's review article published open access under a CC-BY 4.0 licence [47]). The interested reader is referred to [47] and references therein for further details. Furthermore, the essence of this symposium is reflected in the recently formed Sino–German Cooperation Group on Functional Nano-Materials Sciences (FNMS-COOP), which has its emphasis on light–matter interactions with 2D systems and hybrids made of low-dimensional nanomaterials for optoelectronic applications, with groups from Baoding, Beijing and Hangzhou contributing to its aims on the Chinese side, and Frankfurt, Giessen and Marburg on the German side.

1.2 Advances in Functional Nanomaterials Sciences

It is well understood that the investigation of novel and functional materials is of great importance for the optimisation and development of electrical and optical devices (cf. Fig. 1.2). From such devices one does not only expect higher efficiencies, but access

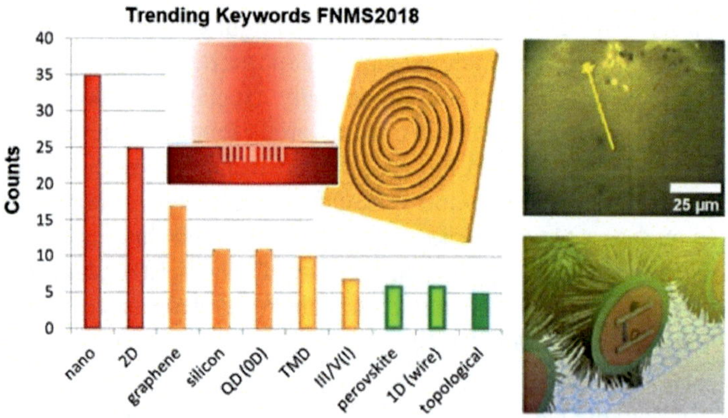

Fig. 1.1 Keyword counts for contributions to the 1st Sino–German Symposium on Functional Nano-Materials Sciences held in 2018 (FNMS2018) [47]. Left and central inset: Sketch of enhanced directional luminescence from a 2D material on a Bragg-grating in-plane optical microcavity designed for improved local in- and outcoupling of light (subject of investigation in [46]). Reproduced with permission under the terms of the CC-BY Creative Commons Attribution 4.0 International Licence (http://creativecommons.org/licenses/by/4.0/). [47] Copyright 2020 The Author(s), published by Wiley-VCH. Upper right: Chemical-vapour-deposition grown monolayer semiconductor optically investigated by the author and co-workers in [48] in comparison with exfoliated counterparts. The arrowed line indicates a line scan for Raman spectroscopy and optical contrast (not shown here). Reproduced under the terms of the CC-BY 3.0 Licence (http://creativecommons.org/licenses/by/3.0/) from Lippert et al., 2D Mater., "Influence of the substrate material on the optical properties of tungsten diselenide monolayers", https://doi.org/10.1088/2053-1583/aa5b21. [48] Copyright 2017 IOP Publishing Ltd. Bottom right: Schematic drawing of an optically-excited colloidal quantum dot on a graphene channel of a homemade field-effect transistor, investigated with regard to Förster energy transfer as a function of the backgate-bias-induced graphene doping. Adapted under the terms of the CC-BY 4.0 Licence (http://creativecommons.org/licenses/by/4.0/). [49] Copyright 2016 The Author(s), published by Springer Nature

to the development of utterly new concepts. These concepts are strongly demanded by modern information processing, quantum or medical technologies and sensing applications.

In this context, a Sino–German Symposium (FNMS2018) was recently organised with topical focus on functional nanomaterials, exhibiting how intensively such systems and materials will be explored in the coming years in China, as well as Germany, among other nations. The content of its 16 scientific sessions covered a wide range of aspects, such as the physics of novel materials, as well as materials engineering, characterisation and applications.

Some of the subjects relevant to such international conferences and workshops are also reflected in the current research landscape of the Philipps-Universität Marburg (UMR), Germany, and the Zhejiang University (ZJU) in Hangzhou, China, which have been the hosts of the two co-chairs of that symposium. In fact, the strategic partnership between the UMR and the ZJU has provided a helpful platform for successful collaborations within the past six years. Typical platforms for interdisciplinary materials science research in Marburg are the local collaborative research centre "Structure and Dynamics of Internal Interfaces" (DFG SFB1083) and Research Training Group "Functionalization of Semiconductors" (DFG GRK1782), which are similarly found with other topical focus at other German universities.

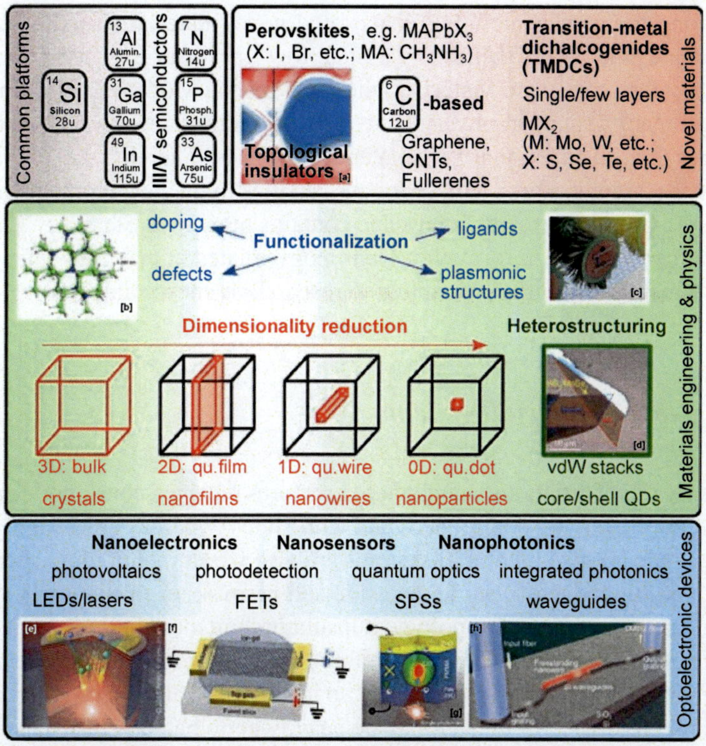

Fig. 1.2 Chart of representative topics highlighted in a review article on advances in functional nanomaterials science arranged in three major sections, and their relationship to each other [47]. For an orientation in the typical optoelectronic materials landscape, common platforms and novel materials are indicated. Schematically, the symbolic reference to elements of the periodic table with atomic number as well as mean rounded mass in atomic weight units u, and a reference to more complex materials such as graphene, carbon nanotubes (CNTs), fullerenes (C_{60} molecules) and their derivatives are given. Among the family of transition-metal dichalcogenides (TMDCs), the most popular 2D semiconductors are indicated. Functionalisation, dimensionality reduction and heterostructuring are typical approaches of materials engineering used to obtain improved optoelectronic and quantum technological devices. qu: quantum; QD: quantum dot; vdW: van der Waals. Applications range for instance from lasers/light-emitting diodes (LEDs), field-effect transistors (FETs) to single-photon sources (SPSs), and more. Adapted with permission under the terms of the CC-BY Creative Commons Attribution 4.0 International Licence (http://creativecommons.org/licenses/by/4.0/). [47] Copyright 2020 The Author(s), published by Wiley-VCH. **a** Reproduced with permission. [50] Copyright 2017 PCCP Owner Societies. **b** Reproduced with permission. [51] Copyright 2016 Wiley-VCH. **c** Adapted under the terms of the CC-BY 4.0 Licence. [49] Copyright 2016 The Author(s), published by Springer Nature. **d** Reproduced with permission. [52] Copyright 2017 Springer Nature. **e** Adapted from the author's original. Copyright 2013 Arash Rahimi-Iman. **f** Reproduced with permission. [53] Copyright 2018 Springer Nature. **g** Reproduced under the terms of the CC-BY 4.0 Licence. [54] Copyright 2017 The Author(s), published by Springer Nature. **h** Reproduced under the terms of the CC-BY 4.0 Licence. [55] Copyright 2017 The Author(s), published by Springer Nature. **c**, **g**, **h** CC-BY 4.0 Licence according to http://creativecommons.org/licenses/by/4.0/ and **a**, **b**, **d**, **f** with DOI web-address in Refs

A strong focus on materials sciences and novel material systems also exists in Hangzhou, China. The ZJU hosts two State Key Laboratories, one of them the State Key Laboratory of Silicon Materials (the other one on Modern Optical Instrumentation), and many groups in physics, chemistry and other departments researching in this wider field of functional nanomaterials sciences. Thereby, the ZJU contributes significantly among other Chinese top universities to the developments in this field. Indeed, the Chinese research landscape contains numerous State Key Laboratories, as well as nanoscience, information-technology-related or materials science research centers which are strongly committed to research on these subjects.

1.2.1 Novel Material Systems

Obviously, 2D materials and their heterostructures received considerable attention in recent years on various conferences and workshops, whereas other highly-attractive conventional and novel material systems are also worth mentioning, such as silicon and III/V semiconductors on the one side and perovskites or materials with special topological properties (e.g. topological insulators) on the other, respectively. Nevertheless, not all of the discovered and tailored material systems that were also subject of the FNMS2018 symposium may end up in commercial devices in the near future, although their long-term impact may be heavily anticipated at this stage.

With the vision of harnessing the remarkable properties of 2D materials, Y. Chai from the Hong Kong Polytechnic University summarised that as the semiconductor industry enters into the post-Moore era, the "heterogeneous integration" becomes a trend. He and his co-workers considered the introduction of 2D materials for future interconnect and transistor applications.

Naturally, the controlled synthesis of 2D materials, such as graphene and MoS_2 (a prominent representative of the TMDC class of 2D semiconductors) remains of great importance, in order to enable electronic and optoelectronic applications of 2D layered and non-layered metal–chalcogenides semiconductors, as well as their heterostructures. This was for instance addressed by Beijing's J. He from the National Center for Nanoscience and Technology, and by M. Xu from the State Key Laboratory of Silicon Materials and ZJU, who highlighted uniform graphene growth with eyes towards the application of 2D materials to solar cells, particularly flexible organic solar cells with graphene-based transparent electrodes.

Intriguing findings related to 2D materials properties also rely on their particular crystal structure. New Dirac fermion states in 2D puckered group-Va elements layers were recently found by Y. Lu and co-workers from the ZJU in Hangzhou. In addition, spontaneous electric polarisation and ferroelectricity in 2D elemental Group-V (As, Sb, and Bi) monolayers with the puckered lattice structure similar to that of phosphorene was highlighted by him.

Towards plasmonics with 2D materials, the interaction between nanoscale metallic entities and 2D materials such as graphene or MoS_2 can be modified in a controlled way and these effects can be probed by Raman spectroscopy, P.J. Klar from the Justus-Liebig-Universität Giessen reported.

In addition, surface plasmonic effects in metallic nanoheterostructures were summarised by G. Bi and C. Cai from the ZJU City College in Hangzhou, who investigated the tuning effects of localised surface-plasmon resonances on properties of the nanosystems.

A class of functionalised materials, which has been also attractive for collective spin excitation and dynamics studies, such as in semiconductor-based spintronic materials, was highlighted by X. Zhang from the State Key Laboratory of Superlattices and Microstructures, Beijing. The associated mechanisms governing their magnetisation dynamics were discussed by her for all-optical ultrafast spin excitation and dynamics of typical semiconductor-based spintronic materials.

While the trend goes towards the exploitation of layered (quantum) materials and functionalised nanocrystals, the material class of perovskites has emerged in the past decade as a promising candidate for modern optoelectronic devices. U. Lemmer from the Karlsruhe Institute of Technology outlined how metal–halide perovskites form a class of printable semiconductors with highly tunable optoelectronic properties, from which his group has realised efficient solar cells and optically pumped lasers using ink-jet printing as the deposition method.

1.2.2 Material Engineering and Physics

Given the importance of material engineering and physics for energy-efficient devices or novel device concepts, several FNMS2018 sessions discussed this subject for materials such as perovskites, graphene and related materials, as well as other layered or nano-sized crystals.

Since perovskites showed a considerable application potential, several investigations regarding fundamentals, synthesis and employment of this material class were presented.

By demonstrating a bright-exciton fine-structure splitting as large as several hundreds of μeV in single perovskite $CsPbI_3$ nanocrystals at the National Laboratory of Solid State Microstructures, Collaborative Innovation Center of Advanced Microstructures, Nanjing University, X. Wang's report implied that their fundamental studies and practical applications have now stepped into the quantum information processing regime.

Furthermore, recent steady-state and time-resolved (TR) photoluminescence (PL) studies on quasi-2D perovskite nanoplatelets, 0D ultrasmall perovskite quantum dots, and 2D layered perovskites revealed unique and excellent optical properties of excitons in solution-processed low-dimensional lead–halide perovskites, as H. He from the State Key Laboratory of Silicon Materials and ZJU in Hangzhou highlighted. This would render solution processing an attractive method with respect to perovskite structure fabrication.

With the goal of employing perovskite materials in photovoltaic applications, also high-quality perovskite films were targeted. Such films were demonstrated by blade coating of lead-acetate-trihydrate sourced precursor solutions under harsh ambient condition. From that, W. Kong and his co-workers obtained perovskite solar cells with a championing photovoltaic efficiency of 16.4% at the Southern University of Science and Technology in Shenzhen, he summarised.

Towards flexible electrodes and tunable surfaces, E.-H. Yang from the Stevens Institute of Technology, USA, explained how nanofabrication of 1D and 2D materials by growth is achieved for graphene/carbon electronics and TMDC heterostructure assembly. In addition, he discussed the implementation of tunable wetting and surface interaction with polymers. His group's studies thereby highlighted the potential for manipulation and control of liquid droplets for various applications including oil separation, water treatment and anti-bacterial surfaces.

Indeed, 2D-materials research delivered great promises for the domain of nano-electronics, but the field of optics may also benefit strongly by the employment of novel materials, e.g. when peculiar properties, such as the tunable nonlinear optical response from 2D materials, become appealing for applications. S. Wu from the Fudan University in Shanghai summarised how atomically thin 2D materials, including graphene and TMDCs, became a new playground for nonlinear optics.

Besides experimental studies, theoretical modeling of band structures and optical transitions has remained indispensable for a better understanding of the optoelectronic properties of homo- and hetero-multilayers based on 2D materials. Representative investigations regarding interlayer excitons had been the subject of researchers working together with T. Stroucken and S.W. Koch at the UMR. Using a fully microscopic theory to investigate TMDC homo- and hetero-multilayer structures, T. Stroucken explained that predictions regarding the (co)existence of intra- and interlayer excitons for various multilayer configurations were successfully made.

Also, the role of the environment in the work with 2D semiconductors matters, as theoretical and experimental works have shown. Both, the impact of the substrate and environment on optical properties of 2D semiconductors were discussed by the author himself (UMR), using 2D WSe_2 and heterostructures comprising WSe_2 and hexagonal-BN (hBN) as primary platform for micro-PL (μ-PL) investigations among other spectroscopic studies. It is generally understood that an interplay of different effects influences resonance energies, linewidths and lifetimes, and that optical properties can be further tailored using photonic microstructures.

Naturally, when envisioning devices and structures based on 2D materials, the study of ultrafast charge transfer in 2D semiconductors and heterostructures becomes very important, as Q. Xiong from the Nanyang Technological University in Singapore highlighted. He explained that the significance of ultrafast charge transfer (on the sub-ps scale) and plasmonic hot carrier injection ($\approx$ ps scale) in 2D semiconductors or heterostructures will further increase. New photophysical properties, such as correlated fluorescence blinking and charge-induced second harmonic generation, reportedly related to these effects, may render these materials appealing for all-optical ultrafast control of nonlinear frequency conversion.

Furthermore, Z. Fang from the Peking University shed light on plasmonic hot electrons that were used to dope 2D materials for the tuning of electro-optical properties with relevance to applications, characterising these ultrathin nanomaterial systems with the help of near-field scanning optical microscopy and dark-field optical microscopy.

Although non of the 2D materials have yet conquered the electronics market, graphene and its derivatives remained in the focus of international research with eyes towards optoelectronic and energy materials. In this context, optical and electronic properties of graphene and graphene nanoribbons were studied by H. Wang at the Max Planck Institute for Polymer Research in Mainz, where these properties were controlled by optical and chemical means.

The interest in carbon nanomaterials was also reflected in the device-oriented studies by H.G. Roskos at the Goethe-Universität Frankfurt, who targeted their employment for terahertz detectors and sensors. Exploration of these materials, such as graphene and vertically-aligned carbon nanotubes (VACNTs), took place with regard to applications at THz frequencies. Firstly, the dielectric properties of VACNTs in the THz/mid-IR range, and secondly, THz detection schemes based on antenna-coupled graphene field-effect transistors were addressed by him.

In addition to graphene and the prominent TMDCs, other 2D materials gained prominence due to their unique monolayer properties. For instance, ultrafast nonlinear optical effects in black phosphorus were studied by J. He from the Central South University in Changsha, China, who identified the saturable absorption properties and ultrafast carrier dynamics in black phosphorus nanosheets and quantum dots suspended in solvent at UV–VIS–IR wavelengths as promising playground for nonlinear optics.

Since layered materials and artificial spacer layers play critical roles in photonic and photoelectronic devices, N. Dai and co-workers from the National Laboratory for Infrared Physics, Chinese Academy of Sciences, Shanghai, introduced an interlayer IR excitation to van-der-Waals (vdW) layered heterostructures, which act as functional nanostructures, and investigated artificial spacer layers in the metasurfaces.

Although various research has been focused on trending material systems, silicon materials have not reached the end of the road yet. Recent work on the inverse design of silicon nanomaterials for highly-efficient light emitting and CMOS-compatible qubits[3] was presented by Beijing's J. Luo from the State Key Laboratory of Superlattices and Microstructures, Chinese Academy of Sciences. Furthermore, a deep understanding of light emission from Si quantum dots was summarised. These nanomaterials were for instance envisioned for optoelectronic and spintronic applications. Additionally, Si nanocrystals were expected to improve existing Si-based technologies and advance the use of Si towards new fields.

In terms of optoelectronic applications, high-performance photodetectors had been developed by heavy-boron (B) doping in Si nanocrystals by X. Pi and coworkers at the State Key Laboratory of Silicon Materials, ZJU. Additionally, he referred to high-performance near-IR LEDs employing these nanocrystals.

Apart from heavy investigations on trend materials, conventional systems comprising of (inorganic–inorganic or organic–organic) semiconductor–semiconductor

[3]Acronym for *quantum bits*, which in contrast to classical bits (binary digits) are in a so-called quantum state (superposition of possible eigen-states). Example qubits are single photons, single electrons or ions, superconducting circuits.

interfaces, hybrids of inorganic–organic materials and molecules on surfaces of metals, or topological insulators were discussed.

For instance, experiments on ultrafast electron dynamics of interfaces between metals and organic semiconductors, surfaces of topological insulators and TMDCs were discussed by U. Höfer from the UMR. Among others, he explained how time-resolved two-photon photoemission served as a characterisation method that combined femtosecond pump–probe techniques with photoelectron spectroscopy. With such method, his group obtained detailed information about the ultrafast dynamics of electrons excited into surface- and interface-specific states of these systems.

Regarding selected-area engineering of correlated vanadium oxides and their nanostructures without any lithography and chemicals, C. Xin from the National Laboratory for Infrared Physics, Chinese Academy of Sciences, Shanghai, discussed how nanoscale selected-area chemical doping became promising for constructing functional materials and devices.

Clearly, numerous nanomaterials systems received considerable attention throughout the symposium. One such system formed by self-prepared crystalline thin organic films and molecular heterostructures was presented as a model system for detailed optoelectronic studies of elementary processes in organic electronic devices by G. Witte from the UMR, whose research is focused on molecular nanostructures.

Not less important remain colloidal quantum dots (cQDs) as model nanoclusters in which quantisation effects strongly determine optical properties. L. Hu from the Jiangnan University in Wuxi outlined how the emission of CdS cQDs could be tuned with silver particles. Giving a hybrid II-VI cQDs silver nanocrystal as an example, the luminescence of the hybrid was shown to be modifiable by tuning the ratio of the two components when mixing the components during synthesis quantitatively.

Complementary to the various other spectroscopic measurement schemes used to obtain most of the information discussed about optoelectronic materials, M. Koch from the UMR also conducted material investigation using THz spectroscopy, a well established technique nowadays. He highlighted THz time-domain spectroscopy as a tool for basic research, as well as for non-destructive testing in the industry, providing several examples for both application fields. For instance, the techniques developed in his group allowed him to take a look at the formation of crystals out of solution, or to analyse arbitrarily shaped samples.

The domain of topological and structural effects was briefly discussed as well. Y.P. Feng from the National University of Singapore shed light on the search for a wider gap topological insulator and the enhancement of charge-to-spin conversion. Via first-principles calculations, layered hexagonal Bi_2S_3 was predicted to be convertible into a topological insulator by pressure. Other calculations indicated that the coexistence of topological surface states and the Rashba effect contributed to the enhanced charge-to-spin conversion in a Bi_2Se_3/Ag/CoFeB system.

Furthermore, investigations on the topological nature of the PbTe crystal and the interface of a CdTe/PbTe heterostructure were summarised by the symposium's co-host, H.Z. Wu from the State Key Laboratory of Silicon Materials and ZJU. Theoretical and experimental results demonstrated that by incorporating Te antisites into the host lattice, PbTe could become a topological crystalline insulator.

Moreover, a novel 2D electron gas (2DEG) of Dirac fermion nature was reported which forms at the interface of CdTe/PbTe heterojunctions.

In addition, C. Guo from the Center for Correlated Matter of the ZJU explained how a novel Weyl semimetal was envisioned through heterostructuring, which hinted at a new path towards the realisation of other exotic topological materials via heterostructure fabrication.

1.2.3 Optoelectronic Devices

Miniaturisation not only promises more compact and energy efficient optical devices but also enables novel device concepts. This is where quantum structures and nanophotonics come into play. It becomes clear that functionalised nanomaterials contribute strongly to the developments in this field.

As a classical example, G. Qian from the State Key Laboratory of Functional Materials for Informatic in Shanghai reported how InAs quantum-dot lasers with high device performances were achieved when epitaxially integrated on a germanium substrate. Grown by gas-source molecular-beam epitaxy (MBE), those quantum-dot laser-diode structures emitting in the spectral range of 1.0–1.3 μm exhibited a very low lasing-threshold current density and high output power.

A truly nanophotonic case was represented by the semiconductor nanowire photonic devices, which were studied for long by L. Tong and his co-workers at the State Key Laboratory of Modern Optical Instrumentation and the ZJU. Recent progress on nanowire optical modulators based on a graphene-coated ZnO nanowire and on an on-chip integrated CdS nanowire were summarised by him, and single-nanowire ultrafast optical correlators based on transverse second-harmonic generation from CdS and CdTe nanowires were reported.

A topical bridge towards quantum technologies was made by nanostructured semiconductors and their applications in classical and quantum communication at the telecommunications wavelength of 1.5 μm. J.P. Reithmaier from the University of Kassel highlighted different types of 1.5-μm quantum-dot gain materials for fibre-based applications that were developed with regard to classical high-performance optoelectronic devices as well as for core elements of long-haul quantum-communication systems.

Entering the nonclassical world, quantum technologies have revealed themselves as an emerging topic, both, generally and within the topic landscape of the symposium.

From the beginning, S. Reitzenstein from the Technische Universität Berlin outlined the advantages of nonclassical light emission from quantum-dot–microlens structures. It was shown how deterministic fabrication of bright quantum-dot-based single-photon sources and on-chip waveguide structures by means of in-situ electron-beam lithography promised better yield and faster implementation in quantum-cryptography schemes. In this context, the optical properties and application potential for quantum networks were discussed.

More quantum dots for photonic quantum technologies were brought up by P. Michler from the University of Stuttgart, whose group at the Center for Integrated Quantum Science and Technology (IQST) and SCoPE demonstrated super-resolving phase measurements based on two-photon N00N states, which were generated by quantum-dot single-photon sources. Moreover, he reported first efforts to perform on-chip quantum sensing. In his group's work, the bosonic nature of light was utilised through the Hong–Ou–Mandel effect on a beam splitter.

In addition to conventional concepts incorporating epitaxially grown quantum dots, the work presented by W. Fang from the State Key Laboratory of Modern Optical Instrumentation and ZJU demonstrated high-purity, electrically-driven, and room-temperature single-photon sources based on colloidal quantum dots. Such CdSe/CdS core/shell quantum dots in solution-processed devices were reported to yield high-purity single photons with a low working voltage.

These quantum technologies of the second generation were preceded by quantum technologies of the first generation, such as LEDs and lasers. Indeed, semiconductor light source technologies have remained very attractive due to their maturity and good integrability in existing technologies.

Based on photonic engineering, G. Xu and his co-workers from the Key Laboratory of Infrared Imaging Materials and Detectors in Shanghai demonstrated a novel cavity concept for THz quantum-cascade lasers, which can operate at elevated (cryogenic) temperatures and exhibit highly efficient power extraction owing to the use of a grating coupler.

While lasers and single-photon sources are highly demanded for current and next-generation applications, LEDs have not lost their appeal as fundamental platform for illumination and light-based technologies. For instance, LEDs based on high-performance solution-processed quantum dots, short QLEDs, have promised application in display technologies. Y. Jin from the State Key Laboratory of Silicon Materials and ZJU reviewed activities associated with QLEDs including material chemistry of charge-transporting layers, mechanism studies and the optimisation of prototype devices.

On the optoelectronic device side, high performance graphene/silicon photodetectors and image sensors were reported by Y. Xu, ZJU, making use of the integration of 2D materials into CMOS device design and fabrication. Numerous examples operating in the UV, near-IR or mid-IR spectral range were highlighted in this context, with the aim to solve the typical problems of silicon-based photodetectors and image sensors.

Similarly, H. Zhu from the Hangzhou Dianzi University and his co-workers focused on improved photodetectors by developing blocked-impurity-band (BIB) THz photodetectors using a periodic metal structure adopted in the design of the BIB device. The demonstration of a superior performance in a new operation mode suggested the development of lower-cost and potentially mass-producible THz detectors.

More exotic optoelectronics were represented by oxide-based neuromorphic transistors for neuromorphic computation, highlighted by Q. Wan from the Nanjing University. Accordingly, artificial synapses and neurons were proposed based on protonconductor gated oxide-based electric-double-layer transistors, for which some important synaptic functions were successfully mimicked. Such works were inspired by the energy-efficient and cognitive computational ability of a biological brain, and building blocks for brain-like computers have been therefore under development.

It can be concluded that all these examples, subjects and investigations together show how interdisciplinary, challenging and important research activities on various quantum structures and functionalised nanomaterials have become over the past decades. Indeed, the evidenced remarkable progress can be understood as a consequence of previous and recent breakthroughs in materials sciences, quantum physics and optoelectronics, strongly motivating further investigations. The following chapters discuss some of the relevant topics in these fields within the frame of this work.

References

1. L. Pavesi, D.J. Lockwood (eds.), *Silicon Photonics III: Systems and Applications*, Topics in Applied Physics (Springer, Berlin Heidelberg, 2016)
2. G. Eisenstein, D. Bimberg, *Green Photonics and Electronics* (Springer, Cham, 2017)
3. S.K. Sharma, K. Ali (eds.), *Solar Cells: From Materials to Device Technology* (Springer, Cham, 2020)
4. C. Brosseau, Emerging technologies of plastic carbon nanoelectronics: a review. Surf. Coat. Technol. **206**, 753–758 (2011)
5. D. Akinwande, N. Petrone, J. Hone, Two-dimensional flexible nanoelectronics. Nat. Commun. **5**, 5678 (2014)
6. Y. Che, H. Chen, H. Gui, J. Liu, B. Liu, C. Zhou, Review of carbon nanotube nanoelectronics and macroelectronics. Semicond. Sci. Technol. **29**, 073001 (2014)
7. J.L. O'Brien, A. Furusawa, J. Vučković, Photonic quantum technologies. Nat. Photonics **3**, 687–695 (2009)
8. M. Razeghi, *Technology of Quantum Devices* (Springer, New York Dordrecht Heidelberg London, 2010)
9. W.P. Schleich, K.S. Ranade, C. Anton, M. Arndt, M. Aspelmeyer, M. Bayer, G. Berg, T. Calarco, H. Fuchs, E. Giacobino, M. Grassl, P. Hänggi, W.M. Heckl, I.-V. Hertel, S. Huelga, F. Jelezko, B. Keimer, J.P. Kotthaus, G. Leuchs, N. Lütkenhaus, U. Maurer, T. Pfau, M.B. Plenio, E.M. Rasel, O. Renn, C. Silberhorn, J. Schiedmayer, D. Schmitt-Landsiedel, K. Schönhammer, A. Ustinov, P. Walther, H. Weinfurter, E. Welzl, R. Wiesendanger, S. Wolf, A. Zeilinger, P. Zoller, Quantum technology: from research to application. Appl. Phys. B **122**, 130 (2016)
10. A. Acín, I. Bloch, H. Buhrman, T. Calarco, C. Eichler, J. Eisert, D. Esteve, N. Gisin, S.J. Glaser, F. Jelezko, S. Kuhr, M. Lewenstein, M.F. Riedel, P.O. Schmidt, R. Thew, A. Wallraff, I. Walmsley, F.K. Wilhelm, The quantum technologies roadmap: a European community view. New J. Phys. **20**, 080201 (2018)
11. K. Vahala, *Optical Microcavities* (World Scientific, 2004)
12. A. Kavokin, G. Malpuech, *Cavity Polaritons* (Academic Press, 2003)
13. B. Deveaud, *The Physics of Semiconductor Microcavities* (WILEY-VCH, 2007)
14. C.F. Klingshirn, *Semiconductor Optics* (Springer, 2012)
15. A.V. Kavokin, J.J. Baumberg, G. Malpuech, F.P. Laussy, *Microcavities*, vol. 1 (Oxford University Press, Oxford, 2017)
16. A. Rahimi-Iman, *Polariton Physics: From Dynamic Bose–Einstein Condensates in Strongly-Coupled Light–Matter Systems to Polariton Lasers*, vol. 229, Springer Series in Optical Sciences (Springer International Publishing, Cham, 2020)

17. R. Dingle, W. Wiegmann, C.H. Henry, Quantum states of confined carriers in very thin $Al_xGa_{1-x}As$-GaAs-$Al_xGa_{1-x}As$ heterostructures. Phys. Rev. Lett. **33**, 827 (1974)
18. M.A. Herman, H. Sitter, *Molecular Beam Epitaxy: Fundamentals and Current Status* (Springer, Berlin Heidelberg, 1996)
19. G.B. Stringfellow, *Organometallic Vapor-Phase Epitaxy: Theory and Practice* (Elsevier Science, 2012)
20. M. Sugawara, *Plasma Etching: Fundamentals and Applications* (Oxford Science Publications, Oxford, 1998)
21. G.P. Agrawal, N.K. Dutta, *Semiconductor Lasers* (Springer, US, 1993)
22. L. Coldren, S. Corzine, *Diode Lasers and Photonic Integrated Circuits* (Wiley, 1995)
23. C. Wilmsen, H. Temkin, L.A. Coldren, *Vertical-Cavity Surface-Emitting Lasers* (Cambridge University Press, Cambridge, 1999)
24. M.I. Amanti, M. Fischer, G. Scalari, M. Beck, J. Faist, Low-divergence single-mode terahertz quantum cascade laser. Nat. Photonics **3**, 586–590 (2009)
25. S. Strauf, F. Jahnke, Single quantum dot nanolaser. Laser Photon. Rev. **5**, 607–633 (2011)
26. A. Rahimi-Iman, Recent advances in VECSELs. J. Opt. **18**, 093003 (2016)
27. M.A. Gaafar, A. Rahimi-Iman, K.A. Fedorova, W. Stolz, E.U. Rafailov, M. Koch, Mode-locked semiconductor disk lasers. Adv. Opt. Photonics **8**, 370–400 (2016)
28. M. Guina, A. Rantamäki, A. Härkönen, Optically pumped VECSELs: review of technology and progress. J. Phys. D: Appl. Phys. **50**, 383001 (2017)
29. E.U. Rafailov, M.A. Cataluna, W. Sibbett, Mode-locked quantum-dot lasers. Nat. Photonics **1**, 395–401 (2007)
30. P. Berini, I. De Leon, Surface plasmon-polariton amplifiers and lasers. Nat. Photonics **6**, 16–24 (2012)
31. C. Schneider, A. Rahimi-Iman, N.Y. Kim, J. Fischer, I.G. Savenko, M. Amthor, M. Lermer, A. Wolf, L. Worschech, V.D. Kulakovskii, I.A. Shelykh, M. Kamp, S. Reitzenstein, A. Forchel, Y. Yamamoto, S. Höfling, An electrically pumped polariton laser. Nature **497**, 348–352 (2013)
32. P. Bhattacharya, B. Xiao, A. Das, S. Bhowmick, J. Heo, Solid state electrically injected exciton-polariton laser. Phys. Rev. Lett. **110**, 206403 (2013)
33. P. Bhattacharya, T. Frost, S. Deshpande, M.Z. Baten, A. Hazari, A. Das, Room temperature electrically injected polariton laser. Phys. Rev. Lett. **112**, 236802 (2014)
34. M. Ramezani, A. Halpin, A.I. Fernez-Domuez, J. Feist, S.R.-K. Rodriguez, F.J. Garcia-Vidal, J.G. Rivas, Plasmon-exciton-polariton lasing. Optica **4**, 31–37 (2017)
35. M.D. Fraser, Coherent exciton-polariton devices. Semicond. Sci. Technol. **32**, 093003 (2017)
36. Y. Yamamoto, A. Imamoglu, *Mesoscopic Quantum Optics* (Wiley, 1999)
37. Y. Yamamoto, F. Tassone, H. Cao, *Semiconductor Cavity Quantum Electrodynamics* (Springer, 2000)
38. S. Reitzenstein, A. Forchel, Quantum dot micropillars. J. Phys. D: Appl. Phys. **43**(3), 033001 (2010)
39. A. Boretti, L. Rosa, A. Mackie, S. Castelletto, Electrically driven quantum light sources. Adv. Opt. Mater. **3**, 1012–1033 (2015)
40. A. Amo, J. Bloch, Exciton-polaritons in lattices: a non-linear photonic simulator. C. R. Phys. **17**, 934–945 (2016)
41. N.Y. Kim, Y. Yamamoto, Exciton-polariton quantum simulators, *Quantum Simulations with Photons and Polaritons*, Quantum Science and Technology (Springer International Publishing, 2017), pp. 91–121
42. J. Loredo, M. Broome, P. Hilaire, O. Gazzano, I. Sagnes, A. Lemaitre, M. Almeida, P. Senellart, A. White, Boson sampling with single-photon Fock states from a bright solid-state source. Phys. Rev. Lett. **118** (2017)
43. P. Senellart, G. Solomon, A. White, High-performance semiconductor quantum-dot single-photon sources. Nat. Nanotechnol. **12**, 1026–1039 (2017)
44. C. Antón, J.C. Loredo, G. Coppola, H. Ollivier, N. Viggianiello, A. Harouri, N. Somaschi, A. Crespi, I. Sagnes, A. Lemaître, L. Lanco, R. Osellame, F. Sciarrino, P. Senellart, Interfacing scalable photonic platforms: solid-state based multi-photon interference in a reconfigurable glass chip. Optica **6**, 1471 (2019)

45. S. Rodt, S. Reitzenstein, T. Heindel, Deterministically fabricated solid-state quantum-light sources. J. Phys. Condens. Matter **32**, 153003 (2020)
46. O. Mey, F. Wall, L.M. Schneider, D. Günder, F. Walla, A. Soltani, H. Roskos, N. Yao, P. Qing, W. Fang, A. Rahimi-Iman, Enhancement of the monolayer tungsten disulfide exciton photoluminescence with a two-dimensional material/air/gallium phosphide in-plane microcavity. ACS Nano **13**, 5259–5267 (2019). https://doi.org/10.1021/acsnano.8b09659
47. A. Rahimi-Iman, Advances in functional nano-materials science. Ann. Phys. **532**(9), 2000015 (2020). https://doi.org/10.1002/andp.202000015
48. S. Lippert, L.M. Schneider, D. Renaud, K.N. Kang, O. Ajayi, J. Kuhnert, M.-U. Halbich, O.M. Abdulmunem, X. Lin, K. Hassoon, S. Edalati-Boostan, Y.D. Kim, W. Heimbrodt, E.-H. Yang, J.C. Hone, A. Rahimi-Iman, Influence of the substrate material on the optical properties of tungsten diselenide monolayers. 2D Mater. **4**, 025045 (2017). https://doi.org/10.1088/2053-1583/aa5b21
49. R. Li, L.M. Schneider, W. Heimbrodt, H. Wu, M. Koch, A. Rahimi-Iman, Gate tuning of Förster resonance energy transfer in a graphene - quantum dot FET photo-detector. Sci. Rep. **6**, 28224 (2016). https://doi.org/10.1038/srep28224
50. M. Yang, Y.Z. Luo, M.G. Zeng, L. Shen, Y.H. Lu, J. Zhou, S.J. Wang, I.K. Sou, Y.P. Feng, Pressure induced topological phase transition in layered Bi_2S_3. Phys. Chem. Chem. Phys. **19**(43), 29372–29380 (2017). https://doi.org/10.1039/C7CP04583B
51. Z. Ni, X. Pi, S. Zhou, T. Nozaki, B. Grandidier, D. Yang, Size-dependent structures and optical absorption of boron-hyperdoped silicon nanocrystals. Adv. Opt. Mater. **4**, 700–707 (2016). https://doi.org/10.1002/adom.201500706
52. W. Xu, W. Liu, J.F. Schmidt, W. Zhao, X. Lu, T. Raab, C. Diederichs, W. Gao, D.V. Seletskiy, Q. Xiong, Correlated fluorescence blinking in two-dimensional semiconductor heterostructures. Nature **541**, 62–67 (2017). https://doi.org/10.1038/nature20601
53. T. Jiang, D. Huang, J. Cheng, X. Fan, Z. Zhang, Y. Shan, Y. Yi, Y. Dai, L. Shi, K. Liu, C. Zeng, J. Zi, J.E. Sipe, Y.-R. Shen, W.-T. Liu, S. Wu, Gate-tunable third-order nonlinear optical response of massless Dirac fermions in graphene. Nat. Photonics **12**, 430–436 (2018). https://doi.org/10.1038/s41566-018-0175-7
54. X. Lin, X. Dai, C. Pu, Y. Deng, Y. Niu, L. Tong, W. Fang, Y. Jin, X. Peng, Electrically-driven single-photon sources based on colloidal quantum dots with near-optimal antibunching at room temperature. Nat. Commun. **8**, 1132 (2017). https://doi.org/10.1038/s41467-017-01379-6
55. B. Chen, H. Wu, C. Xin, D. Dai, L. Tong, Flexible integration of free-standing nanowires into silicon photonics. Nat. Commun. **8**, 20 (2017). https://doi.org/10.1038/s41467-017-00038-0

Chapter 2
Entering a Two-Dimensional Materials World

Abstract The class of two-dimensional materials has emerged as an important field within nanomaterials science, and quantum-materials research has caught the attention of various research communities. With the rise of graphene as a promising ultrathin, electronic and optoelectronic, flexible and robust material, the exploration of other similar van-der-Waals materials in the monolayer regime came into focus about a decade ago. Among them are the semiconducting 2D crystals of the transition-metal dichalcogenide family, the features of which can be typically even improved in combination with the atomically-flat 2D insulator hexagonal boron nitride. Currently, the characterisation, growth and utilisation of monolayers, as well as heterostructures thereof, are heavily targeted owing to their extraordinary properties and the prospects of device miniaturisation, as well as functionality boosts, in different technological domains. This chapter summarises various aspects of this material class with a focus on 2D semiconductors and their application examples with relation to the fields of (opto)electronics, photonics, as well as quantum technologies. An overview is given on the remarkable features of these 2D materials from both a fundamental and application-oriented perspective. A brief discussion of graphene and related materials is embedded. Here, also the role of stacked layered materials as well as heterostructuring is introduced, before subjects such as light–matter interactions and lasing with 2D semiconductors are addressed.

2.1 The Rise of the 2D Materials

2D materials and their heterostructures have received considerable attention in recent years in the wider scientific literature, and they are a highly discussed topic on various conferences and in numerous reviews. After systematic and pioneering studies on single-layer (monolayer) graphite, known as graphene, were performed by Novoselov et al. [1], a vast amount of work on van-der-Waals (vdW) materials targeted the optical, electrical, mechanical and (opto)electronic properties from the very beginning [2], with a strong focus on the monolayer regime [1, 3–7]. With regard to the early experiences and developments in the fields of 2D-materials research, the interested reader is for instance referred to the Nobel lecture on graphene by Novoselov [8].

A. Rahimi-Iman, *Semiconductor Photonics of Nanomaterials and Quantum Structures*, Springer Series in Solid-State Sciences 196,
https://doi.org/10.1007/978-3-030-69352-7_2

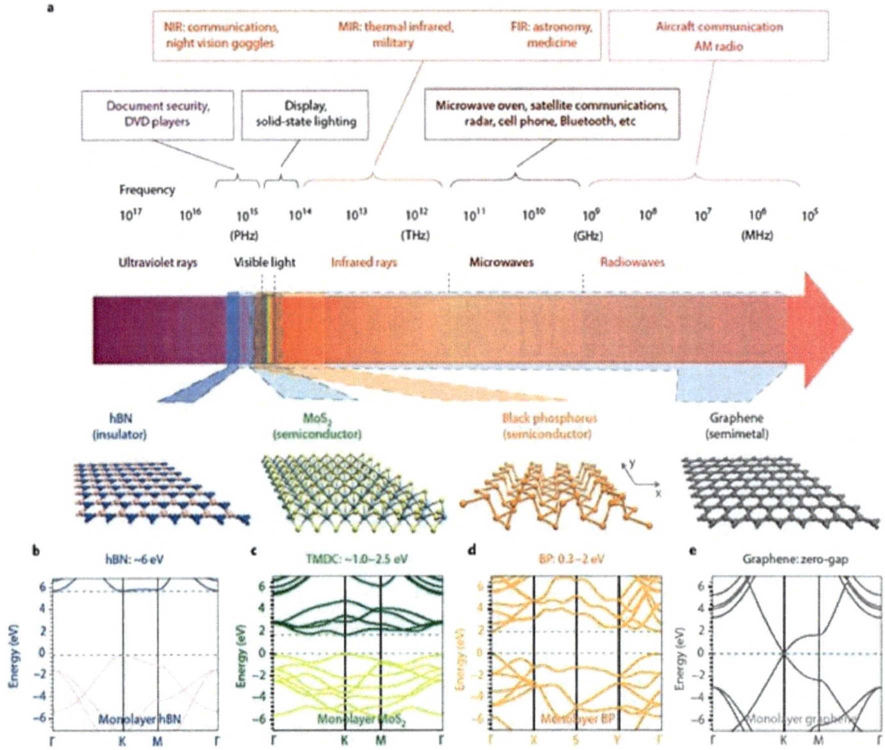

Fig. 2.1 **a** Schematically indicated spectral map of 2D materials and the typical applications for each frequency range. Corresponding to their energy gaps, different materials address different spectral regions, as shown here at the examples of (**b**) hBN, **c** TMDCs (here MoS_2), **d** black phosphorus and **e** graphene (see gap information and band structure diagrams displayed). Large gap van-der-Waals materials act as insulators, whereas gapless materials can be very good conductors. In-between lie medium-sized gaps, which are attractive for optoelectronic applications. Reproduced with permission. [9] Copyright 2014 Springer Nature

Potential Application Reviews and Excitonic Features
Naturally, potential applications of ultrathin 2D materials had come into focus quickly [9–20] (cf. Fig. 2.1). Nonetheless, a lot of fundamentals regarding mechanical, electronic, and optical properties, as well as synthesis, remained for exploration. On top, 2D semiconductors have offered a unique platform for investigations on a vast pool of excitonic species [21]. This goes beyond the neutral excitonic species with peculiar Rydberg-like characteristics [22–25]. Such pool includes trions [26, 27], biexcitons [28] and dark exciton states [29].

With time, more and more aspects of the (monolayer) 2D semiconductor's valley dichroism [30, 31], valley coherence [32], pseudo-spin texture within the light cone [33], their (density and temperature-dependent) dynamics [34–41], radiatively-limited dephasing (temperature-dependent homogeneous broadening) [35], diffusion [42], dissociation [43], orientation [44], selection rules [45], fine-structure [46, 47], exciton or trion g-factors [48–50], electronic dispersion [51], optical dispersion [33,

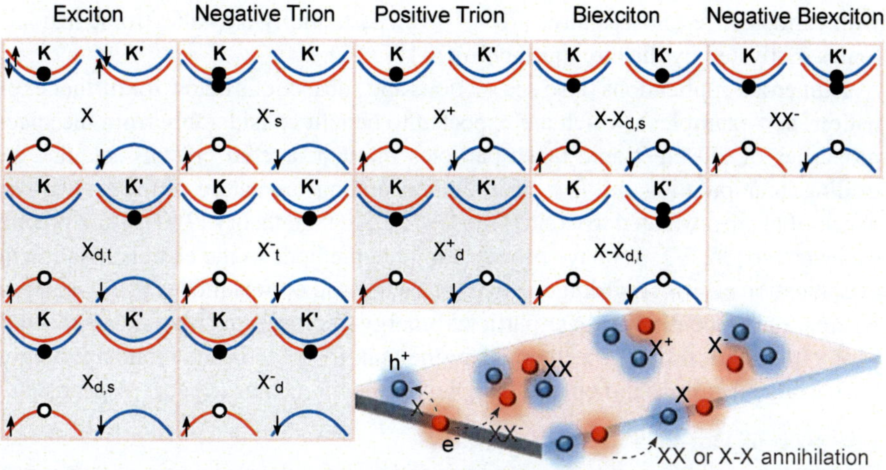

Fig. 2.2 A zoo of excitonic species and so-called multi-particle states across valleys can be obtained in 2D semiconductors owing to the three-fold symmetry and the correspondingly peculiar spin–valley locking. Here, a schematic overview diagram of possible exciton complexes is shown, drawn freely after [41, 54, 55], indicating intra- and intervalley configurations for excitons (X), biexcitons (XX) and their charged analogues, the trions ($X^{\pm}$) and charged biexcitons (XX^{-}). The index d denotes optically-dark states (grey-shaded boxes), s and t denote singlet and triplet state, respectively. Black solid and open circles represent electrons and holes, respectively. By displaying here solely the underlying charge carriers in their different host bands, Coulomb-binding energies are not represented. Here, K and K' represent the two distinct spin-opposite direct gap symmetry points for monolayers. Black arrows indicate spin up/down corresponding to the colour-coded (red/blue) band dispersions. Inset: Sketch of a 2D semiconductor layer with different crystal quasi-particles

52], excitation-detuning-dependent formation of macroscopic polarisation (coherent excitons, also valley-polarised) [33], radiation patterns [47], phonon-assisted modes [53] and the multi-particle states across valleys in 2D semiconductors [54, 55] have been unraveled.

2D Excitons in Quantum Materials

Quantum materials such as monolayer TMDCs are particularly interesting due to the zoo of exciton species hosted in their lattices after optical excitation (see Fig. 2.2), and more importantly due to their strong light–matter interactions and the inherent spin–valley locking [21, 56], which have attracted unmatched attention by the research community. The former peculiar feature of these 2D semiconductors results from their extraordinarily large oscillator strength as well as binding energy, whereas the latter property is a consequence of the symmetry-breaking in the plane of the 2D crystal.

Semiconducting few-layer and monolayer materials with their sharp optical resonances such as WSe_2, which recently were reported close to the homogeneous linewidth limit [7, 52, 57–60], have been extensively studied and envisioned for applications in the weak [61–63] as well as strong light–matter coupling [64–69] regimes. Also, they have been extensively studied for effective nanolaser operation with various different structures [70–74], and particularly for valleytronic nanophotonics motivated by the circular dichroism [17, 75].

Numerous applications (also see sections and subsections below for further explanations and examples), which are expected to benefit considerably from the special properties of 2D electronic quasi-particles in such layered crystals, require controlling, manipulating and first of all understanding the nature of the optical resonances that are attributed to exciton modes [52]. In fact, theory [76–79] and previous experiments [34, 80, 81] have provided unique methods to the characterisation and classification efforts regarding the band structure and optical modes in 2D materials, whereas only recently the quasi-particles' energy–momentum dispersion came under exploration with different techniques in the bulk [51] and in the monolayer regime [52], using electronic and optical mapping of the momentum-space, respectively.

Prospects of Van-der-Waals Materials

The properties of 2D materials in the few- and single-layer regime are widely tunable by doping, strain, external fields and environmental effects [5, 10, 84, 85], owing to the materials' atomic thickness and the strong influence of the surroundings [40, 49, 58, 86–88]. Their size, features and sheet flexibility promises the development of a new generation of nano/micro-devices, in which flexible ultrathin 2D membranes can be integrated [9, 89, 90] and even be stacked layer by layer [56, 91]. Such few- and single-layer-based heterostructures as well as devices benefit from the pronounced vdW interaction between adjacent layers that give rise to high-quality homo- or heterojunction interfaces with little constraints on lattice matching. This encourages production schemes such as vdW epitaxy and offers compatibility with existing semiconductor technologies. Accordingly, novel device concepts on the nanoscale have been envisioned for electronic, photonic and quantum technological applications (Fig. 2.3).

Heterostructuring Possibilities and Novel Twist-Related Phenomena

Leading back the 2D crystals into the 3D world, their out-of-plane (vertical) heterostructures gained unrivaled importance as a means of obtaining new functionality (e.g. through hybridisation, environmental or proximity effects) by building (functionalised) sequences of layers or encapsulated materials in stacks [5, 56, 91, 92]. This was supported by theoretical considerations [93–96]. New design possibilities are furthermore enabled by lateral heterostructuring, offering in-plane interfaces [94, 97], or monolayer preparation with different lateral doping, leading to in-plane pn-junctions (examples in [43, 98]).

Recently, the whole research on vertical heterostructures even received a new twist. This happened, on the one hand, with the emergence of in-plane superlattices of 2D-periodic potential islands and moiré minibands [99–101], as well as moiré states [102–105]. And on the other hand, it happened with the discovery of the extraordinary behaviours of twisted homobilayers. Among those extraordinary behaviours, which were initially theoretically obtained [106] and later experimentally demonstrated, are superconductivity [107, 108], Mott-insulation [109] or emerging ferromagnetism [110] in graphene bilayers. Thus, this new degree of freedom with regard to crystal orientations has been termed in the literature as 'twistronics' [99].

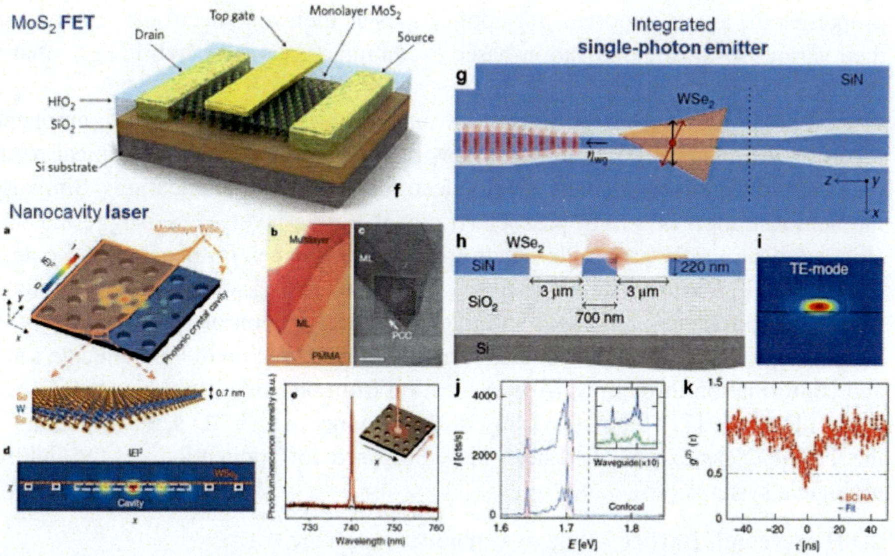

Fig. 2.3 Applications for 2D semiconductors shown at the hand of typical examples with relation to the fields of (nano)electronics, (nano)photonics and quantum (nano)technologies. Bottom left: **a** Microlaser employing a monolayer on a photonic crystal cavity. **b** Micrograph of a TMDC flake on PMMA before the transfer onto the nanofabricated cavity. **c** SEM image of the device. **d** Side-view of the simulated light-field intensity distribution. **e** Polarisation-resolved PL spectra with sketch of the cavity–beam geometry. **a–e** Reproduced with permission. [70] Copyright 2015 Springer Nature. Top left: **f** Artist's interpretation of a field-effect transistor made of MoS_2. Reproduced with permission. [82] Copyright 2011 Springer Nature. Right: Example of a chip-integration approach for WSe_2 quantum emitters based on monolayer flakes on single-mode SiN waveguides (**g**), which were 220 nm thick and processed with surrounding microsized air trenches (**h**) in [83]. **i** Cross-sectional mode profile at 750 nm showing the wave-guided TE-mode. **j** Spectral signatures and **k** quantum emitter's second-order autocorrelation function histogram $g^{(2)}(\tau)$. **g–k** Adapted with permission under the terms of the CC-BY Creative Commons Attribution 4.0 International Licence (http://creativecommons.org/licenses/by/4.0/). [83] Copyright 2019 The Author(s), published by Springer Nature

In fact, the coverage of all these activities in a short topical overview will be out of the scope and hardly complete. Below, an introduction to the rich world of 2D materials is given to provide a profound background for later discussions.

2.2 Fundamentals of 2D Materials

Owing to their reduced dimensionality, remarkable physical properties are found for 2D materials (cf. [56, 91]). The extraordinary aspect ratio between the surface and the volume of such structures exposes them heavily to the influence of their substrate and dielectric environment, while their strong in-plane covalent bonds give them remarkable mechanical strength and flexibility. Their mechanics and mechanical properties [6] are as important for applications as their optoelectronic features and their various forms of polarisation-based excitations (polaritons, hybrid light–matter modes[1]) [111].

Recent studies revealed the high elastic modulus of 2D materials such as graphene, MoS_2 or WS_2 [112], which open up new pathways to ultrathin mechanical membranes for different micro/nano- (opto/electro-) mechanical applications. Similarly, the demonstration of strong piezoelectricity in single-layer MoS_2 suggested possible applications in nanoscale electromechanical devices for sensing and energy harvesting [11]. Presumably, one of the most outstanding possible applications of suspended graphene membranes in biology is the detection of DNA, when drilled nanoholes in that graphene layer enable an indirect estimation of the molecule's size and conformation through DNA translocation from one side of the pin-holed sheet to the other one [113]. Indeed, many of the features, which 2D semiconductors of the TMDC class exhibit, had been also discussed for applications in the field of biological systems [15].

2D Honeycomb Lattice Made of Carbon: Graphene

Graphene, a single sheet of carbon atoms arranged in a hexagonal 2D lattice and highly conductive, is the most studied representative of the class of 2D materials. One can think of a pencil's graphite tip being thinned down to the minimum of one single layer, which is about a quarter of a nanometer thick but, remarkably, exhibits a universal broadband absorption of about $\pi\alpha = 2.3\%$.[2] Correspondingly, graphene's optical contrast on certain substrates is good enough to identify monolayer flakes visually under the microscope. How the remarkable properties could be used for optoelectronic applications is briefly highlighted in the following subsection.

Finding Gaped 2D-Material Systems

However, the lack of a band gap renders graphene unsuitable for many digital electronic and optoelectronic applications. Consequently, significant efforts have been devoted to identifying alternative 2D materials of semiconducting type [91, 116]. In the past decade, several classes of non-carbonic compounds with layered structure have been synthesised, including hBN [117], TMDCs [10], oxides [118, 119], hydroxides [120], and oxychlorides [121]. Remarkably, they address different optical frequencies (indicated for instance in spectral diagrams in [9, 122, 123], see Fig. 2.1). Moreover, they provide a unique pool of vdW materials that has even become attractive for vdW epitaxy (see for instance map of energy gap versus lattice constant in [124], see Fig. 7.1 in Chap. 7).

[1] For 2D materials, the literature discusses species such as plasmon–polaritons, phonon–polaritons, exciton–polaritons, cavity–polaritons, Cooper-pair–polaritons, and magnon–polaritons.

[2] Here, α corresponds to Sommerfeld's (fine-structure) constant, a fundamental natural constant related to the electro-magnetic vacuum field's interaction with charged elementary particles (important for QED)—so-to-say the coupling constant for electromagnetism determining the strength of interactions between charged particles. Thus, it drew the attention of scientists working on graphene, such as K. Novoselov, that graphene's universal absorption (around its Dirac cone/point) was 'quite a good direct measure' of that constant [8].

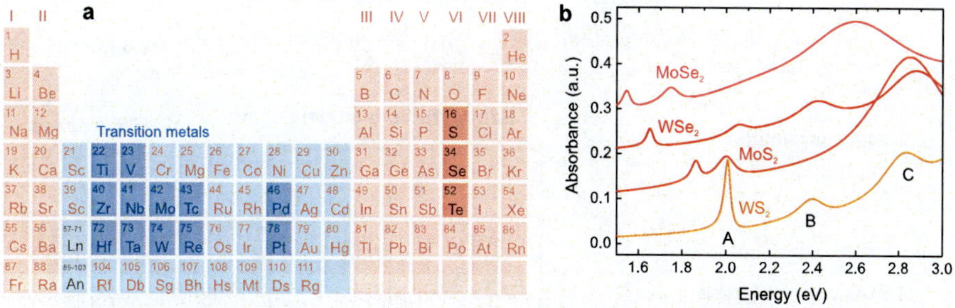

Fig. 2.4 Pool of TMDC materials, with the elements in the periodic table (**a**) highlighted after [114], from which numerous alloys and heterostructures can be produced through synthesis and stacking, respectively. **b** Typical absorption signatures from the four prominent TMDCs (drawn freely after [115], individual spectra off-set against each other arbitrarily for clarity). The two lower-energetic peaks correspond to the A and B exciton of the respective material. The tungsten-based TMDCs show sharper and more pronounced A excitons than their molybdenum-based counterparts, which instead feature a smaller separation between A and B resonances that arise in TMDCs from a considerable spin–orbit coupling

TMDCs, with formula MX_2 (M: transition metal, X: chalcogen, see Fig. 2.4a), are among the most studied non-carbonic compounds. They are usually chemically, thermally and mechanically stable even in the monolayer regime, while some compounds exhibit considerably faster degradation rates than others. The most prominent monolayer ones are MoS_2, WS_2, $MoSe_2$ and WSe_2, with their similar and characteristic spectral features shown in Fig. 2.4b.

Nevertheless, the else semimetal-like graphene has opened up new possibilities in (twisted) bilayer configuration [8, 84], which enabled the formation of sub-bands and controllable gaps [125–128], or even the observation of extraordinary behaviours under "magic" twist angles [107–110].

The TMDC Family—Beyond Graphene

Known as popular 2D semiconductors, the members of the TMDC family exhibit a wide range of different electrical [82, 129–133] and optical [3, 4, 30, 134] properties, depending on the polytype of the TMDCs and the number of transition metal d-electrons. In fact, TMDCs as bulk crystals had been already studied in the past century (cf. [130, 135–137], and other layered semiconductors [138]). Even thinned-down systems (so-to-say few-layer configurations) cleaved with 'Scotch tape' were under investigation in the years after, at that time (see e.g. [139]). Changes in interlayer coupling, quantum confinement, and symmetry breaking lead to dramatic differences in the electronic structure and optical properties of single-layer TMDCs compared with their bulk counterparts.

Given the Dirac cone in the band structure of monolayer graphene, charge carriers in graphene are considered relativistic particles, whereas in the semiconducting counterparts they feature a finite effective mass. The mobility may be also quite high in the plane of a 2D semiconductor, but the formation of bound electron–hole pairs can dominate the optical and charge-transport properties of those. With record

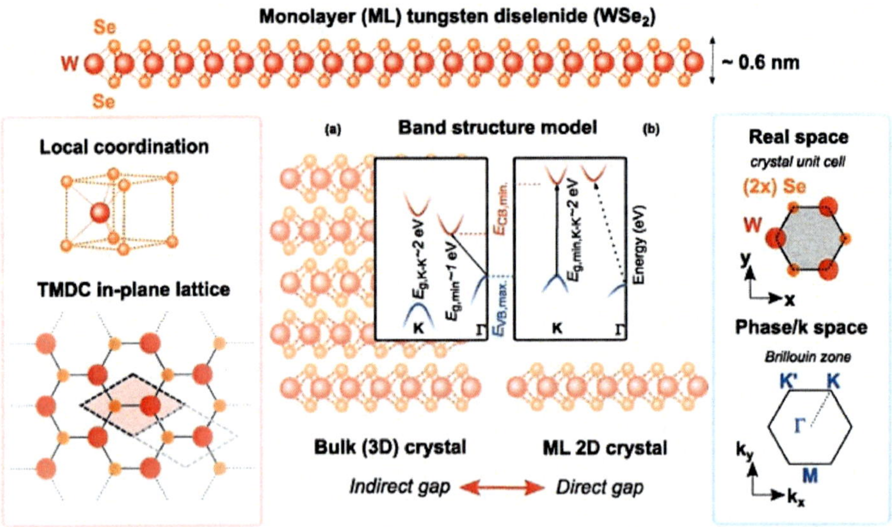

Fig. 2.5 Top: Typical schematic side-view of a monolayer (ML) TMDC with metal atoms sandwiched between chalcogen atom layers, rendering the monolayer semiconductor a covalently-configured three-layer crystalline system. Left box: Sketch of the local coordination (top) in the monolayer crystal lattice with a primitive unit cell (bottom), indicated in black (dashed line). Note the two-element basis. An exemplary translation of that cell to the next lattice point is shown in grey. Also, note that the chalcogen atoms are not in the same plane as the transition metal. Right box: Common representation of a real-space (top) and momentum-space (bottom) elementary unit cell for a monolayer TMDC indicating a three-fold rotation symmetry of a hexagonally-arranged unit cell and of the corresponding Brillouin zone, respectively. Differently coloured balls indicate atoms in the unit cell (in sum containing one W and two Se atoms), and K and K' denote the two distinguishable high-symmetry points on the corners of the hexagonally-shaped Brillouin zone. Note that a three-fold symmetry exists in both the real space and the k space. Central figure with inset: Schematic comparison of a natural multilayer configuration (**a**) with indirect electronic band gap (cf. inset, left) and a monolayer (**b**) with direct gap (cf. inset, right). The inset shows a simplified sketch of the band structure model for bulk and monolayer (Mo/W)(S/Se)$_2$. Arrows mark a momentum-indirect (unlikely) valence-to-conduction-band transition for bulk and a direct gap transition for the monolayer case (K–K). Energy scales according to [4, 93]

mobilities up to 350,000 cm^2/(Vs) [140], graphene has become an ideal candidate for high-mobility devices, while monolayer TMDCs address the demand on the optoelectronics side due to their semiconducting nature (see for instance [16] and references therein).

Structural Properties of TMDCs

One single TMDC layer typically has a thickness of about 0.6–0.7 nm which consists of a hexagonally-packed layer of metal atoms sandwiched between two layers of chalcogen atoms (see Fig. 2.5). However, a three-fold crystal symmetry—and correspondingly k-space symmetry—renders the corners of the hexagonally-shaped 2D Brillouin zone distinguishable, in a way that K and K' valleys are located next to each other. Time-reversal-symmetry exists between these high-symmetry points of the Brillouin zone. From this, the valley degree of freedom gives rise to a pseudo-spin attributed to each valley and makes K and K' valleys separately accessible to oppositely circularly polarised light [30, 31].

Mechanical Exfoliation and Production of Heterostructures

The intralayer MX bonds are mainly covalent in nature, whereas the sandwiched layers are coupled by weak vdW forces, thereby allowing the crystal to be readily cleaved along the layer surface. Thus, mechanical exfoliation is the most common lab-scale approach to deliver isolated (high-quality) monolayers to experiments. Such monolayers can—in a simplified picture—be even stacked in a "LEGO" building-block fashion by means of vdW interactions between adjacent layers to form high-quality homo- or heterojunction interfaces (cf. [56, 91]).

Stacking TMDC monolayers to form different heterostructures offers the unique opportunity to design advanced electronic and optoelectronic devices [16, 56]. Commonly, dry-stamping techniques, wet-transfer or pick-and-lift techniques are employed for the desired vertical heterostructuring [56, 141, 142], whereas chemical-vapour-deposition (CVD) growth can deliver, both, crystalline in-plane as well as lattice-matched out-of-plane heterostructures [97]. Indeed, lateral heterojunctions and pn-junctions are understood to also occur for manually stacked monolayers on other 2D materials at the edge of overlap due to the influence of the environment on the energetics and doping, respectively, in the stacked materials [143].

Confinement and Screening in Monolayers

The strong enhancement of the Coulomb interaction in the monolayer limit leads to an indirect-to-direct band gap transition [3, 4] (see sketched 'band structure model', inset in Fig. 2.5), large exciton binding energy [23, 24] (cf. table of electronic properties of 2D TMDCs in [16]), the abundance of multi-excitons [54, 55] and sensitivity to the surrounding dielectrics [40, 49, 58, 86–88, 144]. Mostly, these drastic changes are attributed to the screening changes from bulk to single-layer crystal. These changes do not only introduce a drastic quantum confinement of charge carriers to one single layer but also expose the excitonic dipole to harsh dielectric changes at the interface (e.g. from monolayer to vacuum/air on two sides in the case of fully suspended flakes) which results in a modified screening for them (the excitons and their higher excitations, cf. [23, 88]).

Band Properties and Interlayer Coupling

In addition, the p-orbital contributions to the electronic band structure originating from the chalcogen atoms (basically sitting at the interfaces) are more prone to variations of the environment than the d-orbital contributions from the transition metal (sandwiched in the central layer). This affects the indirect gap's conduction band minimum at the Σ and the valence band's maximum at the Γ crystal Brillouin-zone symmetry points. In contrast, little happens at the d-orbital dominated K and K' points, at which the direct gap is given in the monolayer regime [4, 145]. Furthermore, a strong spin–orbit coupling leads to a degeneracy lifting of, both, the highest valence and lowest conduction bands, whereas the splitting of the valence band is typically much larger and is the main source of the energy difference between A and B excitons (cf. Fig. 2.4b) corresponding to these spin-opposite optical transitions. Thereby, spin

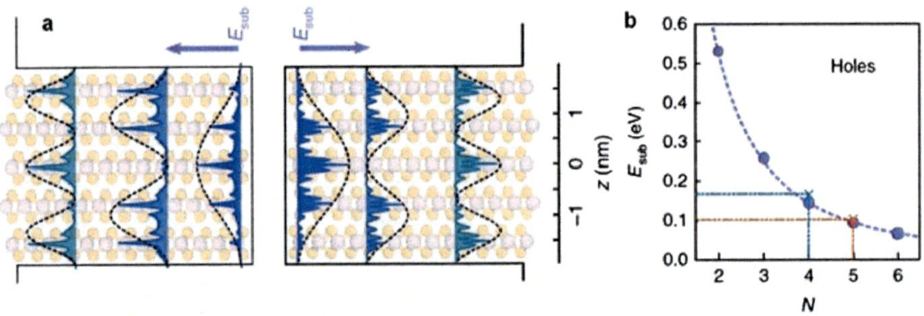

Fig. 2.6 **a** Schematic drawing of a five-layer TMDC with conduction and valence subbands displayed in the common quantum-well representation (left and right part, respectively) with barrier levels encapsulating the well's minimum gap. Purple arrows indicate intersubband transitions. Blueish shaded curves on the subband energy levels display calculated spatial probability functions for charge carriers in the respective bands, with the envelope Bloch wave shown as dotted curves. **b** For various numbers N of layers, the calculated lowest intersubband transitions for holes are compared with the measured valence band transitions for $N = 4$ and 5. Adapted with permission. [148] Copyright 2018 Springer Nature

and valley pseudo-spin degree of freedoms are so-to-say locked [146, 147]. Left (right) circularly polarised light resonant to the A (B) exciton transition will only populate a single valley and vice versa for the other valley, whereas decoherence due to scattering processes may repopulate excitons over time (see for instance [32]).

In most TMDC cases, the bilayer already features a momentum-space indirect exciton I (cf. [3])—the monolayer being a good direct semiconductor. Recently, even the common quantum-well picture and eigen-state formation in out-of-plane direction has been revisited in an investigation that probed (with nanoimaging capabilities) experimentally intersubband transitions for a four- and five-layer TMDC [148]. While the Bloch wave-function of a periodic system—describing the unit cell of a crystal multiplied with a plane envelope wave—does not apply to single layers, for which only standing wave solutions of the envelope function avoid destructive wave-function interference in the out-of-plane direction, the transition from an "atomistic" to a periodic system naturally exhibits the increase of states (formation of bands) with increasing number of layers from the monolayer towards the bulk regime (cf. Fig. 2.6).

Oscillating Polarisation Waves of Coherent Excitons

In fact, due to the lack of out-of-plane dipole–dipole couplings (i.e. hybridisation, in the sense of long-distance exchange interactions) no delocalised bulk excitons (3D exciton–polariton states, to be more precise, cf. [149]) with typically weaker 3D binding energies (oscillator strength) are formed in monolayer TMDCs, but 2D excitons (2D exciton–polaritons, with mere in-plane long-range exchange interactions). This is not only important for the strength of light–matter interactions, but also relevant for the discussion of the energy–momentum dispersion of optically-active resonances in 2D materials (see [33, 52]).

Here, the (complex) dielectric function $\epsilon(\omega)$ of 2D semiconductors (cf. [115]) implies for very good-quality 2D materials with very high oscillator strength and near-homogeneously broadened linewidth (such as achieved by [7, 52, 57–59, 142, 150]) a Reststrahlenbande[3] in the reflection spectrum[4] [149] and, thus, hybrid light–matter states with very low effective masses of their composite particles [33, 52]. These hybrid quasi-particles can be regarded as highly-mobile, delocalised, macroscopically valley-polarised correlated electron–hole pairs dressed with light.

Previously, the link between coherent excitons and exciton–polaritons had been widely disregarded for TMDCs, but this aspect has become more significant when the occurrence of negative permittivities around exciton resonances is justified by the narrow linewidths. These findings and many more leave plenty of room for deeper investigations of this material class. Recently, also the concept of renormalisation of longitudinal-mode energies within the light cone and the degeneracy lifting between longitudinal and transverse exciton modes (previously discussed for III/V quantum-well systems [151, 152]) has been revisited (see [33] and references therein, and Chap. 3). Furthermore, within the light cone, the exciton fine-structure's pseudo-spin and angle-resolved valley polarisation was optically probed in [33].

Impact of the Substrate and Environment on Monolayer TMDCs

Although 2D materials are an ideal (ultrathin) model system in which charge-carriers are strongly confined to the layer of transition metals between the chalcogen layers, they are in fact still not perfect. First of all, many properties are only ideal for a defect- and disorder-free suspended monolayer in vacuum. Besides the fact that most samples are not (strain-free perfectly) suspended, defects in the material spoil various optoelectronic properties, as summarised further below. Furthermore, optoelectronic properties of TMDC monolayers are affected by the dielectric environment (screening) and substrate-induced effects, as early studies indicated.

An interesting overview of the many different values for excitonic energies and linewidths (compared for the four prominent members of the TMDC family) from the early characterisation efforts in the literature is provided as a table with references in [40]—a work concerning the impact of the environment on excitonic resonances following a similar investigation focusing on the influence of the substrate material on optical properties of monolayers [87].

Naturally (well before these works occurred), the questions arose whether the substrates induced doping and strain, what the roles of local roughness and chemical affinities were, and how the band-structure was changed by the local environment.

For instance, it was shown by Lin et al. that the trion/exciton ratio changed in the emission (PL) of monolayer MoS_2 and the peak positions changed as well, when a covering solvant was systematically varied to span a range of solvant dielectric con-

[3]The Reststrahlenbande is a high-reflectivity band (photonic stop band), which is the consequence of the zero-crossings of the real part of ϵ, i.e. for $\mathrm{Re}\{\epsilon\} = \epsilon' < 0$.

[4]Note that spatial dispersion discussed in Chap. 5 of [149] may affect the maximum reflectivity in the spectral region between the transverse and longitudinal exciton, i.e. between their resonances ω_0 and ω_L, respectively.

stants between about 2 and 33 [144]. Their obtained peak position changes indicated binding energy modifications for A and B excitons.

Nevertheless, no significant wavelength shifts for suspended monolayers (initially surrounded by vacuum from both sides) due to a change of the dielectric function of the environment were reported by Rigosi et al.—an effect attributed to band-gap renormalisation simultaneous to the change of binding energy [153].

It was also shown that the PL efficiency in the form of intensity changes could be varied with the help of substrate interactions engineering according to Yu et al. [86], hinting at the role of doping; and effects of the substrate type and the material–substrate bonding on high-temperature behaviour of monolayer WS_2 were discussed by Su et al. [154].

In addition, magneto-optical measurements on monolayer WSe_2 concerning the dielectric environment (on mainly exfoliated monolayer samples, that were 2 out of 3 environment settings) by Stier et al. showed a diamagnetic shift of the monolayer resonance revealing a change of the deduced mean exciton radius[5] by 33% [49]. Correspondingly, binding energy changes were evidenced with altered dielectric environments (also see [88]).

Shortly after, from the comprehensive and comparative work of Lippert et al., which was in preparation in parallel to aforementioned studies in that important time period of excitonic studies in monolayers, it could be concluded that there is a complex interplay between local strain, dielectric environment and doping [87]. Therefor PL, time-resolved PL and Raman measurements were performed—both at room temperature and 10 K at the example of an archetype monolayer semiconductor, that is WSe_2. Moreover, biexcitons were only observable—at cryogenic temperatures—for low-doping and low-strain samples (i.e. with low trion/exciton ratio and low Raman shifts for a certain mode, respectively) (see [87]). In addition, spectral signatures showed that reduced inhomogeneous-broadening and low dephasing support biexciton formation (i.e. a narrow linewidth and pronounced degree of circular polarisation, respectively). It was also shown that room-temperature species exhibited a linearity factor common for bound[6] and uncommon for free excitonic species, which were obtained at low temperatures (below 100 K, also see temperature behaviour of the linearity factor in [40]). Follow-up exciton–exciton annihilation studies using time-resolved PL also gave hints at the Mott density and exciton Bohr radius modifications through the dielectric environment [38].

Thus, these studies inspired further detailed investigations concerning substrate and environment related optoelectronic properties of 2D materials and supported the tailoring of application oriented 2D systems. An example how the substrate landscape

[5]Unfortunately, the reference sample of that study comprised CVD-grown monolayer material, being well known to be inferior/different in quality than their exfoliated counterparts, until very recently.

[6]Here, 'bound' is not referring to Coulomb binding, but meaning bound in the sense of localised (also see examples from 2D research in [28, 155]). Later, the higher-temperature species obtained between about 100 K and room temperature had been viewed in the quasi-particle picture as excitons in a polarisation cloud with phonons, i.e. forming polarons. This was analysed and discussed in [53] and understood as one reason for the vanishing of optical dispersion above 100 K in [52].

can be further utilised to manipulate the absorption and emission behaviour is given in [156] (also see Figs. 2.13, 5.4 and 7.6).

Disorder in 2D Materials

It shall be noted that recent works in the literature have in detail addressed the role of disorder in vdWs monolayers and heterostructures of 2D materials which can obscure intrinsic properties (see [7]): Among the common sources of intrinsic disorder are vacancies, anti-sites, substitutions, edges and grain boundaries. Typically, quality differences exist between mechanically exfoliated 2D material from crystalline bulk material and physical-vapour-deposition (PVD) or CVD grown samples, as the comparison of common point defects in MoS_2 or nanoscopic analysis of graphene indicated. For certain applications, point defects are rather a feature than a bug, when they are optically-active (see for instance [157]) and can be deterministically generated (e.g. by nano-precise ion bombardment [158]).

Also, sources of extrinsic disorder such as strain[7] (recognisable through Raman mode shifts [160], band gap changes [161] or second-harmonic intensity profiles [162]), adsorbates, surface roughness, charged impurities on surfaces and oxidation of the 2D material have to be taken into account. It has been even shown that the band gap of a TMDC can vary locally on the nanoscale between direct and indirect type [7], motivating suspension of monolayers as a remedy.

Moreover, a proven method to improve the monolayer quality for various experiments is given by (few- to multi-layer) hBN encapsulation, as it improves the brightness and linewidth [40, 58], as well as the emission lifetime [40], suppresses adsorbate-related emission features [57, 59] and environmental disorder [42], reduces strain, doping levels [58], as well as exciton–phonon coupling [40] and the effective mass [52]. Furthermore, hBN encapsulation enabled the observation of various exciton complexes and the study of the grey exciton's radiation profile (Fig. 2.7, see [47] and references therein). On the other hand, hBN encapsulation reduces the binding energy and band gap energy compared to suspended monolayers (see [78, 88], also cf. [60]).

The interested reader is referred to [7] for further details on the discussion of disorder.

In conclusion, the extraordinary properties of these materials [5, 56, 111] have already encouraged worldwide endeavours to harness the 2D-materials' features for all kind of nanodevices (see [9, 13, 16] and references therein), but also for "valleytronics" (see [17, 75] and references therein), with graphene and hBN naturally serving as ideal 2D contacts and insulators, respectively.

[7] Strain can indeed affect excitonic modes, as a stretched lattice with increased mean particle distances in the plane can exhibit a band-gap energy reduction (cf. [159], also see comparison WSe_2 grown by CVD with its exfoliated counterpart both on a sapphire substrate and strain-related discussion in [87]).

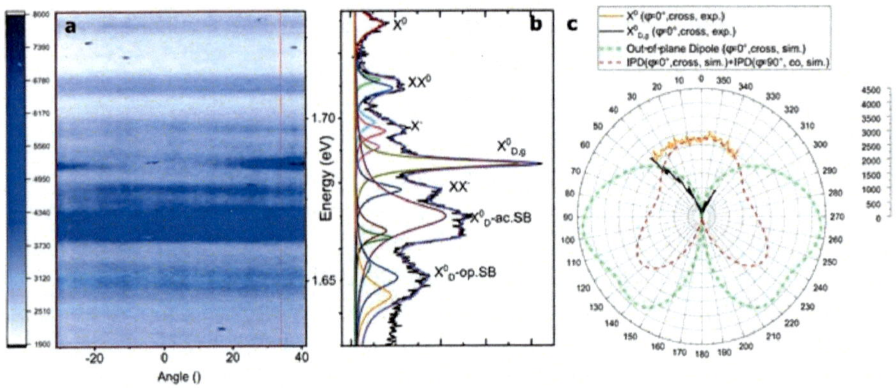

Fig. 2.7 Angle-resolved spectroscopy of an hBN-encapsulated monolayer of WSe_2 exhibiting multiple excitonic features and phonon sidebands, which were attributed in the literature to neutral as well as charged excitons (X and $X^{\pm}$, respectively) and exciton complexes, such as neutral/charged biexcitons (XX/XX^{-}), z-mode excitons ($X_{D,g}$, "grey", i.e. not totally dark) and optical (op.) and acoustic (ac.) dark-exciton (X_D) phonon sidebands (PSB), respectively. **a** Overview spectrum given as contour diagram with energy versus emission angle (false-colour scale: intensity increases towards dark blue). **b** Angle-integrated spectrum (red boxed in **a**) exhibiting different excitonic intensity peaks. **c** Radiation profile of the out-of-plane emitting bright (lateral) exciton and the in-plane emitting 'grey' exciton (a dark exciton with out-of-plane dipole orientation, i.e. z polarisation), obtained under a cross-polarised detection scheme, showing good agreement with the expected profile highlighted by simulated data. Reproduced under the terms of the CC-BY 4.0 Licence (http://creativecommons.org/licenses/by/4.0/). [47] Copyright 2020 The Author(s), published by Springer Nature

2.3 Graphene and Related Materials

A 2D World Opened Up for Use

While the fascination for graphene has continuously grown, graphene and many of its derivatives have not really conquered the electronics market yet. Nevertheless, graphene has become an important platform for 2D-materials research as well as device prototyping, and many review articles such as the one by Novoselov et al. [163] have pointed out a roadmap for graphene, and even for materials beyond graphene [116]. Although popular for its electronic features, it has also become appealing for light–matter interaction studies [164]. Since the beginning of the past decade, graphene has been widely employed and studied for the development of novel electro-optical devices, among others as transparent electrodes [165], for photovoltaic [166] or plasmonic devices [164, 167, 168], in optical modulators [169], and sensing/sensor devices [170–172].

The Rise of Graphene

Based on its remarkable properties such as a broad spectral bandwidth and high electron mobility, high-performance photodetectors were envisaged for graphene [13], and numerous examples indicate the successful use in this regard [173–177], including laser-defined functionalised graphene photodetectors with extraordinary linear

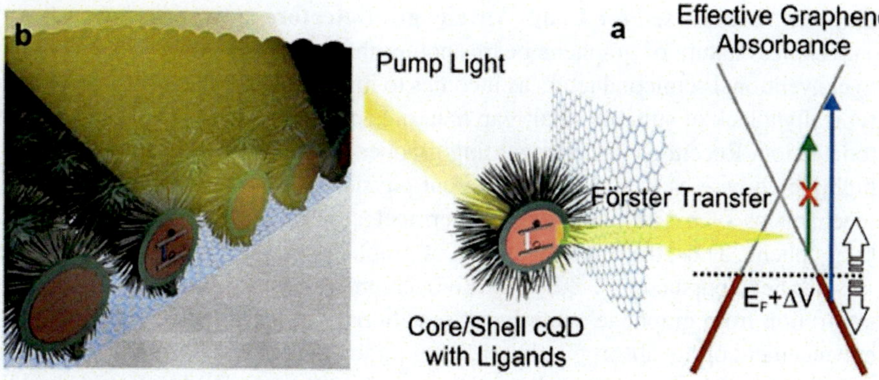

Fig. 2.8 **a** Sketch of the Förster resonant energy transfer (FRET) from an excited cQD exciton mode to an electronically-doped graphene Dirac cone by backgate bias tuning. This configuration was experimentally investigated with regard to backgate-dependent cQD-emission lifetime (i.e. revealing FRET as a gate-tunable loss channel) as well as photodetector responsivity modifications. **b** Schematic drawing of optically-excited cQDs of core/shell type with long ligands on a (laser-defined) graphene channel of a homemade field-effect transistor. **a**, **b** Reproduced under the terms of the CC-BY 4.0 Licence (http://creativecommons.org/licenses/by/4.0/). [191] Copyright 2016 The Author(s), published by Springer Nature

dynamic range [178]. For instance, a monolithic integration of a CMOS integrated circuit with graphene, operating as a high-mobility phototransistor, was recently reported [179]. However, the weak light absorption limits its application for photodetection in the same broad spectral region where it relies on the states of its Dirac cone of a single layer of ultrathin material.

Graphene Photodetection Schemes

In this context, a combination of graphene with an excellent light absorber has been pursued for the fabrication of efficient graphene-based photodetectors. Several approaches including a graphene–semiconductor heterojunction [180–183], a graphene pn-junction [184], quantum-dots–graphene hybrid detectors [185–192], graphene–TMDC–graphene heterostructures [173, 193, 194] and many more were reported recently.

Quantum-Dot Hybrid Systems with Graphene

Among these platforms, the colloidal quantum dots (cQDs) hybrid structure has remained a promising candidate for photodetection schemes owing to the cQDs' wavelength tunablility as well as their efficient light-absorbing capability. Therefore, one aims at combining the optical advantages of cQDs together with the outstanding electrical properties of graphene. Typically, optical power is absorbed by the layer of cQDs on graphene and the excitation energy transferred to the graphene sheet. This can lead to detectors with ultra-high gain and a high external quantum efficiency of up to 25% [189]. In addition, the transfer rate can be manipulated by an applied gate voltage [191, 195] (indicated for Förster resonant energy transfer schematically in Fig. 2.8).

Gapless Nature Used for Long-Wavelength Detectors
The gapless nature of graphene bears another important advantage when compared to conventional semiconductors as it comes to long-wavelength detection, for which typically a lack of suitable small-gap materials prevents efficient and sensitive photodetection. Recent advances regarding graphene-based far-IR detectors are briefly highlighted here as well. Due to different physical phenomena, different detection mechanisms such as a photovoltaic, thermoelectric or bolometric one are enabled by graphene. Thus, the implementation of graphene-based IR and THz photodetectors has been pursued [13, 20, 196]. To overcome limitations imposed by the weak absorption from graphene's interband transitions, many efforts were devoted to the enhancement of the absorption by utilizing microcavities [197], employing antenna coupling [198], patterning graphene into periodic metamaterials [199], and including plasmonic nanostructures on top of the graphene [200]. In fact, plasmonic nanoparticles can even be used in combination with conventional THz antenna materials to boost their performance [201].

2.4 Layered Systems Based on Monolayer Semiconductors

Owing to the transition from indirect to direct semiconductor towards the monolayer regime, TMDCs have received considerable attention as possible building blocks for miniaturised device concepts and energy-efficient optoelectronics. Naturally, vdW epitaxy becomes attractive with regard to the achievement of TMDC heterostructure assembly and the exploitation of the benefits when returning to the 3D world from the 2D plane.

2.4.1 Physics of Transition-Metal Dichalcogenide Heterostructures

Currently, research activities regarding heterostructure fabrication and studies are thriving. A plethora of publications deals with the role of monolayer stacking principles (see for instance [56, 91]) as well as optical and electronic properties of 2D heterostructures—prominently with eyes towards interlayer excitons [40, 202–206] for possible observation of collective behaviour such as condensation and superfluidity [207–209]. Optoelectronic heterostructure properties have been studied in dependence of the band-alignment [93–96], interlayer band-hybridisation [210], dielectric changes [40, 88, 211], phase-space mismatch [212] (generally relevant for vdW stacks, e.g. for twisted graphene bilayers [213]), lifetime modifications [40], moiré potential landscapes [101, 102, 214], twist angles [215], and stacking symmetry [216–219].

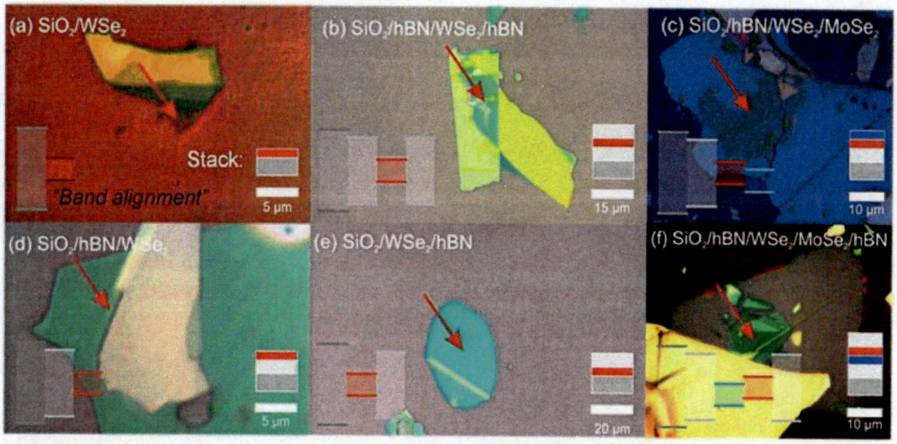

Fig. 2.9 Manual heterostructuring enables rich variations in the stacking sequence as well as interlayer alignment of single-layer and few-layer vdWs materials. **a–f** Example of a stacking variation involving primarily monolayers of WSe_2. Red arrows point at the heterostructure regions. The most common heterostructures are indeed achieved by hBN buffering (**d**), capping (**e**) or encapsulation (**b**), having a noticeable influence on the properties of ultrathin 2D TMDCs, else often directly deposited on substrates (**a**) due to simplicity (common in the early phase of 2D-materials research). Property improvements and modifications are nowadays widely targeted with the help of sophisticated stackings, ideally to achieve "band-gap engineering". Monolayer–monolayer heterostructures such as in (**c**) and (**f**) have become particularly promising due to the achievable band offsets that give rise to charge-transfer states across type-II heterojunctions. While (**a**, **b**, **d**, **e**) clearly address changes of the dielectric environment including in both half-spaces surrounding the monolayer, **c**, **f** show even more effects, using manually incorporated $MoSe_2$ above (**c**) and below (**f**) the WSe_2 specimen, as studied in [40] (published 2017) and many more works of the same time. Remarkably, it can be deduced that the capping of an hBN-supported WSe_2 monolayer by even one single monolayer TMDC (**c**) is enough to achieve a similarly strong impact on the linewidth and energy of excitonic features as the capping with multilayer hBN (**b**). This indicates that the dominant screening effect takes place within the first surrounding layers, while it should be noted that $MoSe_2$ has not only a larger dielectric constant, but can also drain charge carriers (transfer in a unipolar fashion under phase-matching conditions, see [212]) from or feed energy (excitation resonantly transferred, see [220]) to the WSe_2 [40]. Left insets: Sketches of the over-simplified band alignments (CB/VB energy versus z coordinate). Shaded areas between "CB minimum" and "VB maximum" levels indicate the energy gap. Right insets: Stacking configurations for these examples. Colour code: grey, light grey, blue and red correspond to substrate, hBN, $MoSe_2$ and WSe_2, respectively. **a–f** Adapted with permission. [40] Copyright 2017 Elsevier B.V

Alignment of Bands in Heterostructures

Fundamental properties of resulting heterostructures depend crucially on the alignment of electronic bands at the interface (cf. insets in Fig. 2.9), thereby affecting the electron and hole wave-functions. In type-I (type-II) heterostructures, wave-functions are confined in a common (different) spatial region. Type-II heterostructures [210, 221–223] (cf. Fig. 2.9c, f), which can induce spatial charge separation [224, 225], lead to the desired indirect-exciton species [222, 226], as well as interlayer coupling strength and charge-transfer rate, which can be tuned by introducing barriers between the two layers of interest [221].

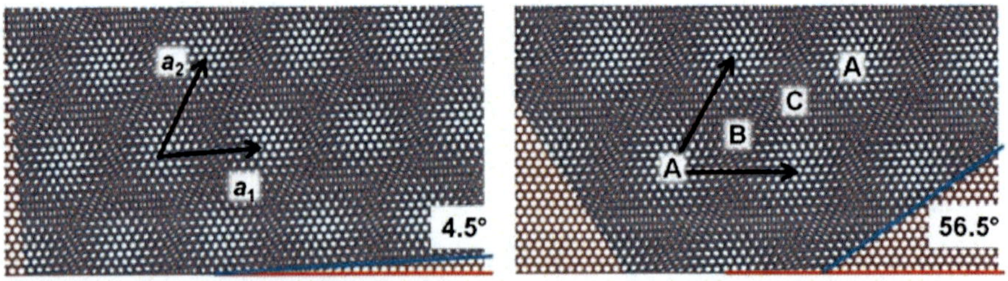

Fig. 2.10 Two schematic hexagonal lattices on top of each other with two different twist angles showing moiré patterns, representing typical scenarios for arbitrarily stacked TMDC heterobilayers. These angles were visually approximated in [228] for a manually-stacked monolayer–monolayer heterostructure in a study of intra- and interlayer excitons in such artificial TMDC heterobilayer with expected moiré landscape. Large superlattices are generally expected for small twists around 0 (aligned) and 60° (anti-aligned for systems with three-fold symmetry). The sketch indicates the translation unit vectors (a_1 and a_2) of the formed supercell and labels high-symmetry points of the heterosystem's supercell according to the nomenclature in [101]

Charge and Energy Transfer
The preservation of spin- and valley-degrees of freedom after charge transfer [210, 224] even renders these heterostructures applicable in the field of valleytronics. In fact, charge transfer in a type-II 2D heterostructure can be accompanied by a Förster-like energy transfer which prevails even with thin interlayer-hBN spacers [220]. In addition, electrical control of the exciton flux in vdWs heterostructures was recently demonstrated [227]. Hereby, existing studies clearly show the attractiveness and potential of heterostructuring for science and applications.

Stacking Sequence and Orientations
It is the choice of 2D materials and their stacking sequence that make 2D heterostructures unique in terms of optoelectronic properties and their use, e.g. for high-performance photodetectors (see e.g. [229, 230]) or efficient photovoltaics (e.g. [98, 231, 232] and references therein). It shall be noted that many of these applications can also benefit from in-plane as well as out-of-plane homojunctions of differently doped monolayers. Furthermore, owing to the emergence of interesting physics in twisted heterobilayers (sketched in Fig. 2.10), which introduce periodic potential landscapes that give rise to moiré minibands [99, 100], new concepts such as twistable electronics with dynamically rotatable heterostructures were recently envisaged [233].

hBN-Based Heterostructures
In addition to conventional monolayer–monolayer heterostructures, heterostructures of monolayers with hBN attracted considerable attention due to their impact on the excitonic signatures [57–59], as well as dynamics and phonon coupling [38, 40]. On the one hand, the emission linewidth considerably narrows when monolayer TMDCs are encapsulated by hBN. On the other hand, the lifetime, exciton–exciton annihilation behaviour and the exciton/trion coupling to phonons is altered, as reported among others in [40]. Moreover, the conditions obtained through hBN encapsulation

have provided access to a measurable optical dispersion (in the meV range, corresponding to effective masses of $< 10^{-3}\, m_0$)[8] of TMDC monolayer resonances [52], and indications of other effects, such as polaron formation (coupling of excitons to the optical-phonon bath) at elevated temperatures above 100 K [52, 53]. Recently, also the role of hBN encapsulation as a nonuniform environment configuration for monolayers on the exciton properties was discussed, showing experimentally that excitons in encapsulated TMDC monolayers exhibit a Rydberg-like behaviour in a fractional dimension between 2D and 3D [60].

Multistacks of TMDCs

Towards vertical structures consisting of multiple 2D materials, double wells of monolayers can be considered as an introduction into this domain, for instance to achieve stronger coupling phenomena in microcavities [65]. Recently, spatially indirect excitons (in the sense of dipolar excitons with electron and hole in separate layers) formed between two hBN-buffered MoS_2 monolayers were shown and compared to their intralayer counterparts [234].

Direct Gap Materials for Heterostructuring

MoS_2 and WS_2 are the most studied materials in the TMDC family. They feature a relatively bright PL spectrum due to their direct band gap in the monolayer limit. On the other hand, $MoSe_2$ and WSe_2 monolayers have smaller band gaps and higher electron mobility compared to their sulfide analogues [235], and exhibit a higher PL intensity compared to MoS_2 monolayers due to a high non-radiative recombination of electrons and holes in MoS_2 [231, 236]. Almost always, higher mobility leads to better device performance rendering $MoSe_2$ and WSe_2 more promising for practical applications, including transistors and optoelectronic detectors. However, an increasing number of layers for these TMDCs induces a transition from a direct to an indirect band gap, thereby reducing the PL intensity significantly [133]. Thus, it is presumed that their direct-gap character can get spoiled within hybridised heterobilayers (i.e. for monolayer–monolayer heterosystem resembling a homobilayer-like situation). In comparison, $MoSe_2$ and $MoTe_2$ retain their direct band gaps in a regime of two layers (tellurides up to a few layers) [237, 238], leading to higher PL intensities beyond the monolayer regime compared to MoS_2 and other TMDCs in monolayer heterostructures. Accordingly, they could become interesting components for heterobilayers.

Telluride-Based Heterostructures

Early considerations revealed that the combination of $MoSe_2$ and $MoTe_2$ monolayers results in type-II heterostructures (see for instance the calculated alignments in [93–96]), which find applications as photovoltaic devices and detectors. In contrast, the combination of WSe_2 and $MoTe_2$ materials is expected to provide a type-I heterostructure (see aforementioned theoretical works) with direct excitons, which has applications in low-threshold diode lasers. Moreover, owing to the aforementioned direct-gap nature of mono-to-few-layer tellurides, they likely retain their direct gap even in heterostructures of bilayer tellurides with other monolayer TMDCs.

[8] Free electron mass m_0, natural constant.

Band Gap Engineering
An incredible effort is going on worldwide to harvest the yield from 2D heterostructuring and the anticipated "band-gap engineering" (for applications) using the existing and known materials at ones disposal. A few of these outcomes are highlighted in a number of timely reviews [7, 56, 98, 205, 239]. One of such heterostructuring endeavours is reflected by one of the author's projects (DFG RA2841/5-1), in which 2D semiconductor type-I and type-II heterostructures incorporating $MoTe_2$ are primarily investigated.

Harnessing Post-TMDCs—Golden GaTe Example
Novel nanomaterials commonly are based on existing bulk crystalline systems, which have been later rediscovered in the nanoscale as interesting candidates for optoelectronics with 2D materials. Similar to TMDCs, group-III monochalcogenides exhibit interesting properties as layered materials. One example is GaTe, which remarkably behaves oppositely to TMDCs when thinned down to a monolayer. It becomes an indirect semiconductor as monolayer due to a change in the crystal structure (transition from a monoclinic to a hexagonal phase) [240], whereas as few-layer system it features a direct gap [241]. Thereby, it offers direct-gap emission for a wide range of thicknesses. When decorated with gold nanoparticles, which efficiently grow on facets of multi-layered GaTe 2D crystals due to the defect density on the material surface, GaTe has been shown to act as a very sensitive probe for certain aromatic molecules and could become a promising candidate for surface-enhanced Raman spectroscopy [242].

Twist Angle and Moiré Superlattices
Nevertheless, the world of heterostructure studies with 2D materials is far more rich than the mere variability of material combinations. Twist angles are known to play a major role and promise unprecedented properties from monolayer–monolayer stacks. While phase-space mismatch introduced by an interlayer twist has a direct impact on charge-transfer processes [212], more changes to the system arise from moiré potential landscapes originating from the lattice mismatch and/or twist angles between adjacent layers [99, 100, 107–110]. Thereby, periodic nanoislands are formed acting as perfect superlattices of quantum dots with supercell symmetry points affecting both interlayer spacings and energy scales locally [101, 102]. Thus, the recently reported moiré exciton states unravel interesting optical and electronic properties of bilayer systems [102–105, 214]. Current signatures for testbed heterostructures stacked and probed in Marburg indicate strong oscillator strengths of new excitonic species in possible moiré landscapes via optical absorption spectra [228] and very long PL lifetimes (see Figs. 5.11 and 5.9, respectively), triggering further investigations. So far, the observation of energy-split bilayer inter- and intralayer excitons has emerged as a spectral fingerprint for the formation of such superlattices as discussed in the literature.

High-Symmetry Bilayer Configurations
In addition to arbitrary twist angles, high-symmetry stacking configurations [19, 243] open up new possibilities to alter the electronic band structure of homobilayers

[216, 217, 244], enabling to obtain spin–layer or spin–valley locking scenarios for the naturally (AA') [219, 245] and artificially (AB)[9] stacked TMDCs, respectively [219]. A recent study by the author's team with the help of collaboration partners showed among other spectral signatures that a noticeable difference of circular polarisation anisotropy is evidenced for selectively CVD-grown AA' and AB configured homobilayers (corresponding schematics and band structure calculations displayed in Fig. 2.11), with the latter exhibiting a higher value [219]; for hBN-supported AA' and AB flakes, for which dephasing induced by defects and doping is reduced (cf. [7]), a more pronounced linear and circular polarisation anisotropy is obtained [219]. Furthermore, their theory predicted a lifted degeneracy between layers, observable in reflection-contrast experiments. These distinct bilayer cases could be for instance interesting for future 'valleytronic' or even 'layertronic' device schemes.

2.5 Photonics and Optoelectronics of 2D Semiconductor TMDCs

Besides the recent progress in applications of 2D semiconductor TMDCs with an emphasis on strong excitonic effects in monolayers and heterostructures [21, 205], the mere electronic and optical properties of 2D materials, as well as the spin- and valley-dependent properties, caught the attention of various research communities worldwide (see for instance the activities described in [16, 17, 111, 246]). It is worth noting that, while most of the 2D research takes place sailing the graphene 'flagship', various accompanying 2D 'catamarans' have been recently explored and have shown a strong potential for next generation optoelectronics and materials sciences (see for instance [5, 7, 12, 18, 20, 56, 92, 98, 232, 239]).

Optoelectronic Application Potential

Optoelectronic applications are understood to gain significantly from the research on 2D materials in the future and have still remained largely untapped. The following examples may emphasise the great potential they exhibit. Naturally, the high carrier mobilities render 2D semiconductors ideal candidates for efficient photodetector and transistor schemes. Furthermore, the direct gap nature, pronounced light–matter interactions and the controllable optoelectronic properties clearly promote the use in semiconductor photonic devices. In addition, complex excitonic species, such as charged excitons (trions) and exciton molecules (e.g. biexcitons), can enrich the electrical transport of photoexcitation [247].

Benefits for Photonic 2D Devices

Remarkably, the strong exciton binding energies and the correspondingly large exciton oscillator strength in TMDCs can enable room-temperature strong light–matter

[9] Note that in the literature, the use and definition of bilayer abbreviations may not be consistent, so that care has to be taken when comparing different works referring to AA, AA', AB and other stacking situations.

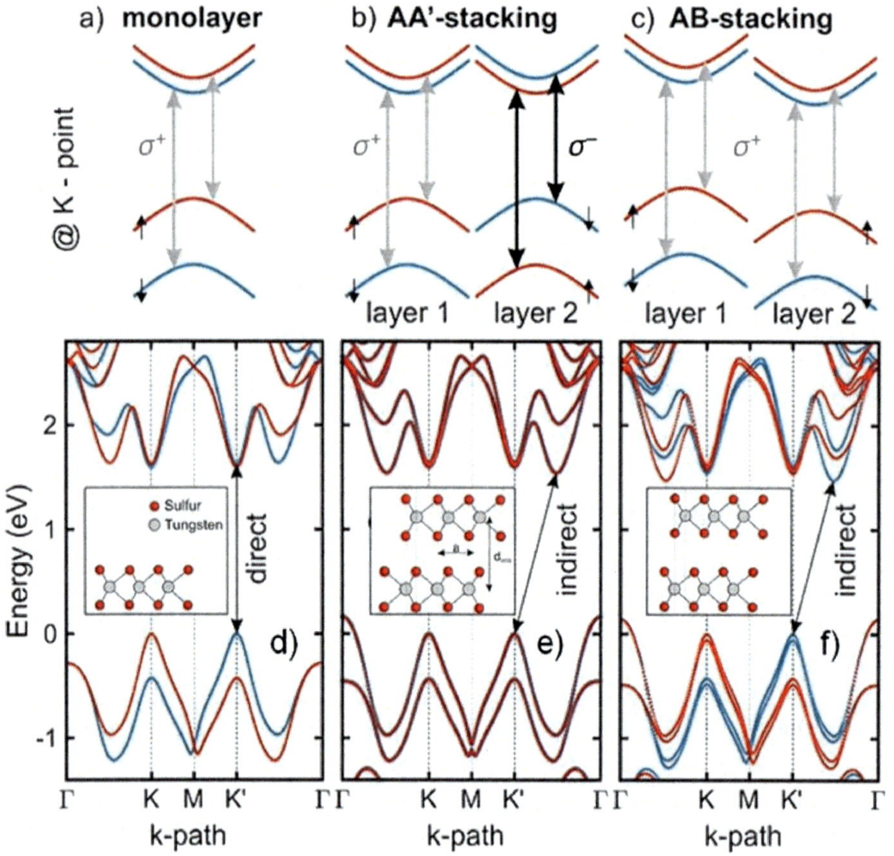

Fig. 2.11 WS_2 in different highly-symmetric stacking configurations, the naturally occurring AA'-stacked type of bilayer and an artificially-stacked AB type. Sketch of the valley-selective optical transitions (circularly polarised light, $\sigma^{\pm}$, represented by grey/black vertical arrows) at the K point between valence and conduction bands for the single-layer (**a**), AA' (**b**) and AB (**c**) bilayer configurations. Spin orientations up (red) and down (blue) are associated with each band. While here only the behaviour at the K point is described, the spin situation and polarisation selectivity at the K' point behaves simply oppositely. The calculated band structures of the three systems (**a–c**) are depicted in (**d–f**) in the same order. Arrows indicate direct/indirect gaps. Insets depict schematically the stacking alignment with in-plane constant a and mean interlayer (tungsten–tungsten) distance d_{WW}. **a–f** and insets adapted with permission. [219] Copyright 2019 American Chemical Society

coupling [64–66] and promise the achievement of exciton and polariton condensation at relatively high temperatures (up to room temperature for polaritons) [207, 248, 249]. These systems can become ideal testbeds for many-body phenomena, e.g. in optical microcavities, while others exploit 2D-material microcavities for second-harmonic generation [250, 251]. Moreover, the valley- and spin-dependent properties of TMDCs are attractive for potential applications in information processing [16, 17, 246], promising novel concepts that enable optical "valleytronics"—using principles of spintronics [252]—such as 'spin-and-valley laser' (making use of valley- and spin-dependent optical selection rules). These novel device schemes promise for instance efficient transmission of information over long distances via polarised photons.

Overall, it is clear that the integration of TMDCs into photonic structures and circuits becomes increasingly important. This is also benefiting from the generally much-relaxed lattice matching constraints compared to other semiconductor crystalline systems regarding deposition of materials onto each other (e.g. onto silicon, silica or polymer substrates). Besides the employment of mere monolayer and bilayer materials as saturable absorbers [253–255], which can be used for (passive) ultrafast modulation of intensities, TMDCs have become attractive as active material (see below) or in connection with waveguides [256–259], photonic crystals [11, 260], nanoresonators [261] and in-plane grating/cavity elements [156], promising advances in integrated photonics [9].

Roadmap Towards 2D Photonics

In order to pave the way for these concepts to become practical for applications, current challenges have to be addressed, where strong interaction between materials sciences and physics will be key to a successful development of this field:

Firstly, scalable production of 2D materials with high quality and crystallinity is still a problem, although several methods have been developed to grow, transfer, as well as isolate monolayers and continuously improved to deliver the desired tailor-made 2D crystals [5, 7].

Secondly, the improvement of PL quantum yield (QY) (cf. Fig. 2.12a, also see for instance [142, 156, 262]) is important for optical structures that incorporate TMDCs for light-emitting devices [14], polariton physics [248, 249] (cf. Fig. 2.12b, c) and ultralow threshold lasing [70–72, 74, 263] (cf. Fig. 2.12d).

Thirdly, the rich exciton physics of TMDCs has to be further explored [21], understood and utilised.

Nevertheless, a plethora of studies can be enabled on the lab scale, as the following examples of 2D–microcavity research reflect.

2D-Materials–Grating Structures for Integrated Photonics

Recently, circular grating structures with central half-wavelength cavity gap, also referred to as circular in-plane distributed-Bragg-reflector-based optical microcavity (CIDBROM, similarly possible in linear, parallel grating configuration, as PIDBROM), have been investigated with the aim to deliver strong in-plane confinement of light [156]. Such optical confinement was achieved by the lateral height-profile modulation in a dielectric substrate (see Fig. 2.13). Simultaneously good out-of-plane confinement was provided according to simple optical interference considerations as a consequence of the air-trench depths (also see the Supporting Information of [156]). This approach, which used air–GaP CIDBROMs in combination with monolayer WS_2 on buffering thin hBN, not only aims at the Purcell enhancement and directionality of output offered by the radial-symmetric "bull's-eye" Bragg grating structure (also studied for different wavelength regions as well as applications in the literature [264–268]). It also aims at the maximisation of light–matter interaction in terms of optical in-coupling and out-coupling in the central region of the designed ring pattern on top of the grating with its high refractive index contrast.

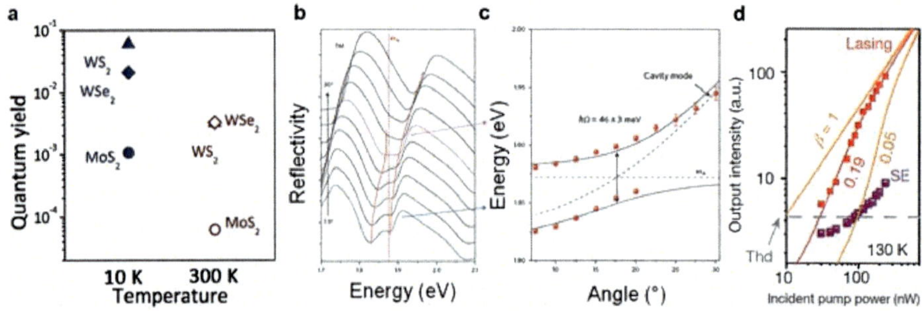

Fig. 2.12 Examples of 2D photonics, which have made use of optical microcavities in combination with monolayer TMDCs. **a** Quantum yield as a function of temperature. Data shown for 300 and 10 K for three prominent TMDC materials. Reproduced with permission. [72] (Supporting Information) Copyright 2015 Springer Nature. **b** Typical detuning series in waterfall representation of reflectivity spectra recorded from a 2D–microcavity system in the strong light–matter coupling regime. The anticrossing behaviour is a signature of the establishment of (cavity–)polariton modes, hybrids of light and matter, also known as excitons dressed with photons. Accordingly, the extracted mode energies with characteristic Rabi splitting (here of 46 meV at a finite angle situation) are displayed in (**c**). **b**, **c** Adapted with permission. [64] Copyright 2014 Springer Nature. **d** Double-logarithmic input–output curve for a 2D microlaser (red) in comparison to the material's spontaneous emission (SE, violet symbols). Reproduced with permission. [70] Copyright 2015 Springer Nature

Using scattering-type optical scanning near-field microscopy (s-SNOM, see raster-scanned image in Fig. 2.13d), reflection-contrast measurements and luminescence (μ-PL) investigations, the successful demonstration of a markedly enhanced out-of-plane emission from such monolayer structure indicates the effectiveness of this CIDBROM structure; such structure indeed served as a model system for in-plane confinement and enhanced vertical PL extraction, as well as local, spectrally-sensitive absorption enhancement [156].

These experiments were performed with eyes towards future waveguide-coupled on-chip applications, for example, for optical integrated circuitry and valleytronics. Both, micro/nano-sized 2D LEDs and 2D photovoltaics may benefit from these established interference-based optical confinement approaches. The author's ongoing investigations aim at the demonstration of tailored emission control of out-of-plane dipole resonances (as a stronger confinement is predicted by simulated fields for out-of-plane than in-plane fields, see Supporting Information of [156]), such as interlayer excitons in heterostructures and z-mode (out-of-plane dipole, "grey") excitons in monolayers, as well as at improved absorption/light–matter-interaction capabilities, e.g. for photovoltaics or nonlinear optics.

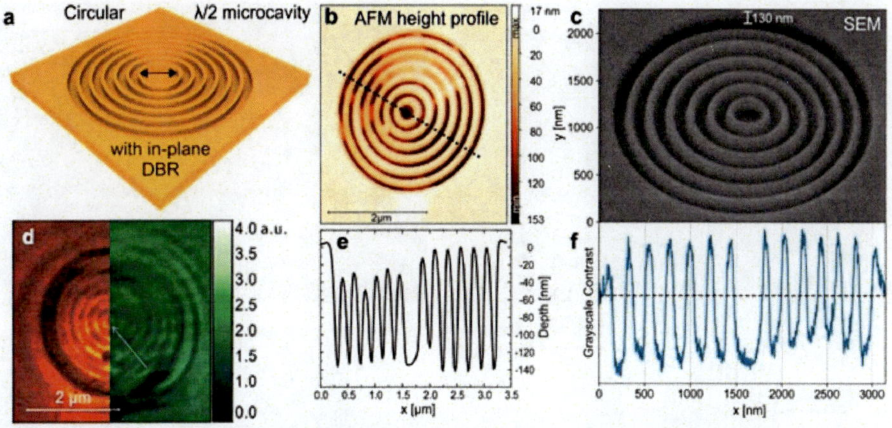

Fig. 2.13 **a** Sketch of a circular in-plane distributed-Bragg-reflector-based optical microcavity. **b** Atomic-force micrograph and **c** scanning-electron micrograph of a patterned substrate. **e** and **f** show corresponding cross sections to **b** and **c**, respectively. **d** Two s-SNOM measurements (false-colour intensity maps) at different probe-laser wavelengths manually combined in one diagram, with the near-IR wavelength off-resonant with regard to the structure's design range shown in red (left) and the in-range green wavelength data displayed in green (right). The white arrow indicates the central part of the ring system. Remarkably, the profiles show opposite radial intensity modulation, which is in line with the design considerations. **a–f** Adapted with permission. [156] Copyright 2019 American Chemical Society

2.5.1 *Strong Light–Matter Interaction and Lasing with 2D Materials*

The room-temperature observation of strong light–matter coupling and nonlinearities in that regime in practical emitter–resonator systems has required the use of large-band-gap materials, which have promised correspondingly high binding energies of Wannier-Mott excitons [269], or alternatively Frenkel excitons provided by organic materials, which are strongly localised and naturally deliver a high oscillator strength [270, 271]. In this context, 2D materials offer a new and unique testbed for light–matter coupling experiments at elevated temperatures, as pioneering work showed [64].

The Wave of Hybrid Light–Matter States

The strong interest in TMDCs is explained by their unique properties as nearly perfect quantum wells with exceptionally high binding energies of strongly confined excitons. This has led to further breakthroughs in this domain, particularly using tunable and open cavity designs [65, 66, 69] (example shown in Fig. 2.14). Consecutively, many reports followed in an effort to explore this field deeper in recent times [67, 68, 272]. Up-to-date, different cavity designs have been discussed and different TMDCs employed, ranging from WS_2 to $MoSe_2$ [249]. Naturally, these microcavities can be also designed to work with 2D perovskites [273]. While commonly the Fabry–Pérot-type planar microcavities are used to obtain cavity–polaritons, some

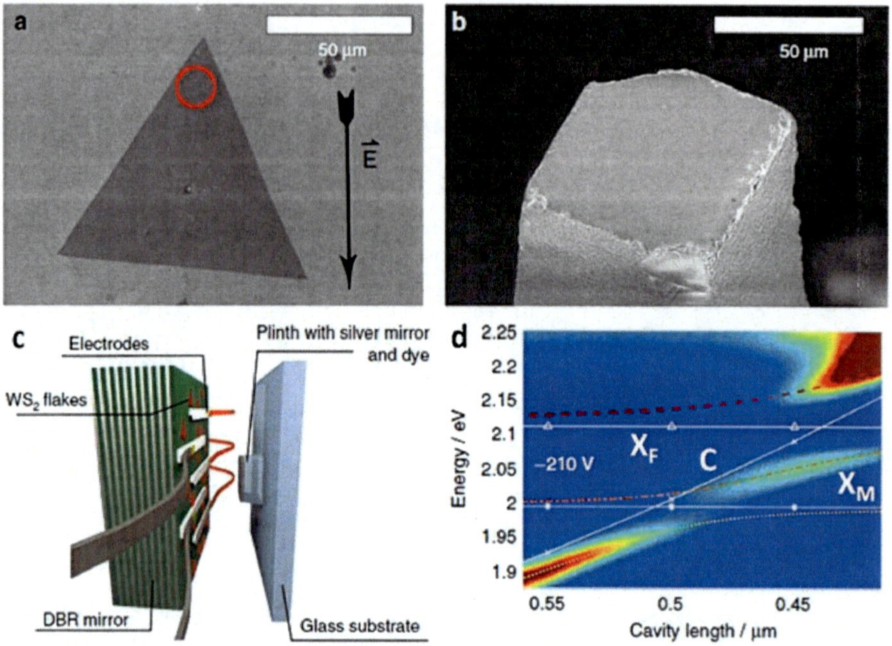

Fig. 2.14 Light–matter coupling experiment employing Mott–Wannier (M) excitons (X) in an inorganic 2D semiconductor simultaneously with Frenkel (F) excitons in an organic material, achieving hybridisation of up to three modes (resulting in hybrid polaritons). A CVD-grown monolayer WS_2 flake (**a**) is employed on a dielectric mirror (**c**). Electrodes enable exciton energy tuning by lateral electric fields. The tunable microcavity system is completed by an opposing silver mirror (top facet of a plinth, shown in (**b**)). **d** Spectral signatures of the hybrid Frenkel–Mott polaritons obtained with lateral bias of −210 V in transmission geometry as a function of the cavity length. **a**–**d** Reproduced under the terms of the CC-BY 4.0 Licence (http://creativecommons.org/licenses/by/4.0/). [69] Copyright 2017 The Author(s), published by Springer Nature

examples employ strongly-confined light fields in plasmonic nanocavities with 2D materials [67, 274].

Flexible and Tunable Coupling Experiments

With regard to flexibility in the coupling situation and a continuous tunability of the strength of light–matter interaction around the exceptional point of the system, work with 2D-materials–open-cavity concepts has remained promising and was further explored in this context [275]. Recently, also fibre-tip based microcavities gained popularity [276] which may offer new possibilities in combination with 2D materials when it comes to low-mode-volume cavities with tunable 2D-membrane positions [275].

Towards Valley Polaritons and Their Condensates

Particularly the domain of valley-polarisation has become a hot topic for polariton research, given access to valley-selective excitation and detection schemes and the outlook for “optical valleytronics”. Recent reports showed pronounced circular polarisation of 2D polaritons up to room temperature and explored optical control of polaritons in their microcavities [272, 277–280]. Still, further optimisation of these systems is needed to obtain polariton condensates, the demonstration of which is basically imminent given all the recent achievements in 2D-materials-based microcavities.

Ultralow-Threshold Lasing

An equally exciting research direction arose from the efforts to achieve ultralow-threshold lasing with 2D semiconductors. Here, the Purcell effect for a gain medium coupled to an optical micro- or nanocavity can significantly enhance the spontaneous emission rate, which has important consequences for the lasing threshold [281]. Ultralow-threshold optically-pumped lasing has been reported recently in a coupled monolayer-WSe_2–photonic-crystal high-quality[10] (high-Q) nanocavity system under a continuous-wave (CW) pumping for a wide temperature range below 250 K [70] (see Fig. 2.15 (top left box)). Also, room-temperature CW lasing from oxygen-plasma-treated few-layer MoS_2 coupled to an optical ring resonator was reported, making use of high-Q whispering-gallery modes of the microcavity [71] (see Fig. 2.15 (bottom box)). Similarly, lasing in a high-Q optical ring resonator coupled to monolayer WS_2 was reported under pulsed optical excitation at low temperature [72] (see Fig. 2.15 (top right box)). However, the lasing thresholds in these studies drastically differ and the unambiguous demonstration of a transition to lasing is still pending. A comprehensive nanolaser demonstration with profound threshold characterisation can be for instance achieved with the help of photon statistics measurements and coherence studies.

Rich Photonics—from Optical Gain to Quantum Emitters

Although gain has been predicted by the study of population inversion in WS_2 [81] (that is in the regime of an uncorrelated electron–hole plasma), ultralow-threshold lasers usually exploit the Purcell effect [282] with excitonic species in quantum structures [283, 284]. These species can range from excitons, trions and biexcitons to bound excitons and exciton–polaritons in TMDCs depending on the temperature range, sample properties and environment (cf. [40, 47, 52, 87]). Even defect states can play a significant role with regard to TMDC emission, as their use for single-photon emission with 2D materials reveals [285–289]. These examples strongly motivate to explore the rich application potentials in the field of (nano)photonics.

Optomechanical Experiments Hardly Touched

Ultimately, beyond the employment of 2D–cavity systems for optoelectronic coupling, 2D-membrane mechanical resonators and miniature high-finesse optical cavities can enable sophisticated (quantum-)optomechanical experiments—a hardly explored domain with huge demand for fundamental research [290].

[10]The quality of a resonator or microcavity is expressed by $Q = E/\gamma$, the quality factor, with mode energy E and mode linewidth γ in full-width at half-maximum (FWHM).

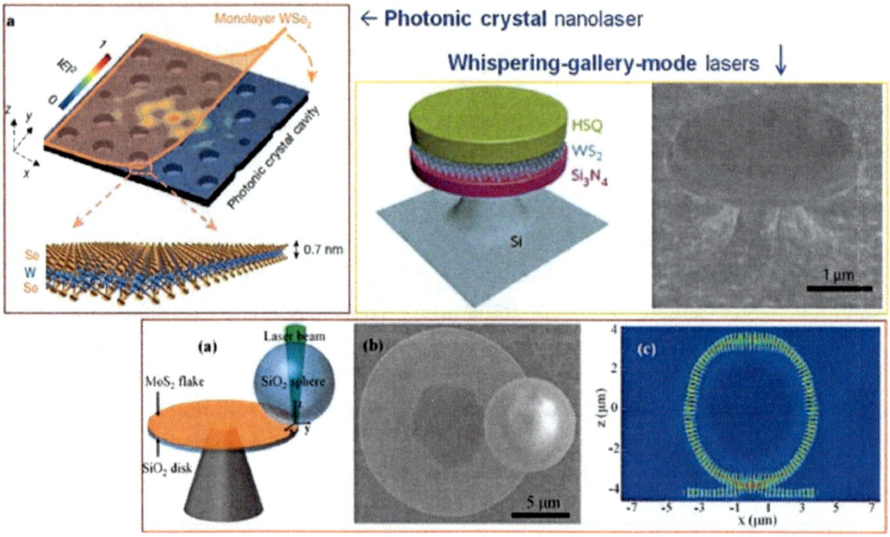

Fig. 2.15 Three early examples of microcavity lasers based on 2D-semiconductor gain regions. Top left box: A photonic-crystal cavity with monolayer, also see Fig. 2.3. Reproduced with permission. [70] Copyright 2015 Springer Nature. Top right box: Sketch of a whispering-gallery-mode laser based on a processed Si_3N_4 microdisk resonator on silicon substrate (left) and corresponding SEM image of the actual device (right). The 2D semiconductor deposited onto the disk is capped by a transparent medium (HSQ: hydrogen silsesquioxane) for an optimised out-of-plane mode distribution profile, i.e. for the maximisation of the laser mode's electric field in the plane of the 2D gain medium. Reproduced with permission. [72] Copyright 2015 Springer Nature. Bottom box: **a** Sketch of a microlaser based on a combination of two whispering-gallery-mode resonators (a sphere on a disk) with monolayer gain material deposited onto the disk. **b** Top-view SEM image of the device. **c** Calculated mode pattern in a cross-sectional view on the sphere (circular part) attached to the disk (flat part). As this image shows, this approach allows to maximise the laser mode's electric field at the position of the gain material in-between the two attached cavities at the intersection owing to the joint mode pattern. Moreover, owing to the different free-spectral ranges (FSR) of the two resonators, a strong mode reduction is achieved, since the standing-wave pattern must match both resonator conditions. Reproduced with permission. [71] Copyright 2015 American Chemical Society

References

1. K.S. Novoselov, A.K. Geim, S.V. Morozov, D. Jiang, Y. Zhang, S.V. Dubonos, I.V. Grigorieva, A.A. Firsov, Electric field effect in atomically thin carbon films. Science **306**, 666–669 (2004)
2. K.S. Novoselov, D. Jiang, F. Schedin, T.J. Booth, V.V. Khotkevich, S.V. Morozov, A.K. Geim, Two-dimensional atomic crystals. Proc. Natl. Acad. Sci. **102**, 10451–10453 (2005)
3. K.F. Mak, C. Lee, J. Hone, J. Shan, T.F. Heinz, Atomically thin MoS_2: a new direct-gap semiconductor. Phys. Rev. Lett. **105**, 136805 (2010)
4. A. Splendiani, L. Sun, Y. Zhang, T. Li, J. Kim, C.-Y. Chim, G. Galli, F. Wang, Emerging photoluminescence in monolayer MoS_2. Nano Lett. **10**, 1271–1275 (2010)
5. Z. Lin, A. McCreary, N. Briggs, S. Subramanian, K. Zhang, Y. Sun, X. Li, N. J. Borys, H. Yuan, S. K. Fullerton-Shirey, A. Chernikov, H. Zhao, S. McDonnell, A.M. Lindenberg, K. Xiao, B.J. LeRoy, M. Drndić, J.C.M. Hwang, J. Park, M. Chhowalla, R.E. Schaak, A. Javey, M.C. Hersam, J. Robinson, M. Terrones, 2D materials advances: from large scale synthesis and

controlled heterostructures to improved characterization techniques, defects and applications. 2D Mater **3**, 042001 (2016)
6. D. Akinwande, C.J. Brennan, J.S. Bunch, P. Egberts, J.R. Felts, H. Gao, R. Huang, J.S. Kim, T. Li, Y. Li, K.M. Liechti, N. Lu, H.S. Park, E.J. Reed, P. Wang, B.I. Yakobson, T. Zhang, Y.W. Zhang, Y. Zhou, Y. Zhu, A review on mechanics and mechanical properties of 2D materials–Graphene and beyond. Extrem. Mech. Lett. **13**, 42–77 (2017)
7. D. Rhodes, S.H. Chae, R. Ribeiro-Palau, J. Hone, Disorder in van der Waals heterostructures of 2D materials. Nat. Mater. **18**, 541–549 (2019)
8. K.S. Novoselov, Graphene: materials in the Flatland (Nobel lecture). Angew. Chem. Int. Ed. Engl. **50**, 6986–7002 (2011)
9. F. Xia, H. Wang, D. Xiao, M. Dubey, A. Ramasubramaniam, Two-dimensional material nanophotonics. Nat. Photonics **8**, 899–907 (2014). https://www.nature.com/articles/nphoton.2014.271
10. Q.H. Wang, K. Kalantar-Zadeh, A. Kis, J.N. Coleman, M.S. Strano, Electronics and optoelectronics of two-dimensional transition metal dichalcogenides. Nat. Nanotechnol. **7**, 699–712 (2012)
11. W. Wu, L. Wang, Y. Li, F. Zhang, L. Lin, S. Niu, D. Chenet, X. Zhang, Y. Hao, T.F. Heinz, J. Hone, Z.L. Wang, Piezoelectricity of single-atomic-layer MoS_2 for energy conversion and piezotronics. Nature **514**, 470–474 (2014)
12. D. Jariwala, V.K. Sangwan, L.J. Lauhon, T.J. Marks, M.C. Hersam, Emerging device applications for semiconducting two-dimensional transition metal dichalcogenides. ACS Nano **8**, 1102–1120 (2014)
13. F.H.L. Koppens, T. Mueller, P. Avouris, A.C. Ferrari, M.S. Vitiello, M. Polini, Photodetectors based on graphene, other two-dimensional materials and hybrid systems. Nat. Nanotechnol. **9**, 780–793 (2014)
14. F. Withers, O.D. Pozo-Zamudio, A. Mishchenko, A.P. Rooney, A. Gholinia, K. Watanabe, T. Taniguchi, S.J. Haigh, A.K. Geim, A.I. Tartakovskii, K.S. Novoselov, Light-emitting diodes by band-structure engineering in van der Waals heterostructures. Nat. Mater. **14**, 301–306 (2015)
15. K. Kalantar-zadeh, J.Z. Ou, T. Daeneke, M.S. Strano, M. Pumera, S.L. Gras, Two-dimensional transition metal dichalcogenides in biosystems. Adv. Funct. Mater. **25**, 5086–5099 (2015)
16. K.F. Mak, J. Shan, Photonics and optoelectronics of 2D semiconductor transition metal dichalcogenides. Nat. Photonics **10**, 216–226 (2016)
17. J.R. Schaibley, H. Yu, G. Clark, P. Rivera, J.S. Ross, K.L. Seyler, W. Yao, X. Xu, Valleytronics in 2D materials. Nat. Rev. Mater. **1**, 16055 (2016)
18. H. Tian, M.L. Chin, S. Najmaei, Q. Guo, F. Xia, H. Wang, M. Dubey, Optoelectronic devices based on two-dimensional transition metal dichalcogenides. Nano Res. **9**(6), 1543–1560 (2016)
19. S. Wu, Y. Du, S. Sun, Transition metal dichalcogenide based nanomaterials for rechargeable batteries. Chem. Eng. J. **307**, 189–207 (2017)
20. M. Wang, E.-H. Yang, THz applications of 2D materials: graphene and beyond. Nano-Struct. Nano-Objects **15**, 107–113 (2018)
21. G. Wang, A. Chernikov, M.M. Glazov, T.F. Heinz, X. Marie, T. Amand, B. Urbaszek, Colloquium: excitons in atomically thin transition metal dichalcogenides. Rev. Mod. Phys. **90**, 21001 (2018)
22. K. He, N. Kumar, L. Zhao, Z. Wang, K.F. Mak, H. Zhao, J. Shan, Tightly bound excitons in monolayer WSe_2. Phys. Rev. Lett. **113**, 026803 (2014)
23. A. Chernikov, T.C. Berkelbach, H.M. Hill, A. Rigosi, Y. Li, O.B. Aslan, D.R. Reichman, M.S. Hybertsen, T.F. Heinz, Exciton binding energy and nonhydrogenic Rydberg series in monolayer WSe_2. Phys. Rev. Lett. **113**, 076802 (2014)
24. Z. Ye, T. Cao, K. OBrien, H. Zhu, X. Yin, Y. Wang, S.G. Louie, X. Zhang, Probing excitonic dark states in single-layer tungsten disulphide. Nature **513**, 214–218 (2014)
25. H.M. Hill, A.F. Rigosi, C. Roquelet, A. Chernikov, T.C. Berkelbach, D.R. Reichman, M.S. Hybertsen, L.E. Brus, T.F. Heinz, Observation of excitonic Rydberg states in monolayer MoS_2 and WS_2 by photoluminescence excitation spectroscopy. Nano Lett. **15**, 2992–2997 (2015)

26. K.F. Mak, K. He, C. Lee, G.H. Lee, J. Hone, T.F. Heinz, J. Shan, Tightly bound trions in monolayer MoS_2. Nat. Mater. **12**, 207–211 (2013)
27. J.S. Ross, S. Wu, H. Yu, N.J. Ghimire, A.M. Jones, G. Aivazian, J. Yan, D.G. Mandrus, D. Xiao, W. Yao, X. Xu, Electrical control of neutral and charged excitons in a monolayer semiconductor. Nat. Commun. **4**, 1–6 (2013)
28. Y. You, X.-X. Zhang, T.C. Berkelbach, M.S. Hybertsen, D.R. Reichman, T.F. Heinz, Observation of biexcitons in monolayer WSe_2. Nat. Phys. **11**, 477–481 (2015)
29. X.-X. Zhang, Y. You, S.Y.F. Zhao, T.F. Heinz, Experimental evidence for dark excitons in monolayer WSe_2. Phys. Rev. Lett. **115**, 257403 (2015)
30. K.F. Mak, K. He, J. Shan, T.F. Heinz, Control of valley polarization in monolayer MoS_2 by optical helicity. Nat. Nanotechnol. **7**, 494–498 (2012)
31. T. Cao, G. Wang, W. Han, H. Ye, C. Zhu, J. Shi, Q. Niu, P. Tan, E. Wang, B. Liu, J. Feng, Valley-selective circular dichroism of monolayer molybdenum disulphide. Nat. Commun. **3**, 1–5 (2012)
32. A.M. Jones, H. Yu, N.J. Ghimire, S. Wu, G. Aivazian, J.S. Ross, B. Zhao, J. Yan, D.G. Mandrus, D. Xiao, W. Yao, X. Xu, Optical generation of excitonic valley coherence in monolayer WSe_2. Nat. Nanotechnol. **8**, 634–638 (2013)
33. L.M. Schneider, S.S. Esdaille, D.A. Rhodes, K. Barmak, J.C. Hone, A. Rahimi-Iman, Optical dispersion of valley-hybridised coherent excitons with momentum-dependent valley polarisation in monolayer semiconductor. 2D Mater. **8**, 015009 (2020)
34. C. Poellmann, P. Steinleitner, U. Leierseder, P. Nagler, G. Plechinger, M. Porer, R. Bratschitsch, C. Schüller, T. Korn, R. Huber, Resonant internal quantum transitions and femtosecond radiative decay of excitons in monolayer WSe_2. Nat. Mater. **14**, 889–893 (2015)
35. T. Jakubczyk, V. Delmonte, M. Koperski, K. Nogajewski, C. Faugeras, W. Langbein, M. Potemski, J. Kasprzak, Radiatively limited dephasing and exciton dynamics in MoS_2 monolayers revealed with four-wave mixing microscopy. Nano Lett. **16**, 5333–5339 (2016)
36. C. Ruppert, A. Chernikov, H.M. Hill, A.F. Rigosi, T.F. Heinz, The role of electronic and phononic excitation in the optical response of monolayer WS_2 after ultrafast excitation. Nano Lett. **17**, 644–651 (2017)
37. P. Steinleitner, P. Merkl, P. Nagler, J. Mornhinweg, C. Schüller, T. Korn, A. Chernikov, R. Huber, Direct observation of ultrafast exciton formation in a monolayer of WSe_2. Nano Lett. **17**, 1455–1460 (2017)
38. L.M. Schneider, S. Lippert, J. Kuhnert, D. Renaud, K.N. Kang, O. Ajayi, M.-U. Halbich, O.M. Abdulmunem, X. Lin, K. Hassoon, S. Edalati-Boostan, Y.D. Kim, W. Heimbrodt, E.H. Yang, J.C. Hone, A. Rahimi-Iman, The impact of the substrate material on the optical properties of 2D WSe_2 monolayers. Semiconductors **52**, 565–571 (2018). https://doi.org/10.1134/S1063782618050275
39. L.M. Schneider, J. Kuhnert, S. Lippert, D. Renaud, W. Heimbrodt, O. Ajayi, Y.D. Kim, J.C. Hone, K.N. Kang, E.-H. Yang, A. Rahimi-Iman, Density-dependent excitonic properties and dynamics in 2d heterostructures consisting of boron nitride and monolayer or few-layer tungsten diselenide, in *Micro- and Nanotechnology Sensors, Systems, and Applications X*, vol. 10639, p. 106390V, International Society for Optics and Photonics (2018)
40. L.M. Schneider, S. Lippert, J. Kuhnert, O. Ajayi, D. Renaud, S. Firoozabadi, Q. Ngo, R. Guo, Y.D. Kim, W. Heimbrodt, J.C. Hone, A. Rahimi-Iman, The influence of the environment on monolayer tungsten diselenide photoluminescence. Nano-Struct. Nano-Objects **15**, 84–97 (2018), online 2017. https://doi.org/10.1016/j.nanoso.2017.08.009
41. J. Fu, J.M.R. Cruz, F. Qu, Valley dynamics of different trion species in monolayer WSe_2. Appl. Phys. Lett. **115**, 082101 (2019)
42. J. Zipfel, M. Kulig, R. Perea-Causín, S. Brem, J.D. Ziegler, R. Rosati, T. Taniguchi, K. Watanabe, M.M. Glazov, E. Malic, A. Chernikov, Exciton diffusion in monolayer semiconductors with suppressed disorder. Phys. Rev. B **101**(11), 1–20 (2020)
43. M. Massicotte, F. Vialla, P. Schmidt, M.B. Lundeberg, S. Latini, S. Haastrup, M. Danovich, D. Davydovskaya, K. Watanabe, T. Taniguchi, V.I. Falko, K.S. Thygesen, T.G. Pedersen, F.H.L. Koppens, Dissociation of two-dimensional excitons in monolayer WSe_2. Nat. Commun. **9**, 1633 (2018)

44. J.A. Schuller, S. Karaveli, T. Schiros, K. He, S. Yang, I. Kymissis, J. Shan, R. Zia, Orientation of luminescent excitons in layered nanomaterials. Nat. Nanotechnol. **8**, 271–276 (2013)
45. G. Wang, C. Robert, M.M. Glazov, F. Cadiz, E. Courtade, T. Amand, D. Lagarde, T. Taniguchi, K. Watanabe, B. Urbaszek, X. Marie, In-plane propagation of light in transition metal dichalcogenide monolayers: optical selection rules. Phys. Rev. Lett. **119**, 047401 (2017)
46. A. Steinhoff, M. Florian, A. Singh, K. Tran, M. Kolarczik, S. Helmrich, A.W. Achtstein, U. Woggon, N. Owschimikow, F. Jahnke, X. Li, Biexciton fine structure in monolayer transition metal dichalcogenides. Nat. Phys. **14**, 1199–1204 (2018)
47. L.M. Schneider, S.S. Esdaille, D.A. Rhodes, K. Barmak, J.C. Hone, A. Rahimi-Iman, Direct measurement of the radiative pattern of bright and dark excitons and exciton complexes in encapsulated tungsten diselenide. Sci. Rep. **10**, 8091 (2020). https://doi.org/10.1038/s41598-020-64838-z
48. G. Wang, L. Bouet, M.M. Glazov, T. Amand, E.L. Ivchenko, E. Palleau, X. Marie, B. Urbaszek, Magneto-optics in transition metal diselenide monolayers. 2D Mater. **2**(3), 034002 (2015)
49. A.V. Stier, N.P. Wilson, G. Clark, X. Xu, S.A. Crooker, Probing the influence of dielectric environment on excitons in monolayer WSe_2: insight from high magnetic fields. Nano Lett. **16**, 7054–7060 (2016)
50. J. Kuhnert, A. Rahimi-Iman, W. Heimbrodt, Magneto photoluminescence measurements of tungsten disulphide monolayers. J. Phys.: Condens. Matter **29**, 08LT02 (2017)
51. R. Schuster, Y. Wan, M. Knupfer, B. Büchner, Nongeneric dispersion of excitons in the bulk of WSe_2. Phys. Rev. B **94**, 085201 (2016)
52. L.M. Schneider, S.S. Esdaille, D.A. Rhodes, K. Barmak, J.C. Hone, A. Rahimi-Iman, Shedding light on exciton's nature in monolayer quantum material by optical dispersion measurements. Opt. Express **27**, 37131–37149 (2019). https://doi.org/10.1364/OE.27.037131
53. S. Brem, A. Ekman, D. Christiansen, F. Katsch, M. Selig, C. Robert, X. Marie, B. Urbaszek, A. Knorr, E. Malic, Phonon-assisted photoluminescence from indirect excitons in monolayers of transition-metal dichalcogenides. Nano Lett. **20**(4), 2849–2856 (2020)
54. S.-Y. Chen, T. Goldstein, T. Taniguchi, K. Watanabe, J. Yan, Coulomb-bound four- and five-particle intervalley states in an atomically-thin semiconductor. Nat. Commun. **9**, 1–8 (2018)
55. M. Barbone, A.R.-P. Montblanch, D.M. Kara, C. Palacios-Berraquero, A.R. Cadore, D.D. Fazio, B. Pingault, E. Mostaani, H. Li, B. Chen, K. Watanabe, T. Taniguchi, S. Tongay, G. Wang, A.C. Ferrari, M. Atatüre, Charge-tuneable biexciton complexes in monolayer WSe_2. Nat. Commun. **9**, 1–6 (2018)
56. K.S. Novoselov, A. Mishchenko, A. Carvalho, A.H. Castro Neto, 2D materials and van der Waals heterostructures. Science **353**, aac9439 (2016)
57. F. Cadiz, E. Courtade, C. Robert, G. Wang, Y. Shen, H. Cai, T. Taniguchi, K. Watanabe, H. Carrere, D. Lagarde, M. Manca, T. Amand, P. Renucci, S. Tongay, X. Marie, B. Urbaszek, Excitonic linewidth approaching the homogeneous limit in MoS_2-based van der Waals heterostructures. Phys. Rev. X **7**, 021026 (2017)
58. O.A. Ajayi, J.V. Ardelean, G.D. Shepard, J. Wang, A. Antony, T. Taniguchi, K. Watanabe, T.F. Heinz, S. Strauf, X.Y. Zhu, J.C. Hone, Approaching the intrinsic photoluminescence linewidth in transition metal dichalcogenide monolayers. 2D Mater. **4**, 031011 (2017)
59. J. Wierzbowski, J. Klein, F. Sigger, C. Straubinger, M. Kremser, T. Taniguchi, K. Watanabe, U. Wurstbauer, A.W. Holleitner, M. Kaniber, K. Müller, J.J. Finley, Direct exciton emission from atomically thin transition metal dichalcogenide heterostructures near the lifetime limit. Sci. Rep. **7**, 12383 (2017)
60. M.R. Molas, A.O. Slobodeniuk, K. Nogajewski, M. Bartos, L. Bala, A. Babinski, K. Watanabe, T. Taniguchi, C. Faugeras, M. Potemski, Energy spectrum of two-dimensional excitons in a nonuniform dielectric medium. Phys. Rev. Lett. **123**, 136801 (2019)
61. X. Gan, Y. Gao, K. Fai Mak, X. Yao, R.J. Shiue, A. Van Der Zande, M.E. Trusheim, F. Hatami, T.F. Heinz, J. Hone, D. Englund, Controlling the spontaneous emission rate of monolayer MoS_2 in a photonic crystal nanocavity. Appl. Phys. Lett. **103**, 181119 (2013)
62. G.M. Akselrod, T. Ming, C. Argyropoulos, T.B. Hoang, Y. Lin, X. Ling, D.R. Smith, J. Kong, M.H. Mikkelsen, Leveraging nanocavity harmonics for control of optical processes in 2D semiconductors. Nano Lett. **15**, 3578–3584 (2015)

63. S. Wu, S. Buckley, A.M. Jones, J.S. Ross, N.J. Ghimire, J. Yan, D.G. Mandrus, W. Yao, F. Hatami, J. Vučković, A. Majumdar, X. Xu, Control of two-dimensional excitonic light emission via photonic crystal. 2D Mater. **1**, 011001 (2014)
64. X. Liu, T. Galfsky, Z. Sun, F. Xia, E.C. Lin, Y.H. Lee, S. Kéna-Cohen, V.M. Menon, Strong light-matter coupling in two-dimensional atomic crystals. Nat. Photonics **9**, 30–34 (2014). https://doi.org/10.1038/nphoton.2014.304
65. S. Dufferwiel, S. Schwarz, F. Withers, A.A. Trichet, F. Li, M. Sich, O. Del Pozo-Zamudio, C. Clark, A. Nalitov, D.D. Solnyshkov, G. Malpuech, K.S. Novoselov, J.M. Smith, M.S. Skolnick, D.N. Krizhanovskii, A.I. Tartakovskii, Exciton-polaritons in van der Waals heterostructures embedded in tunable microcavities. Nat. Commun. **6**, 8579 (2015)
66. L.C. Flatten, Z. He, D.M. Coles, A.A. Trichet, A.W. Powell, R.A. Taylor, J.H. Warner, J.M. Smith, Room-Temperature exciton-polaritons with two-dimensional WS_2. Sci. Rep. **6**, 33134 (2016)
67. S. Wang, S. Li, T. Chervy, A. Shalabney, S. Azzini, E. Orgiu, J.A. Hutchison, C. Genet, P. Samorì, T.W. Ebbesen, Coherent coupling of WS_2 monolayers with metallic photonic nanostructures at room temperature. Nano Lett. **16**(7), 4368–74 (2016)
68. N. Lundt, S. Klembt, E. Cherotchenko, S. Betzold, O. Iff, A.V. Nalitov, M. Klaas, C.P. Dietrich, A.V. Kavokin, S. Höfling, C. Schneider, Room-temperature Tamm-plasmon exciton-polaritons with a WSe_2 monolayer. Nat. Commun. **7**, 13328 (2016)
69. L.C. Flatten, D.M. Coles, Z. He, D.G. Lidzey, R.A. Taylor, J.H. Warner, J.M. Smith, Electrically tunable organic-inorganic hybrid polaritons with monolayer WS_2. Nat. Commun. **8**, 14097 (2017). https://doi.org/10.1038/ncomms14097
70. S. Wu, S. Buckley, J.R. Schaibley, L. Feng, J. Yan, D.G. Mandrus, F. Hatami, W. Yao, J. Vučković, A. Majumdar, X. Xu, Monolayer semiconductor nanocavity lasers with ultralow thresholds. Nature **520**, 69–72 (2015). https://doi.org/10.1038/nature14290
71. O. Salehzadeh, M. Djavid, N.H. Tran, I. Shih, Z. Mi, Optically pumped two-dimensional MoS_2 lasers operating at room-temperature. Nano Lett. **15**(8), 5302–5306 (2015). https://doi.org/10.1021/acs.nanolett.5b01665
72. Y. Ye, Z.J. Wong, X. Lu, X. Ni, H. Zhu, X. Chen, Y. Wang, X. Zhang, Monolayer excitonic laser. Nat. Photonics **9**, 733–737 (2015). https://doi.org/10.1038/nphoton.2015.197
73. Y. Li, J. Zhang, D. Huang, H. Sun, F. Fan, J. Feng, Z. Wang, C.Z. Ning, Room-temperature continuous-wave lasing from monolayer molybdenum ditelluride integrated with a silicon nanobeam cavity. Nat. Nanotechnol. **12**(10), 987–992 (2017)
74. J. Shang, C. Cong, Z. Wang, N. Peimyoo, L. Wu, C. Zou, Y. Chen, X.Y. Chin, J. Wang, C. Soci, W. Huang, T. Yu, Room-temperature 2D semiconductor activated vertical-cavity surface-emitting lasers. Nat. Commun. **8**(1), 543 (2017)
75. K.F. Mak, D. Xiao, J. Shan, Light-valley interactions in 2D semiconductors. Nat. Photonics **12**, 451–460 (2018)
76. T.C. Berkelbach, M.S. Hybertsen, D.R. Reichman, Theory of neutral and charged excitons in monolayer transition metal dichalcogenides. Phys. Rev. B **88**, 045318 (2013)
77. D. Van Tuan, B. Scharf, I. Žutić, H. Dery, Marrying excitons and plasmons in monolayer transition-metal dichalcogenides. Phys. Rev. X **7**, 041040 (2017)
78. I.C. Gerber, X. Marie, Dependence of band structure and exciton properties of encapsulated WSe_2 monolayers on the hBN-layer thickness. Phys. Rev. B **98**(24), 28–31 (2018)
79. T. Deilmann, K.S. Thygesen, Finite-momentum exciton landscape in mono- and bilayer transition metal dichalcogenides. 2D Mater. **6**, 035003 (2019)
80. R. Suzuki, M. Sakano, Y.J. Zhang, R. Akashi, D. Morikawa, A. Harasawa, K. Yaji, K. Kuroda, K. Miyamoto, T. Okuda, K. Ishizaka, R. Arita, Y. Iwasa, Valley-dependent spin polarization in bulk MoS_2 with broken inversion symmetry. Nat. Nanotechnol. **9**(8), 611–617 (2014)
81. A. Chernikov, C. Ruppert, H.M. Hill, A.F. Rigosi, T.F. Heinz, Population inversion and giant bandgap renormalization in atomically thin WS2 layers. Nat. Photonics **9**, 466–470 (2015)
82. B. Radisavljevic, A. Radenovic, J. Brivio, V. Giacometti, A. Kis, Single-layer MoS_2 transistors. Nat. Nanotechnol. **6**, 147–150 (2011). https://doi.org/10.1038/nnano.2010.279

83. F. Peyskens, C. Chakraborty, M. Muneeb, D. Van Thourhout, D. Englund, Integration of single photon emitters in 2D layered materials with a silicon nitride photonic chip. Nat. Commun. **10**, 4435 (2019). https://doi.org/10.1038/s41467-019-12421-0
84. A.H. Castro Neto, F. Guinea, N.M.R. Peres, K.S. Novoselov, A.K. Geim, The electronic properties of graphene. Rev. Mod. Phys. **81**, 109–162 (2009)
85. A.K. Geim, Graphene: status and prospects. Science **324**, 1530–1534 (2009)
86. Y. Yu, Y. Yu, C. Xu, Y.Q. Cai, L. Su, Y. Zhang, Y.W. Zhang, K. Gundogdu, L. Cao, Engineering substrate interactions for high luminescence efficiency of transition-metal dichalcogenide monolayers. Adv. Funct. Mater. **26**, 4733–4739 (2016)
87. S. Lippert, L.M. Schneider, D. Renaud, K.N. Kang, O. Ajayi, J. Kuhnert, M.-U. Halbich, O.M. Abdulmunem, X. Lin, K. Hassoon, S. Edalati-Boostan, Y.D. Kim, W. Heimbrodt, E.-H. Yang, J.C. Hone, A. Rahimi-Iman, Influence of the substrate material on the optical properties of tungsten diselenide monolayers. 2D Mater. **4**, 025045 (2017). http://dx.doi.org/10.1088/2053-1583/aa5b21
88. L. Meckbach, T. Stroucken, S.W. Koch, Influence of the effective layer thickness on the ground-state and excitonic properties of transition-metal dichalcogenide systems. Phys. Rev. B **97**, 035425 (2018)
89. F. Bonaccorso, Z. Sun, T. Hasan, A.C. Ferrari, Graphene photonics and optoelectronics. Nat. Photonics **4**, 611–622 (2010)
90. A.N. Grigorenko, M. Polini, K.S. Novoselov, Graphene plasmonics. Nat. Photonics **6**(11), 749–758 (2012)
91. A.K. Geim, I.V. Grigorieva, Van der Waals heterostructures. Nature **499**, 419 (2013)
92. D. Jariwala, T.J. Marks, M.C. Hersam, Mixed-dimensional van der Waals heterostructures. Nat. Mater. **16**, 170–181 (2017)
93. H. Terrones, F.L-Ur, M. Terrones, Novel hetero-layered materials with tunable direct band gaps by sandwiching different metal disulfides and diselenides. Sci. Rep. **3**, 1–7 (2013)
94. J. Kang, S. Tongay, J. Zhou, J. Li, J. Wu, Band offsets and heterostructures of two-dimensional semiconductors. Appl. Phys. Lett. **102**, 012111 (2013)
95. C. Gong, H. Zhang, W. Wang, L. Colombo, R.M. Wallace, K. Cho, Band alignment of two-dimensional transition metal dichalcogenides: application in tunnel field effect transistors. Appl. Phys. Lett. **103**, 053513 (2013)
96. V.O. Özçelik, J.G. Azadani, C. Yang, S.J. Koester, T. Low, Band alignment of two-dimensional semiconductors for designing heterostructures with momentum space matching. Phys. Rev. B **94**, 035125 (2016)
97. Y. Gong, J. Lin, X. Wang, G. Shi, S. Lei, Z. Lin, X. Zou, G. Ye, R. Vajtai, B.I. Yakobson, H. Terrones, M. Terrones, B.K. Tay, J. Lou, S.T. Pantelides, Z. Liu, W. Zhou, P.M. Ajayan, Vertical and in-plane heterostructures from WS_2/MoS_2 monolayers. Nat. Mater. **13**, 1135–1142 (2014)
98. D. Jariwala, A.R. Davoyan, J. Wong, H.A. Atwater, Van der Waals materials for atomically-thin photovoltaics: promise and outlook. ACS Photon. **4**, 2962–2970 (2017)
99. B. Hunt, J.D. Sanchez-Yamagishi, A.F. Young, M. Yankowitz, B.J. LeRoy, K. Watanabe, T. Taniguchi, P. Moon, M. Koshino, P. Jarillo-Herrero, R.C. Ashoori, Massive Dirac fermions and Hofstadter butterfly in a van der Waals heterostructure. Science **340**(6139), 1427–1430 (2013)
100. C.R. Dean, L. Wang, P. Maher, C. Forsythe, F. Ghahari, Y. Gao, J. Katoch, M. Ishigami, P. Moon, M. Koshino, T. Taniguchi, K. Watanabe, K.L. Shepard, J. Hone, P. Kim, Hofstadter's butterfly and the fractal quantum Hall effect in moiré superlattices. Nature **497**, 598–602 (2013)
101. H. Yu, G.-B. Liu, J. Tang, X. Xu, W. Yao, Moiré excitons: from programmable quantum emitter arrays to spin-orbitcoupled artificial lattices. Sci. Adv. **3**, e1701696 (2017)
102. K.L. Seyler, P. Rivera, H. Yu, N.P. Wilson, E.L. Ray, D.G. Mandrus, J. Yan, W. Yao, X. Xu, Signatures of moiré-trapped valley excitons in $MoSe_2/WSe_2$ heterobilayers. Nature **567**, 66–70 (2019)

103. K. Tran, G. Moody, F. Wu, X. Lu, J. Choi, K. Kim, A. Rai, D.A. Sanchez, J. Quan, A. Singh, J. Embley, A. Zepeda, M. Campbell, T. Autry, T. Taniguchi, K. Watanabe, N. Lu, S.K. Banerjee, K.L. Silverman, S. Kim, E. Tutuc, L. Yang, A.H. MacDonald, X. Li, Evidence for moiré excitons in van der Waals heterostructures. Nature **567**, 71–75 (2019)
104. E.M. Alexeev, D.A. Ruiz-Tijerina, M. Danovich, M.J. Hamer, D.J. Terry, P.K. Nayak, S. Ahn, S. Pak, J. Lee, J.I. Sohn, M.R. Molas, M. Koperski, K. Watanabe, T. Taniguchi, K.S. Novoselov, R.V. Gorbachev, H.S. Shin, V.I. Falko, A.I. Tartakovskii, Resonantly hybridized excitons in moiré superlattices in van der Waals heterostructures. Nature **567**, 81–86 (2019)
105. X. Lu, P. Stepanov, W. Yang, M. Xie, M.A. Aamir, I. Das, C. Urgell, K. Watanabe, T. Taniguchi, G. Zhang, A. Bachtold, A.H. MacDonald, D.K. Efetov, Superconductors, orbital magnets and correlated states in magic-angle bilayer graphene. Nature **574**, 653–657 (2019)
106. R. Bistritzer, A.H. MacDonald, Moire bands in twisted double-layer graphene. Proc. Natl. Acad. Sci. **108**, 12233–12237 (2011)
107. Y. Cao, V. Fatemi, S. Fang, K. Watanabe, T. Taniguchi, E. Kaxiras, P. Jarillo-Herrero, Unconventional superconductivity in magic-angle graphene superlattices. Nature **556**, 43–50 (2018)
108. M. Yankowitz, S. Chen, H. Polshyn, Y. Zhang, K. Watanabe, T. Taniguchi, D. Graf, A.F. Young, C.R. Dean, Tuning superconductivity in twisted bilayer graphene. Science **363**, 1059–1064 (2019)
109. Y. Cao, V. Fatemi, A. Demir, S. Fang, S.L. Tomarken, J.Y. Luo, J.D. Sanchez-Yamagishi, K. Watanabe, T. Taniguchi, E. Kaxiras, R.C. Ashoori, P. Jarillo-Herrero, Correlated insulator behaviour at half-filling in magic-angle graphene superlattices. Nature **556**, 80 (2018)
110. A.L. Sharpe, E.J. Fox, A.W. Barnard, J. Finney, K. Watanabe, T. Taniguchi, M.A. Kastner, D. Goldhaber-Gordon, Emergent ferromagnetism near three-quarters filling in twisted bilayer graphene. Science **365**, 605–608 (2019)
111. D.N. Basov, M.M. Fogler, F.J. García De Abajo, Polaritons in van der Waals materials. Science, **354**, aag1992 (2016)
112. K. Liu, J. Wu, Mechanical properties of two-dimensional materials and heterostructures. J. Mater. Res. **31**, 832–844 (2016)
113. S. Garaj, W. Hubbard, A. Reina, J. Kong, D. Branton, J.A. Golovchenko, Graphene as a subnanometre trans-electrode membrane. Nature **467**, 190–193 (2010)
114. E. Gibney, 2D or not 2D. Nature **522**, 274 (2015)
115. Y. Li, A. Chernikov, X. Zhang, A. Rigosi, H.M. Hill, A.M. van der Zande, D.A. Chenet, E.-M. Shih, J. Hone, T.F. Heinz, Measurement of the optical dielectric function of monolayer transition-metal dichalcogenides: MoS_2, $MoSe_2$, WS_2, WSe_2. Phys. Rev. B **90**, 205422 (2014)
116. S.Z. Butler, S.M. Hollen, L. Cao, Y. Cui, J.A. Gupta, H.R. Gutiérrez, T.F. Heinz, S.S. Hong, J. Huang, A.F. Ismach, E. Johnston-Halperin, M. Kuno, V.V. Plashnitsa, R.D. Robinson, R.S. Ruoff, S. Salahuddin, J. Shan, L. Shi, M.G. Spencer, M. Terrones, W. Windl, J.E. Goldberger, Progress, challenges, and opportunities in two-dimensional materials beyond graphene. ACS Nano **7**, 2898–2926 (2013)
117. D. Golberg, Y. Bando, Y. Huang, T. Terao, M. Mitome, C. Tang, C. Zhi, Boron nitride nanotubes and nanosheets. ACS Nano **4**, 2979–2993 (2010)
118. J.M. Rondinelli, S.J. May, J.W. Freeland, Control of octahedral connectivity in perovskite oxide heterostructures: An emerging route to multifunctional materials discovery. MRS Bull. **37**, 261–270 (2012)
119. J. Chakhalian, A.J. Millis, J. Rondinelli, Whither the oxide interface. Nat. Mater. **11**, 92–94 (2012)
120. F. Zhang, X. Xiang, F. Li, X. Duan, Layered double hydroxides as catalytic materials: recent development. Catal. Surv. Asia **12**, 253 (2008)
121. J. Jiang, K. Zhao, X. Xiao, L. Zhang, Synthesis and facet-dependent photoreactivity of BiOCl single-crystalline nanosheets. J. Am. Chem. Soc. **134**, 4473–4476 (2012)
122. A. Castellanos-Gomez, Why all the fuss about 2D semiconductors? Nat. Photonics **10**, 202–204 (2016)
123. J.Y. Lee, J.-H. Shin, G.-H. Lee, C.-H. Lee, Two-dimensional semiconductor optoelectronics based on van der Waals heterostructures. Nanomaterials **6**, 193 (2016)

124. P. Gupta, A.A. Rahman, S. Subramanian, S. Gupta, A. Thamizhavel, T. Orlova, S. Rouvimov, S. Vishwanath, V. Protasenko, M.R. Laskar, H.G. Xing, D. Jena, A. Bhattacharya, Layered transition metal dichalcogenides: promising near-lattice-matched substrates for GaN growth. Sci. Rep. **6**, 1–8 (2016). https://doi.org/10.1038/srep23708
125. T. Ohta, A. Bostwick, T. Seyller, K. Horn, E. Rotenberg, Controlling the electronic structure of bilayer graphene. Science **313**, 951 (2006)
126. E.V. Castro, K.S. Novoselov, S.V. Morozov, N.M.R. Peres, J.M.B.L. dos Santos, J. Nilsson, F. Guinea, A.K. Geim, A.H.C. Neto, Biased bilayer graphene: semiconductor with a gap tunable by the electric field effect. Phys. Rev. Lett. **99**, 216802 (2007)
127. J.B. Oostinga, H.B. Heersche, X. Liu, A.F. Morpurgo, L.M.K. Vandersypen, Gate-induced insulating state in bilayer graphene devices. Nat. Mater. **7**, 151–157 (2008)
128. A.B. Kuzmenko, E. van Heumen, D. van der Marel, P. Lerch, P. Blake, K.S. Novoselov, A.K. Geim, Infrared spectroscopy of electronic bands in bilayer graphene. Phys. Rev. B **79**, 115441 (2009)
129. R. Fivaz, E. Mooser, Electron-phonon interaction in semiconducting layer structures. Phys. Rev. **136**, A833–A836 (1964)
130. J.A. Wilson, A.D. Yoffe, The transition metal dichalcogenides discussion and interpretation of the observed optical, electrical and structural properties. Adv. Phys. **18**, 193–335 (1969)
131. R.L. Withers, J.A. Wilson, An examination of the formation and characteristics of charge-density waves in inorganic materials with special reference to the two- and one-dimensional transition-metal chalcogenides. J. Phys. C: Solid State Phys. **19**, 4809–4845 (1986)
132. T. Shishidou, A.J. Freeman, R. Asahi, Effect of GGA on the half-metallicity of the itinerant ferromagnet CoS_2. Phys. Rev. B **64**, 180401 (2001)
133. J.T. Ye, Y.J. Zhang, R. Akashi, M.S. Bahramy, R. Arita, Y. Iwasa, Superconducting dome in a gate-tuned band insulator. Science **338**, 1193 (2012)
134. N. Kumar, S. Najmaei, Q. Cui, F. Ceballos, P.M. Ajayan, J. Lou, H. Zhao, Second harmonic microscopy of monolayer MoS_2. Phys. Rev. B **87**, 161403 (2013)
135. A.R. Beal, W.Y. Liang, Excitons in 2H-WSe_2 and 3R-WS_2. J. Phys. C Solid State Phys. **9**, 2459–2466 (1976)
136. H. Meinhold, G. Weiser, Modulation spectroscopy on MoS_2, and $MoSe_2$. Phys. Status Solidi B **73**(1), 105–115 (1976)
137. D.G. Mead, J.C. Irwin, Long wavelength optic phonons in WSe_2. Can. J. Phys. **55**, 379–382 (1977)
138. R.A. Taylor, J.F. Rayn, Time-resolved exciton photoluminescence in GaSe and GaTe. J. Phys. C: Solid State Phys. **20**, 6175–6187 (1987)
139. K.K. Kam, B.A. Parkinson, Detailed photocurrent spectroscopy of the semiconducting group VIB transition metal dichalcogenides. J. Phys. Chem. **86**, 463–467 (1982)
140. L. Banszerus, M. Schmitz, S. Engels, J. Dauber, M. Oellers, F. Haupt, K. Watanabe, T. Taniguchi, B. Beschoten, C. Stampfer, Ultrahigh-mobility graphene devices from chemical vapor deposition on reusable copper. Sci. Adv. **1**, e1500222 (2015)
141. A. Castellanos-Gomez, M. Buscema, R. Molenaar, V. Singh, L. Janssen, H.S.J. van der Zant, G.A. Steele, Deterministic transfer of two-dimensional materials by all-dry viscoelastic stamping. 2D Mater. **1**, 011002 (2014)
142. D. Edelberg, D. Rhodes, A. Kerelsky, B. Kim, J. Wang, A. Zangiabadi, C. Kim, A. Abhinandan, J. Ardelean, M. Scully, D. Scullion, L. Embon, R. Zu, E.J.G. Santos, L. Balicas, C. Marianetti, K. Barmak, X. Zhu, J. Hone, A.N. Pasupathy, Approaching the intrinsic limit in transition metal diselenides via point defect control. Nano Lett. **19**, 4371–4379 (2019)
143. A. Raja, A. Chaves, J. Yu, G. Arefe, H.M. Hill, A.F. Rigosi, T.C. Berkelbach, P. Nagler, C. Schüller, T. Korn, C. Nuckolls, J. Hone, L.E. Brus, T.F. Heinz, D.R. Reichman, A. Chernikov, Coulomb engineering of the bandgap and excitons in two-dimensional materials. Nat. Commun. **8**, 15251 (2017)
144. Y. Lin, X. Ling, L. Yu, S. Huang, A.L. Hsu, Y.-H. Lee, J. Kong, M.S. Dresselhaus, T. Palacios, Dielectric screening of excitons and trions in single-layer MoS_2. Nano Lett. **14**, 5569–5576 (2014)

145. A. Kuc, N. Zibouche, T. Heine, Influence of quantum confinement on the electronic structure of the transition metal sulfide TS_2. Phys. Rev. B **83**, 245213 (2011)
146. D. Xiao, G.-B. Liu, W. Feng, X. Xu, W. Yao, Coupled spin and valley physics in monolayers of MoS_2 and other group-VI dichalcogenides. Phys. Rev. Lett. **108**, 196802 (2012)
147. X. Xu, W. Yao, D. Xiao, T.F. Heinz, Spin and pseudospins in layered transition metal dichalcogenides. Nat. Phys. **10**, 343–350 (2014)
148. P. Schmidt, F. Vialla, S. Latini, M. Massicotte, K.-J. Tielrooij, S. Mastel, G. Navickaite, M. Danovich, D.A. Ruiz-Tijerina, C. Yelgel, V. Falko, K.S. Thygesen, R. Hillenbrand, F.H.L. Koppens, Nano-imaging of intersubband transitions in van der Waals quantum wells. Nat. Nanotechnol. **13**, 1035–1041 (2018). https://doi.org/10.1038/s41565-018-0233-9
149. C.F. Klingshirn, *Semiconductor Optics* (Springer, Berlin, 2012)
150. F.A. Cevallos, S. Guo, H. Heo, G. Scuri, Y. Zhou, J. Sung, T. Taniguchi, K. Watanabe, P. Kim, H. Park, R.J. Cava, Liquid salt transport growth of single crystals of the layered dichalcogenides MoS_2 and WS_2. Cryst. Growth Des. **19**, 5762–5767 (2019)
151. E. Hanamura, Rapid radiative decay and enhanced optical nonlinearity of excitons in a quantum well. Phys. Rev. B **38**, 1228–1234 (1988)
152. L.C. Andreani, F. Bassani, Exchange interaction and polariton effects in quantum-well excitons. Phys. Rev. B **41**, 7536–7544 (1990)
153. A.F. Rigosi, H.M. Hill, Y. Li, A. Chernikov, T.F. Heinz, Probing interlayer interactions in transition metal dichalcogenide heterostructures by optical spectroscopy: MoS_2/WS_2 and $MoSe_2/WSe_2$. Nano Lett. **15**, 5033–5038 (2015)
154. L. Su, Y. Yu, L. Cao, Y. Zhang, Effects of substrate type and material-substrate bonding on high-temperature behavior of monolayer WS_2. Nano Res. **8**, 2686–2697 (2015)
155. T. Godde, D. Schmidt, J. Schmutzler, M. Assmann, J. Debus, F. Withers, E.M. Alexeev, O. Del Pozo-Zamudio, O.V. Skrypka, K.S. Novoselov, M. Bayer, A.I. Tartakovskii, Exciton and trion dynamics in atomically thin $MoSe_2$ and WSe_2: effect of localization. Phys. Rev. B **94**, 165301 (2016)
156. O. Mey, F. Wall, L.M. Schneider, D. Günder, F. Walla, A. Soltani, H. Roskos, N. Yao, P. Qing, W. Fang, A. Rahimi-Iman, Enhancement of the monolayer tungsten disulfide exciton photoluminescence with a two-dimensional material/air/gallium phosphide in-plane microcavity. ACS Nano **13**, 5259–5267 (2019). https://doi.org/10.1021/acsnano.8b09659
157. S. Barja, S. Refaely-Abramson, B. Schuler, D.Y. Qiu, A. Pulkin, S. Wickenburg, H. Ryu, M.M. Ugeda, C. Kastl, C. Chen, C. Hwang, A. Schwartzberg, S. Aloni, S.-K. Mo, D. Frank Ogletree, M.F. Crommie, O.V. Yazyev, S.G. Louie, J.B. Neaton, A. Weber-Bargioni, Identifying substitutional oxygen as a prolific point defect in monolayer transition metal dichalcogenides. Nat. Commun. **10**, 3382 (2019)
158. J. Klein, M. Lorke, M. Florian, F. Sigger, L. Sigl, S. Rey, J. Wierzbowski, J. Cerne, K. Müller, E. Mitterreiter, P. Zimmermann, T. Taniguchi, K. Watanabe, U. Wurstbauer, M. Kaniber, M. Knap, R. Schmidt, J.J. Finley, A.W. Holleitner, Site-selectively generated photon emitters in monolayer MoS_2 via local helium ion irradiation. Nat. Commun. **10**, 2755 (2019)
159. S.B. Desai, G. Seol, J.S. Kang, H. Fang, C. Battaglia, R. Kapadia, J.W. Ager, J. Guo, A. Javey, Strain-induced indirect to direct bandgap transition in multilayer WSe_2. Nano Lett. **14**, 4592–4597 (2014)
160. R. Zhang, V. Koutsos, R. Cheung, Elastic properties of suspended multilayer WSe_2. Appl. Phys. Lett. **108**(4), 042104 (2016)
161. R. Schmidt, I. Niehues, R. Schneider, M. Drüppel, T. Deilmann, M. Rohlfing, S.M. de Vasconcellos, A. Castellanos-Gomez, R. Bratschitsch, Reversible uniaxial strain tuning in atomically thin WSe_2. 2D Mater. **3**, 021011 (2016)
162. L. Mennel, M.M. Furchi, S. Wachter, M. Paur, D.K. Polyushkin, T. Mueller, Optical imaging of strain in two-dimensional crystals. Nat. Commun. **9**, 516 (2018)
163. K.S. Novoselov, V.I. Falko, L. Colombo, P.R. Gellert, M.G. Schwab, K. Kim, A roadmap for graphene. Nature **490**, 192–200 (2012)
164. F.H. Koppens, D.E. Chang, F.J. García De Abajo, Graphene plasmonics: a platform for strong light-matter interactions. Nano Lett. **11**, 3370–3377 (2011)

165. K.S. Kim, Y. Zhao, H. Jang, S.Y. Lee, J.M. Kim, K.S. Kim, J.-H. Ahn, P. Kim, J.-Y. Choi, B.H. Hong, Large-scale pattern growth of graphene films for stretchable transparent electrodes. Nature **457**, 706–710 (2009)
166. A. Pospischil, M.M. Furchi, T. Mueller, Solar-energy conversion and light emission in an atomic monolayer p-n diode. Nat. Nanotechnol. **9**, 257–261 (2014)
167. Y. Liu, R. Cheng, L. Liao, H. Zhou, J. Bai, G. Liu, L. Liu, Y. Huang, X. Duan, Plasmon resonance enhanced multicolour photodetection by graphene. Nat. Commun. **2**, 1–7 (2011)
168. T. Low, P. Avouris, Graphene plasmonics for terahertz to mid-infrared applications. ACS Nano **8**, 1086–1101 (2014)
169. M. Liu, X. Yin, E. Ulin-Avila, B. Geng, T. Zentgraf, L. Ju, F. Wang, X. Zhang, A graphene-based broadband optical modulator. Nature **474**, 64–67 (2011)
170. S. Jiang, T. Shi, X. Zhan, S. Xi, H. Long, B. Gong, J. Li, S. Cheng, Y. Huang, Z. Tang, Scalable fabrication of carbon-based MEMS/NEMS and their applications: a review. J. Micromech. Microeng. **25**, 113001 (2015)
171. C.J. Wan, L.Q. Zhu, Y.H. Liu, P. Feng, Z.P. Liu, H.L. Cao, P. Xiao, Y. Shi, Q. Wan, Proton-conducting graphene oxide-coupled neuron transistors for brain-inspired cognitive systems. Adv. Mater. **28**, 3557–3563 (2016)
172. C.J. Wan, Y.H. Liu, P. Feng, W. Wang, L.Q. Zhu, Z.P. Liu, Y. Shi, Q. Wan, Flexible metal oxide/graphene oxide hybrid neuromorphic transistors on flexible conducting graphene substrates. Adv. Mater. **28**, 5878–5885 (2016)
173. M. Massicotte, P. Schmidt, F. Vialla, K.G. Schädler, A. Reserbat-Plantey, K. Watanabe, T. Taniguchi, K.J. Tielrooij, F.H.L. Koppens, Picosecond photoresponse in van der Waals heterostructures. Nat. Nanotechnol. **11**, 42–46 (2016)
174. T. Yu, F. Wang, Y. Xu, L. Ma, X. Pi, D. Yang, Graphene coupled with silicon quantum dots for high-performance bulk-silicon-based Schottky-junction photodetectors. Adv. Mater. **28**, 4912–4919 (2016)
175. Z. Ni, L. Ma, S. Du, Y. Xu, M. Yuan, H. Fang, Z. Wang, M. Xu, D. Li, J. Yang, W. Hu, X. Pi, D. Yang, Plasmonic silicon quantum dots enabled high-sensitivity ultrabroadband photodetection of graphene-based hybrid phototransistors. ACS Nano **11**, 9854–9862 (2017)
176. A. Ali, K. Shehzad, H. Guo, Z. Wang, P. Wang, A. Qadir, W. Hu, T. Ren, B. Yu, Y. Xu, High-performance, flexible graphene/ultra-thin silicon ultra-violet image sensor, in *2017 IEEE International Electron Devices Meeting*, pp. 8.6.1–8.6.4. IEEE (2017)
177. X. Wan, Y. Xu, H. Guo, K. Shehzad, A. Ali, Y. Liu, J. Yang, D. Dai, C.-T. Lin, L. Liu, H.-C. Cheng, F. Wang, X. Wang, H. Lu, W. Hu, X. Pi, Y. Dan, J. Luo, T. Hasan, X. Duan, X. Li, J. Xu, D. Yang, T. Ren, B. Yu, A self-powered high-performance graphene/silicon ultraviolet photodetector with ultra-shallow junction: breaking the limit of silicon?, npj 2D Mater. Appl. **1**, 4 (2017)
178. A. De Sanctis, G.F. Jones, D.J. Wehenkel, F. Bezares, F.H.L. Koppens, M.F. Craciun, S. Russo, Extraordinary linear dynamic range in laser-defined functionalized graphene photodetectors. Sci. Adv. **3**(5), e1602617 (2017)
179. S. Goossens, G. Navickaite, C. Monasterio, S. Gupta, J.J. Piqueras, R. Pérez, G. Burwell, I. Nikitskiy, T. Lasanta, T. Galn, E. Puma, A. Centeno, A. Pesquera, A. Zurutuza, G. Konstantatos, F. Koppens, Broadband image sensor array based on graphene-cmos integration. Nat. Photonics **11**, 366–371 (2017)
180. C.-H. Liu, Y.-C. Chang, T.B. Norris, Z. Zhong, Graphene photodetectors with ultra-broadband and high responsivity at room temperature. Nat. Nanotechnol. **9**, 273–278 (2014)
181. S. Darbari, V. Ahmadi, P. Afzali, Y. Abdi, M. Feda, Reduced graphene oxide/ZnO hybrid structure for high-performance photodetection. J. Nanopart. Res. **16**, 2798 (2014)
182. S. Liu, Q. Liao, S. Lu, Z. Zhang, G. Zhang, Y. Zhang, Strain modulation in graphene/ZnO nanorod film Schottky junction for enhanced photosensing performance. Adv. Funct. Mater. **26**(9), 1347–1353 (2016)
183. X. An, F. Liu, Y.J. Jung, S. Kar, Tunable graphene silicon heterojunctions for ultrasensitive photodetection. Nano Lett. **13**, 909–916 (2013)

184. C.O. Kim, S. Kim, D.H. Shin, S.S. Kang, J.M. Kim, C.W. Jang, S.S. Joo, J.S. Lee, J.H. Kim, S.-H. Choi, E. Hwang, High photoresponsivity in an all-graphene p n vertical junction photodetector. Nat. Commun. **5**, 1–7 (2014)
185. Y.Q. Huang, R.J. Zhu, N. Kang, J. Du, H.Q. Xu, Photoelectrical response of hybrid graphene-PbS quantum dot devices. Appl. Phys. Lett. **103**, 143119 (2013)
186. D. Zhang, L. Gan, Y. Cao, Q. Wang, L. Qi, X. Guo, Understanding charge transfer at PbS-decorated graphene surfaces toward a tunable photosensor. Adv. Mater. **24**(20), 2715–2720 (2012)
187. S. Kaniyankandy, S. Rawalekar, H.N. Ghosh, Ultrafast charge transfer dynamics in photoexcited CdTe quantum dot decorated on graphene. J. Phys. Chem. C **116**, 16271–16275 (2012)
188. C.-W. Chiang, G. Haider, W.-C. Tan, Y.-R. Liou, Y.-C. Lai, R. Ravindranath, H.-T. Chang, Y.-F. Chen, Highly stretchable and sensitive photodetectors based on hybrid graphene and graphene quantum dots. ACS Appl. Mater. Interfaces. **8**, 466–471 (2016)
189. G. Konstantatos, M. Badioli, L. Gaudreau, J. Osmond, M. Bernechea, F.P. G.D. Arquer, F. Gatti, F.H.L. Koppens, Hybrid graphene–quantum dot phototransistors with ultrahigh gain. Nat. Nanotechnol. **7**, 363–368 (2012)
190. K.J. Tielrooij, L. Orona, A. Ferrier, M. Badioli, G. Navickaite, S. Coop, S. Nanot, B. Kalinic, T. Cesca, L. Gaudreau, Q. Ma, A. Centeno, A. Pesquera, A. Zurutuza, H. de Riedmatten, P. Goldner, F.J. G. de Abajo, P. Jarillo-Herrero, F.H.L. Koppens, Electrical control of optical emitter relaxation pathways enabled by graphene. Nat. Phys. **11**, 281–287 (2015)
191. R. Li, L.M. Schneider, W. Heimbrodt, H. Wu, M. Koch, A. Rahimi-Iman, Gate tuning of Förster resonance energy transfer in a graphene - quantum dot FET photo-detector. Sci. Rep. **6**, 28224 (2016). https://doi.org/10.1038/srep28224
192. D. Kufer, T. Lasanta, M. Bernechea, F.H.L. Koppens, G. Konstantatos, Interface engineering in hybrid quantum dot-2D phototransistors. ACS Photon. **3**, 1324–1330 (2016)
193. W.J. Yu, Y. Liu, H. Zhou, A. Yin, Z. Li, Y. Huang, X. Duan, Highly efficient gate-tunable photocurrent generation in vertical heterostructures of layered materials. Nat. Nanotechnol. **8**, 952–958 (2013)
194. H. Jeong, H.M. Oh, S. Bang, H.J. Jeong, S.-J. An, G.H. Han, H. Kim, S.J. Yun, K.K. Kim, J.C. Park, Y.H. Lee, G. Lerondel, M.S. Jeong, Metal-insulator-semiconductor diode consisting of two-dimensional nanomaterials. Nano Lett. **16**, 1858–1862 (2016)
195. D. Spirito, S. Kudera, V. Miseikis, C. Giansante, C. Coletti, R. Krahne, UV light detection from CdS nanocrystal sensitized graphene photodetectors at kHz frequencies. J. Phys. Chem. C **119**, 23859–23864 (2015)
196. A. Tredicucci, M.S. Vitiello, Device concepts for graphene-based terahertz photonics. IEEE J. Sel. Topics Quantum Electron. **20**, 130–138 (2014)
197. M. Furchi, A. Urich, A. Pospischil, G. Lilley, K. Unterrainer, H. Detz, P. Klang, A.M. Andrews, W. Schrenk, G. Strasser, T. Mueller, Microcavity-integrated graphene photodetector. Nano Lett. **12**, 2773–2777 (2012)
198. A. Zak, M.A. Andersson, M. Bauer, J. Matukas, A. Lisauskas, H.G. Roskos, J. Stake, Antenna-integrated 0.6 THz FET direct detectors based on CVD graphene. Nano Lett. **14**, 5834–5838 (2014)
199. M. Freitag, T. Low, W. Zhu, H. Yan, F. Xia, P. Avouris, Photocurrent in graphene harnessed by tunable intrinsic plasmons. Nat. Commun. **4**, 1–8 (2013)
200. T. Echtermeyer, S. Milana, U. Sassi, A. Eiden, M. Wu, E. Lidorikis, A. Ferrari, Surface plasmon polariton graphene photodetectors. Nano Lett. **16**, 8–20 (2016)
201. O. Abdulmunem, K. Hassoon, M. Gaafar, A. Rahimi-Iman, J.C. Balzer, TiN nanoparticles for enhanced THz generation in TDS systems. J. Infrared Milli. Terahz Waves **38**, 1206–1214 (2017). https://doi.org/10.1007/s10762-017-0412-z
202. H. Chen, X. Wen, J. Zhang, T. Wu, Y. Gong, X. Zhang, J. Yuan, C. Yi, J. Lou, P.M. Ajayan, W. Zhuang, G. Zhang, J. Zheng, Ultrafast formation of interlayer hot excitons in atomically thin MoS_2/WS_2 heterostructures. Nat. Commun. **7**, 12512 (2016)
203. B. Miller, A. Steinhoff, B. Pano, J. Klein, F. Jahnke, A. Holleitner, U. Wurstbauer, Long-lived direct and indirect interlayer excitons in van der Waals heterostructures. Nano Lett. **17**, 5229–5237 (2017)

204. P. Nagler, G. Plechinger, M.V. Ballottin, A. Mitioglu, S. Meier, N. Paradiso, C. Strunk, A. Chernikov, P.C.M. Christianen, C. Schüller, T. Korn, Interlayer exciton dynamics in a dichalcogenide monolayer heterostructure. 2D Mater. **4**, 025112 (2017)
205. U. Wurstbauer, B. Miller, E. Parzinger, A.W. Holleitner, Lightmatter interaction in transition metal dichalcogenides and their heterostructures. J. Phys. D Appl. Phys. **50**, 173001 (2017)
206. S. Dietl, S. Wang, D. Schuh, W. Wegscheider, J.P. Kotthaus, A. Pinczuk, A.W. Holleitner, U. Wurstbauer, Collective electronic excitation in a trapped ensemble of photogenerated dipolar excitons and free holes revealed by inelastic light scattering. Phys. Rev. B **95**, 085312 (2017)
207. M.M. Fogler, L.V. Butov, K.S. Novoselov, High-temperature superfluidity with indirect excitons in van der Waals heterostructures. Nat. Commun. **5**, 4555 (2014)
208. X. Liu, K. Watanabe, T. Taniguchi, B.I. Halperin, P. Kim, Quantum Hall drag of exciton condensate in graphene. Nat. Phys. **13**, 746–750 (2017)
209. Z. Wang, D.A. Rhodes, K. Watanabe, T. Taniguchi, J.C. Hone, J. Shan, K.F. Mak, Evidence of high-temperature exciton condensation in two-dimensional atomic double layers. Nature **574**, 76–80 (2019)
210. N.R. Wilson, P.V. Nguyen, K. Seyler, P. Rivera, A.J. Marsden, Z.P. Laker, G.C. Constantinescu, V. Kandyba, A. Barinov, N.D. Hine, X. Xu, D.H. Cobden, Determination of band offsets, hybridization, and exciton binding in 2D semiconductor heterostructures. Sci. Adv. **3**, e1601832 (2017)
211. M. Florian, M. Hartmann, A. Steinhoff, J. Klein, A.W. Holleitner, J.J. Finley, T.O. Wehling, M. Kaniber, C. Gies, The dielectric impact of layer distances on exciton and trion binding energies in van der Waals heterostructures. Nano Lett. **18**, 2725–2732 (2018)
212. H. Zhu, J. Wang, Z. Gong, Y.D. Kim, J. Hone, X.-Y. Zhu, Interfacial charge transfer circumventing momentum mismatch at two-dimensional van der Waals heterojunctions. Nano Lett. **17**, 3591–3598 (2017)
213. J.D. Sanchez-Yamagishi, J.Y. Luo, A.F. Young, B.M. Hunt, K. Watanabe, T. Taniguchi, R.C. Ashoori, P. Jarillo-Herrero, Helical edge states and fractional quantum Hall effect in a graphene electronhole bilayer. Nat. Nanotechnol. **12**, 118–122 (2017)
214. C. Jin, E. C. Regan, A. Yan, M. Iqbal Bakti Utama, D. Wang, S. Zhao, Y. Qin, S. Yang, Z. Zheng, S. Shi, K. Watanabe, T. Taniguchi, S. Tongay, A. Zettl, F. Wang, Observation of moiré excitons in WSe_2/WS_2 heterostructure superlattices. Nature **567**, 76–80 (2019)
215. J. Kunstmann, F. Mooshammer, P. Nagler, A. Chaves, F. Stein, N. Paradiso, G. Plechinger, C. Strunk, C. Schüller, G. Seifert, D.R. Reichman, T. Korn, Momentum-space indirect interlayer excitons in transition-metal dichalcogenide van der Waals heterostructures. Nat. Phys. **14**, 801–805 (2018)
216. K. Liu, L. Zhang, T. Cao, C. Jin, D. Qiu, Q. Zhou, A. Zettl, P. Yang, S.G. Louie, F. Wang, Evolution of interlayer coupling in twisted molybdenum disulfide bilayers. Nat. Commun. **5**, 4966 (2014)
217. M. Xia, B. Li, K. Yin, G. Capellini, G. Niu, Y. Gong, W. Zhou, P.M. Ajayan, Y.H. Xie, Spectroscopic signatures of AA' and AB stacking of chemical vapor deposited bilayer MoS_2. ACS Nano **9**(12), 12246–12254 (2015)
218. A.A. Puretzky, L. Liang, X. Li, K. Xiao, B.G. Sumpter, V. Meunier, D.B. Geohegan, Twisted $MoSe_2$ bilayers with variable local stacking and interlayer coupling revealed by low-frequency raman spectroscopy. ACS Nano **10**(2), 2736–2744 (2016)
219. L.M. Schneider, J. Kuhnert, S. Schmitt, W. Heimbrodt, U. Huttner, L. Meckbach, T. Stroucken, S.W. Koch, S. Fu, X. Wang, K. Kang, E.-H. Yang, A. Rahimi-Iman, Spin-layer and spin-valley locking in CVD-grown AA'- and AB-stacked tungsten-disulfide bilayers. J. Phys. Chem. C **123**, 21813–21821 (2019). https://doi.org/10.1021/acs.jpcc.9b07213
220. D. Kozawa, A. Carvalho, I. Verzhbitskiy, F. Giustiniano, Y. Miyauchi, S. Mouri, A.H. Castro Neto, K. Matsuda, G. Eda, Evidence for fast interlayer energy transfer in $MoSe_2/WS_2$ heterostructures. Nano Lett. **16**, 4087–4093 (2016)
221. S. Tongay, W. Fan, J. Kang, J. Park, U. Koldemir, J. Suh, D.S. Narang, K. Liu, J. Ji, J. Li, R. Sinclair, J. Wu, Tuning interlayer coupling in large-area heterostructures with CVD-grown MoS_2 and WS_2 monolayers. Nano Lett. **14**, 3185–3190 (2014)

222. P. Rivera, J.R. Schaibley, A.M. Jones, J.S. Ross, S. Wu, G. Aivazian, P. Klement, K. Seyler, G. Clark, N.J. Ghimire, J. Yan, D.G. Mandrus, W. Yao, X. Xu, Observation of long-lived interlayer excitons in monolayer $MoSe_2$-WSe_2 heterostructures. Nat. Commun. **6**, 6242 (2015)
223. H.M. Hill, A.F. Rigosi, K.T. Rim, G.W. Flynn, T.F. Heinz, Band alignment in MoS_2/WS_2 transition metal dichalcogenide heterostructures probed by scanning tunneling microscopy and spectroscopy. Nano Lett. **16**, 4831–4837 (2016)
224. J.R. Schaibley, P. Rivera, H. Yu, K.L. Seyler, J. Yan, D.G. Mandrus, T. Taniguchi, K. Watanabe, W. Yao, X. Xu, Directional interlayer spin-valley transfer in two-dimensional heterostructures. Nat. Commun. **7**, 13747 (2016)
225. J.S. Ross, P. Rivera, J. Schaibley, E. Lee-Wong, H. Yu, T. Taniguchi, K. Watanabe, J. Yan, D. Mandrus, D. Cobden, W. Yao, X. Xu, Interlayer exciton optoelectronics in a 2D heterostructure p-n junction. Nano Lett. **17**, 638–643 (2017)
226. P. Rivera, K.L. Seyler, H. Yu, J.R. Schaibley, J. Yan, D.G. Mandrus, W. Yao, X. Xu, Valley-polarized exciton dynamics in a 2D semiconductor heterostructure. Science **351**, 688–691 (2016)
227. D. Unuchek, A. Ciarrocchi, A. Avsar, K. Watanabe, T. Taniguchi, A. Kis, Room-temperature electrical control of exciton flux in a van der Waals heterostructure. Nature **560**, 340–344 (2018)
228. M. Shah, L.M. Schneider, A. Rahimi-Iman, Observation of intralayer and interlayer excitons in monolayered WSe_2/WS_2 heterostructure. Semiconductors **53**(16), 2140–2146 (2019). https://doi.org/10.1134/S1063782619120273
229. X. Zhou, N. Zhou, C. Li, H. Song, Q. Zhang, X. Hu, L. Gan, H. Li, J. Lü, J. Luo, J. Xiong, T. Zhai, Vertical heterostructures based on $SnSe_2/MoS_2$ for high performance photodetectors. 2D Mater. **4**, 025048 (2017)
230. H. Xue, Y. Dai, W. Kim, Y. Wang, X. Bai, M. Qi, K. Halonen, H. Lipsanen, Z. Sun, High photoresponsivity and broadband photodetection with a band-engineered $WSe_2/SnSe_2$ heterostructure. Nanoscale **11**(7), 3240–3247 (2019)
231. M.M. Furchi, A. Pospischil, F. Libisch, J. Burgdörfer, T. Mueller, Photovoltaic effect in an electrically tunable van der Waals heterojunction. Nano Lett. **14**, 4785–4791 (2014)
232. M. Ye, D. Zhang, Y.K. Yap, Recent advances in electronic and optoelectronic devices based on two-dimensional transition metal dichalcogenides. Electron. **6**(2), 43 (2017)
233. R. Ribeiro-Palau, C. Zhang, K. Watanabe, T. Taniguchi, J. Hone, C.R. Dean, Twistable electronics with dynamically rotatable heterostructures. Science **361**, 690–693 (2018)
234. E.V. Calman, M.M. Fogler, L.V. Butov, S. Hu, A. Mishchenko, A.K. Geim, Indirect excitons in van der Waals heterostructures at room temperature. Nat. Commun. **9**, 1895 (2018)
235. X. Wang, Y. Gong, G. Shi, W.L. Chow, K. Keyshar, G. Ye, R. Vajtai, J. Lou, Z. Liu, E. Ringe, B.K. Tay, P.M. Ajayan, Chemical vapor deposition growth of crystalline monolayer $MoSe_2$. ACS Nano **8**, 5125–5131 (2014)
236. H. Fang, C. Battaglia, C. Carraro, S. Nemsak, B. Ozdol, J.S. Kang, H.A. Bechtel, S.B. Desai, F. Kronast, A.A. Unal, G. Conti, C. Conlon, G.K. Palsson, M.C. Martin, A.M. Minor, C.S. Fadley, E. Yablonovitch, R. Maboudian, A. Javey, Strong interlayer coupling in van der Waals heterostructures built from single-layer chalcogenides. Proc. Natl. Acad. Sci. **111**, 6198–6202 (2014)
237. S. Tongay, J. Zhou, C. Ataca, K. Lo, T.S. Matthews, J. Li, J.C. Grossman, J. Wu, Thermally driven crossover from indirect toward direct bandgap in 2D semiconductors: $MoSe_2$ versus MoS_2. Nano Lett. **12**, 5576–5580 (2012)
238. I.G. Lezama, A. Arora, A. Ubaldini, C. Barreteau, E. Giannini, M. Potemski, A.F. Morpurgo, Indirect-to-direct band gap crossover in few-layer $MoTe_2$. Nano Lett. **15**, 2336–2342 (2015)
239. Y. Liu, N.O. Weiss, X. Duan, H.-C. Cheng, Y. Huang, X. Duan, Van der Waals heterostructures and devices. Nat. Rev. Mater. **1**, 16042 (2016)
240. Q. Zhao, T. Wang, Y. Miao, F. Ma, Y. Xie, X. Ma, Y. Gu, J. Li, J. He, B. Chen, S. Xi, L. Xu, H. Zhen, Z. Yin, J. Li, J. Ren, W. Jie, Thickness-induced structural phase transformation of layered gallium telluride. Phys. Chem. Chem. Phys. **18**(28), 18719–18726 (2016)

241. O. D. Pozo-Zamudio, S. Schwarz, M. Sich, I.A. Akimov, M. Bayer, R.C. Schofield, E.A. Chekhovich, B.J. Robinson, N.D. Kay, O.V. Kolosov, A.I. Dmitriev, G.V. Lashkarev, D.N. Borisenko, N.N. Kolesnikov, A.I. Tartakovskii, Photoluminescence of two-dimensional GaTe and GaSe films. 2D Mater. **2**, 035010 (2015)
242. P. Lu, J. Lang, Z. Weng, A. Rahimi-Iman, H. Wu, Hybrid structure of 2D layered GaTe with Au nanoparticles for ultrasensitive detection of aromatic molecules. ACS Appl. Mater. Interfaces. **10**, 1356–1362 (2018). https://doi.org/10.1021/acsami.7b14121
243. D. Voiry, A. Mohite, M. Chhowalla, Phase engineering of transition metal dichalcogenides. Chem. Soc. Rev. **44**, 2702–2712 (2015)
244. T. Jiang, H. Liu, D. Huang, S. Zhang, Y. Li, X. Gong, Y.-R. Shen, W.-T. Liu, S. Wu, Valley and band structure engineering of folded MoS_2 bilayers. Nat. Nanotechnol. **9**, 825–829 (2014)
245. A.M. Jones, H. Yu, J.S. Ross, P. Klement, N.J. Ghimire, J. Yan, D.G. Mandrus, W. Yao, X. Xu, Spin-layer locking effects in optical orientation of exciton spin in bilayer WSe_2. Nat. Phys. **10**, 130–134 (2014)
246. X. Liu, M.C. Hersam, 2D materials for quantum information science. Nat. Rev. Mater. **4**, 669–684 (2019)
247. D. Sanvitto, F. Pulizzi, A.J. Shields, P.C. Christianen, S.N. Holmes, M.Y. Simmons, D.A. Ritchie, J.C. Maan, M. Pepper, Observation of charge transport by negatively charged excitons. Science **294**(5543), 837–839 (2001)
248. C. Schneider, M.M. Glazov, T. Korn, S. Höfling, B. Urbaszek, Two-dimensional semiconductors in the regime of strong light-matter coupling. Nat. Commun. **9**, 2695 (2018)
249. A. Rahimi-Iman, *Technological Realization of Polariton Systems*. Springer Series in Optical Sciences, vol. 229, pp. 139–166. Springer International Publishing, Cham (2020)
250. J.K. Day, M.-H. Chung, Y.-H. Lee, V.M. Menon, Microcavity enhanced second harmonic generation in 2D MoS_2. Opt. Mater. Express **6**, 2360 (2016)
251. F. Yi, M. Ren, J.C. Reed, H. Zhu, J. Hou, C.H. Naylor, A.T.C. Johnson, R. Agarwal, E. Cubukcu, Optomechanical enhancement of doubly resonant 2D optical nonlinearity. Nano Lett. **16**, 1631–1636 (2016)
252. I. Žuti, J. Fabian, S. Das Sarma, Spintronics: fundamentals and applications. Rev. Mod. Phys. **76**, 323–410 (2004)
253. J. Du, Q. Wang, G. Jiang, C. Xu, C. Zhao, Y. Xiang, Y. Chen, S. Wen, H. Zhang, Ytterbium-doped fiber laser passively mode locked by few-layer Molybdenum Disulfide (MoS_2) saturable absorber functioned with evanescent field interaction. Sci. Rep. **4**, 6346 (2014)
254. K. Wu, X. Zhang, J. Wang, X. Li, J. Chen, WS_2 as a saturable absorber for ultrafast photonic applications of mode-locked and Q-switched lasers. Opt. Express **23**, 11453–11461 (2015)
255. Z. Luo, D. Wu, B. Xu, H. Xu, Z. Cai, J. Peng, J. Weng, S. Xu, C. Zhu, F. Wang, Z. Sun, H. Zhang, Two-dimensional material-based saturable absorbers: towards compact visible-wavelength all-fiber pulsed lasers. Nanoscale **8**(2), 1066–1072 (2016)
256. G. Wei, T.K. Stanev, D. Czaplewski, I.W. Jung, N.P. Stern, Interfacing monolayer MoS_2 with silicon-nitride integrated photonics, in *Advanced Photonics 2015*, p. IM4A.3. Optical Society of America (2015)
257. H. Chen, V. Corboliou, A.S. Solntsev, D.-Y. Choi, M.A. Vincenti, D. de Ceglia, C. de Angelis, Y. Lu, D.N. Neshev, Enhanced second-harmonic generation from two-dimensional $MoSe_2$ on a silicon waveguide. Light Sci. Appl. **6**, e17060–e17060 (2017)
258. M. Blauth, M. Jürgensen, G. Vest, O. Hartwig, M. Prechtl, J. Cerne, J.J. Finley, M. Kaniber, Coupling single photons from discrete quantum emitters in WSe_2 to lithographically defined plasmonic slot waveguides. Nano Lett. **18**, 6812–6819 (2018)
259. G.Q. Ngo, A. George, R.T.K. Schock, A. Tuniz, E. Najafidehaghani, Z. Gan, N.C. Geib, T. Bucher, H. Knopf, C. Neumann, T. Lühder, S. Warren-Smith, H. Ebendorff-Heidepriem, T. Pertsch, M.A. Schmidt, A. Turchanin, F. Eilenberger, Scalable functionalization of optical fibers using atomically thin semiconductors (2020). arxiv:2005.04201
260. Y.J. Noori, Y. Cao, J. Roberts, C. Woodhead, R. Bernardo-Gavito, P. Tovee, R.J. Young, Photonic crystals for enhanced light extraction from 2D materials. ACS Photon. **3**, 2515–2520 (2016)

261. A. Castellanos-Gomez, R. van Leeuwen, M. Buscema, H.S.J. van der Zant, G.A. Steele, W.J. Venstra, Single-layer MoS_2 mechanical resonators. Adv. Mater. **25**, 6719–6723 (2013)
262. M. Amani, D.-H. Lien, D. Kiriya, J. Xiao, A. Azcatl, J. Noh, S.R. Madhvapathy, R. Addou, S. KC, M. Dubey, K. Cho, R.M. Wallace, S.-C. Lee, J.-H. He, J. W. Ager, X. Zhang, E. Yablonovitch, A. Javey, Near-unity photoluminescence quantum yield in MoS_2. Science **350**, 1065–1068 (2015)
263. H. Fang, J. Liu, H. Li, L. Zhou, L. Liu, J. Li, X. Wang, T.F. Krauss, Y. Wang, 1305 nm few-layer $MoTe_2$ -on-silicon laser-like emission. Laser Photon. Rev. **12**, 1800015 (2018)
264. H.J. Lezec, A. Degiron, E. Devaux, R.A. Linke, L. Martin-Moreno, F.J. Garcia-Vidal, T.W. Ebbesen, Beaming light from a subwavelength aperture. Science **297**(5582), 820–822 (2002)
265. J. Scheuer, A. Yariv, Circular photonic crystal resonators. Phys. Rev. E **70**, 036603 (2004)
266. J. Scheuer, W.M. Green, G.A. DeRose, A. Yariv, Lasing from a circular Bragg nanocavity with an ultrasmall modal volume. Appl. Phys. Lett. **86**, 251101 (2005)
267. C.-M. Chiu, H.-W. Chen, Y.-R. Huang, Y.-J. Hwang, W.-J. Lee, H.-Y. Huang, C.-K. Sun, All-terahertz fiber-scanning near-field microscopy. Opt. Lett. **34**(9), 1327–1329 (2009)
268. F. Prins, D.K. Kim, J. Cui, E. De Leo, L.L. Spiegel, K.M. McPeak, D.J. Norris, Direct patterning of colloidal quantum-dot thin films for enhanced and spectrally selective outcoupling of emission. Nano Lett. **17**(3), 1319–1325 (2017)
269. G. Christmann, R. Butté, E. Feltin, J.-F. Carlin, N. Grandjean, Room temperature polariton lasing in a GaN/AlGaN multiple quantum well microcavity. Appl. Phys. Lett. **93**, 051102 (2008)
270. J.D. Plumhof, T. Stöferle, L. Mai, U. Scherf, R.F. Mahrt, Room-temperature Bose–Einstein condensation of cavity exciton-polaritons in a polymer. Nat. Mater. **13**, 247 (2014). Published online Dec. 2013
271. K.S. Daskalakis, S.A. Maier, R. Murray, S. Kéna-Cohen, Nonlinear interactions in an organic polariton condensate. Nat. Mater. **13**(3), 271–278 (2014)
272. Y.J. Chen, J.D. Cain, T.K. Stanev, V.P. Dravid, N.P. Stern, Valley-polarized exciton-polaritons in a monolayer semiconductor. Nat. Photonics **11**, 431–435 (2017)
273. G. Lanty, A. Bréhier, R. Parashkov, J.S. Lauret, E. Deleporte, Strong exciton-photon coupling at room temperature in microcavities containing two-dimensional layered perovskite compounds. New J. Phys. **10**, 065007 (2008)
274. M.E. Kleemann, R. Chikkaraddy, E.M. Alexeev, D. Kos, C. Carnegie, W. Deacon, A.C. De Pury, C. Große, B. De Nijs, J. Mertens, A.I. Tartakovskii, J.J. Baumberg, Strong-coupling of WSe_2 in ultra-compact plasmonic nanocavities at room temperature. Nat. Commun. **8**, 1296 (2017)
275. F. Wall, O. Mey, L.M. Schneider, A. Rahimi-Iman, Continuously-tunable light-matter coupling in optical microcavities with 2D semiconductors. Sci. Rep. **10**, 8303 (2020). https://doi.org/10.1038/s41598-020-64909-1
276. P. Qing, J. Gong, X. Lin, N. Yao, W. Shen, A. Rahimi-Iman, W. Fang, L. Tong, A simple approach to fiber-based tunable microcavity with high coupling efficiency. Appl. Phys. Lett. **114**, 021106 (2019). https://doi.org/10.1063/1.5083011
277. Z. Sun, J. Gu, A. Ghazaryan, Z. Shotan, C.R. Considine, M. Dollar, B. Chakraborty, X. Liu, P. Ghaemi, S. Kéna-Cohen, V.M. Menon, Optical control of roomerature valley polaritons. Nat. Photonics **11**, 491–496 (2017)
278. S. Dufferwiel, T.P. Lyons, D.D. Solnyshkov, A.A. Trichet, F. Withers, S. Schwarz, G. Malpuech, J.M. Smith, K.S. Novoselov, M.S. Skolnick, D.N. Krizhanovskii, A.I. Tartakovskii, Valley-addressable polaritons in atomically thin semiconductors. Nat. Photonics **11**, 497–501 (2017)
279. N. Lundt, C. Schneider, P. Nagler, A. Nalitov, S. Klembt, M. Wurdack, S. Stoll, T.H. Harder, S. Betzold, V. Baumann, A.V. Kavokin, C. Schüller, T. Korn, S. Höfling, Valley polarized relaxation and upconversion luminescence from Tamm-plasmon trion-polaritons with a $MoSe_2$ monolayer. 2D Mater. **4**, 025096 (2017)
280. N. Lundt, S. Stoll, P. Nagler, A. Nalitov, S. Klembt, S. Betzold, J. Goddard, E. Frieling, A.V. Kavokin, C. Schüller, T. Korn, S. Höfling, C. Schneider, Observation of macroscopic

valley-polarized monolayer exciton-polaritons at room temperature. Phys. Rev. B **96**, 241403 (2017)
281. H. Yokoyama, Physics and device applications of optical microcavities. Science **256**, 66–70 (1992)
282. E.M. Purcell, Spontaneous emission probabilities at radio frequencies. Phys. Rev. **69**, 681 (1946)
283. S. Noda, Seeking the ultimate nanolaser. Science **314**, 260 (2006)
284. S. Reitzenstein, A. Forchel, Quantum dot micropillars. J. Phys. D Appl. Phys. **43**(3), 033001 (2010)
285. C. Chakraborty, L. Kinnischtzke, K.M. Goodfellow, R. Beams, A.N. Vamivakas, Voltage-controlled quantum light from an atomically thin semiconductor. Nat. Nanotechnol. **10**, 507–511 (2015)
286. Y.-M. He, G. Clark, J.R. Schaibley, Y. He, M.-C. Chen, Y.-J. Wei, X. Ding, Q. Zhang, W. Yao, X. Xu, C.-Y. Lu, J.-W. Pan, Single quantum emitters in monolayer semiconductors. Nat. Nanotechnol. **10**, 497–502 (2015)
287. M. Koperski, K. Nogajewski, A. Arora, V. Cherkez, P. Mallet, J.-Y. Veuillen, J. Marcus, P. Kossacki, M. Potemski, Single photon emitters in exfoliated WSe_2 structures. Nat. Nanotechnol. **10**, 503–506 (2015)
288. A. Srivastava, M. Sidler, A.V. Allain, D.S. Lembke, A. Kis, A. Imamoglu, Optically active quantum dots in monolayer WSe_2. Nat. Nanotechnol. **10**, 491–496 (2015)
289. P. Tonndorf, R. Schmidt, R. Schneider, J. Kern, M. Buscema, G.A. Steele, A. Castellanos-Gomez, H.S.J. van der Zant, S. Michaelis de Vasconcellos, R. Bratschitsch, Single-photon emission from localized excitons in an atomically thin semiconductor. Optica **2**, 347 (2015)
290. I. Wilson-Rae, J.C. Sankey, H.L. Offerhaus, Editorial for special issue on nano-optomechanics. J. Opt. **19**(8), 80401 (2017)

Chapter 3
Light–Matter Interactions for Photonic Applications

Abstract Photonic applications of nanomaterials and quantum structures heavily rely on light–matter interactions and effective utilisation of their remarkable optical or optoelectronic properties. Typically, excitons in solids are coupled in one form or another to propagating or confined electromagnetic waves, i.e. either weakly or strongly, and interact with different kind of resonances present in their host medium. In addition, a tailored density-of-states for the electronic or the optical system through confinement potentials is commonly exploited to alter the coupling or dissipation behaviours of charge carriers or photons. Moreover, hybridisation of modes can occur which gives access to favourable properties for bosonic quasi-particles in solids. Many of these modifications and adjustments can be harnessed in photonic devices and fundamental studies of light–matter coupled oscillators, which can even undergo a phase transition towards a Bose–Einstein-like condensate in a solid-state platform. Light is shed on matter excitations in semiconductors in this chapter, with a focus on their interactions with photonic modes and peculiarities in monolayer materials. Furthermore, cavity–polaritons are discussed with eyes towards condensation phenomena and the influence of external fields. Particularly, polariton experiments involving terahertz waves are motivated.

3.1 Where Strong Interactions with Light Matters

Since the discovery of cavity–polaritons in 1992 [1],[1] microcavities continuously gained popularity amongst the quantum optics community and led to remarkable breakthroughs with eyes toward photonic applications and the exploration of novel physics [2–9], as they allow for the modification of the mode structure relative to that in free space and the strong spatial confinement of photons [10–12]. The former affects the spontaneous emission rate of light emitted from a structure, as well as its radiation pattern and direction [10, 13], whereas the latter enables a coher-

[1] Here, one typically uses this expression when excitons (or exciton–polaritons that represent coherent exciton modes in matter) in quantum structures are coupled to an optical cavity mode.

A. Rahimi-Iman, *Semiconductor Photonics of Nanomaterials and Quantum Structures*, Springer Series in Solid-State Sciences 196,
https://doi.org/10.1007/978-3-030-69352-7_3

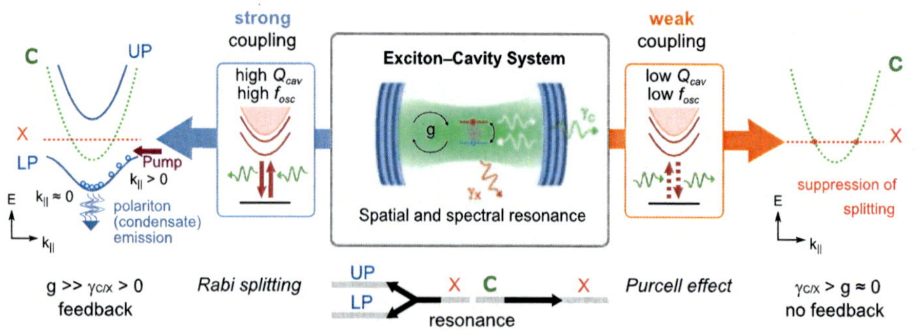

Fig. 3.1 Schematic diagram of two distinct light–matter coupling regimes for emitter–cavity systems. The optical microcavity (light field) and the embedded quantum emitter (two-level system) are indicated in the centre. When the cavity mode C and the exciton mode X are resonant, they can couple either strongly (left panel) or weakly (right panel). The boxes at the sides represent the exciton quasi-particle picture in which cavity light interacts with matter (coupling strength indicated). For low oscillator strength (f_{osc}) and low resonator quality factor Q_{cav}, the coupling strength is typically not enough to achieve a reversible energy transfer within the lifetime of cavity photons prior to their leakage through one of the resonator mirrors. If the coupling strength g is considerably larger than the decay rates for field γ_{C} and emitter γ_{X}, Rabi oscillations between the field and emitter populations with frequency 2Ω occur and new eigen-states are formed, the polariton modes (energy split by $2g = 2\hbar\Omega = \Delta E_{\mathrm{LP/UP}}$, with Ω the Rabi frequency). Thus, for the two regimes, different spectral (and temporal) features are obtained: polariton energy–momentum dispersion branches with anticrossing behaviour (usually damped oscillations due to decay channels), or a Purcell effect with crossing modes (and irreversible decay, whereas the spontaneous emission rate is enhanced/suppressed for resonant/off-resonant detunings, respectively)

ent energy exchange between the emitter and photon states with pronounced spatial and spectral resonance (see Fig. 3.1)[2]—promising various applications [14, 15]. Microcavity-based single-photon sources and nanolasers were a subject of intense studies in the last two decades [12, 16–27] and have left their mark on modern quantum cryptography concepts and light source achievements [28–31].

From Light-Dressed Matter States to Polariton Lasers

Intriguing light–matter states were demonstrated in the single emitter regime [22, 32, 33], and (actively altered) Rabi oscillations have been even monitored in the transient polariton luminescence [34, 35] or transient cavity reflection [36] with ultrafast spectroscopy techniques, which furthermore allow the manipulation of the quantum states via ultrashort optical [34, 35] or THz pulses [36]. Particularly, the light effective mass of polaritons in high-quality microcavities gave rise to the observation of dynamic BEC in solids [37–40] (also see [41–46] on the condensation of polaritons, and related discussions on this subject [47]), from which electrically-driven polariton lasers were later obtained [48–50]. More about polariton lasers is found for instance in the foundational papers [51, 52] or in newer literature [7, 53–58]. In addition,

[2]Ultimately, at the quantum limit, coupling of a single exciton with the vacuum light field in an empty cavity is obtained, leading to a vacuum Rabi splitting between the new eigen-states of the strongly-coupled system.

the physics of superfluids were studied at cryogenic temperatures exploiting condensation phenomena in solids [59–62]. The availability of (novel) semiconducting materials with large excitonic binding energies has even enabled the observation of room-temperature polariton condensation and superfluidity [63–66].

Prospects for Strong Light–Matter Coupling in Applications

The amount of applications proposed for light–matter coupled systems is ever rising, with leading scientists suggesting microcavity-based devices for optical circuits [67–70] (with increased mobility compared to their electronic counterparts), terahertz generation [71–76] and bosonic cascade lasers [77]. Even qubits based on polariton Rabi oscillators have been proposed [78]. Furthermore, also in the domain of quantum technologies, although much more fundamental in nature, a recent experiment demonstrated entanglement of a polariton with a single photon [79]. Towards on-chip photonic integration of devices with ultralow lasing thresholds, even plasmon–exciton–polariton lasers have been reported at room temperature using plasmonic structures with organic materials [80]. Such polariton lasers harness the active metamaterial planar technologies and lower thresholds than conventional lasers based on stimulated-scattering effects.

Thus, the wealth of microcavity physics is evident [15, 46, 81] and can be conveniently studied using well-established optical-spectroscopy techniques [82].

Other Forms of Polariton Systems

In addition to conventional optical cavities, metamaterials nanocavities could also be employed to achieve coupling in wavelength ranges beyond the infrared in the form of terahertz polaritons with systems comprising metallic resonators on top of semiconductor quantum-well structures, in which the intersubband transitions of semiconductor heterostructures can couple to the enhanced light field in the metamaterial nanocavity, as used for instance in [83]. Polaritons, which are coupled-oscillator systems, can be obtained with quantum-well or organic-molecule excitons in microcavities, as exciton–polaritons in bulk crystals [84–87], in systems with phonons [88–92], polarons [93, 94] as well as surface-plasmons [92, 95–97]. The latter is particularly highlighted for the class of 2D materials in [98], and surely there are many more examples not mentioned here (a wider overview is further given in [81]). Strikingly, beyond weak and strong coupling, one can for instance with strongly-confined light fields and organic molecules also obtain coupling strengths comparable to the transition frequencies in the light–matter system, or even larger. The corresponding regime is therefore referred to as the ultra-strong coupling or even deep-strong coupling regime, respectively [99, 100].

Tunable Polariton Microcavities

In recent years, polariton studies at room temperature with organic and inorganic high-binding-energy semiconducting materials gained from various tunable-cavity concepts. The material-independent continuous adjustability of cavity resonances could be for instance utilised for polariton-based chemistry, which can alter reaction pathways and rates inside cavities by precisely adjusting polariton energy levels or light–matter coupling situations (see for instance [102]). In addition, tunable cavities

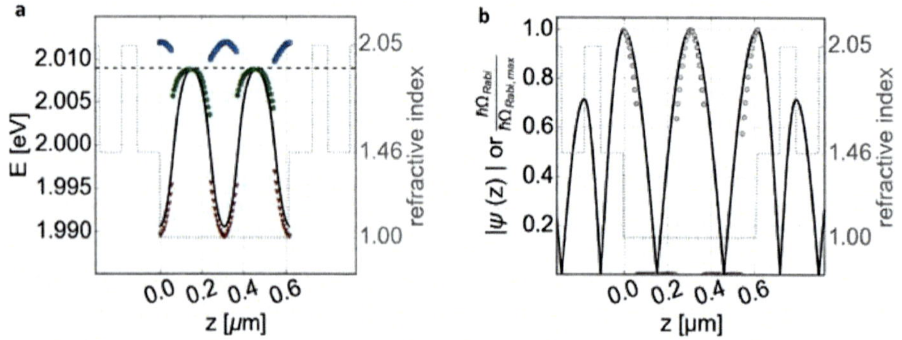

Fig. 3.2 Modelling data for tunable exciton–photon coupling in an open Fabry–Pérot-type optical microcavity with movable 2D semiconductor sheet. The structure is sketched in [101]. Spatial translation of the monolayer WS_2 in this theoretical work along the standing wave electric-field profile in z direction results in different coupling strengths which is evidenced by an extractable mode splitting. **a** Energies of the system (left axis): Blue squares and red triangles indicate the eigen-energies of the strongly-coupled system (upper and lower polariton states, respectively) for each sheet distance z to the left mirror, green circles correspondingly indicate the central peak energy attributed to the weakly-coupled system, according to calculated model reflectivity spectra. The black dashed line marks the 2D-exciton's energy and the solid black line the modelled cavity's resonance corresponding to an uncoupled system, i.e. $E_{res} = \hbar\omega_{cav}$, for which the imaginary part of the refractive index of WS_2 was neglected. **b** Absolute relative electric field strength (cavity photon probability amplitude) calculated for the empty cavity (black line), and the coupled-mode splitting $\Delta E = \hbar\Omega_{Rabi}$ (dots) obtained from calculated spectra and normalised to its maximum (left axis). The refractive index modulation of the empty DBR–DBR cavity with air spacer is displayed for clarity (grey dotted line, right axis).

such as the one in [103] (cf. Fig. 7.7) are not only interesting for single-photon sources and nanolasers, but promising for the investigation of the transition between weak and strong coupling (see Fig. 3.2), which may become possible by smoothly tuning the coupling across the 'exceptional point' [101], which links the two regimes. Interesting phenomena are expected at that specific point where only one complex solution exists for the coupled-oscillator system [104–106].

Quanta of Polarisation in Action

Indeed, polariton physics is not limited to cavity–polaritons. The mere presence of bulk excitons freely propagating through the host lattice of a solid as a polarisation wave manifests the existence of exciton–polaritons [87], the quanta of polarisation in the quasi-particle picture. These examples render the subject of matter excitations highly interesting and rich of physics and phenomena. More about polariton physics with an emphasis on Bose–Einstein condensation and polariton lasing can be found in [9].

3.1.1 Basics of Light–Matter Systems

Obviously, light–matter interaction goes beyond semiconductor physics and can be quantum-mechanically very fundamental, as atom–cavity physics shows. To give an idea of the formalism (after [107]), the example of a two-level atom interacting with a single field mode shall be highlighted. Therefor, one typically considers the interaction in dipole approximation, giving the prominent Hamiltonian $H_{\mathrm{I}} = -e\mathbf{r} \cdot \mathbf{E}(\mathbf{r})$, with spatial vector $\mathbf{r}$ and electric field vector $\mathbf{E}$. For stationary fields, this can be rewritten as

$$H_{\mathrm{I}} = \hbar g(\sigma_+ + \sigma_-)(f + f^\dagger), \tag{3.1}$$

with $\sigma_\pm$ the Pauli spin matrices corresponding to atom excitation (+) and de-excitation (−), f and $f^\dagger$ the photon annihilation and creation operator, respectively, and g the coupling strength. This is the fundamental light–matter interaction term, which implies two energy conserving and two non-conserving processes (see e.g. [107]). In dipole and rotating wave approximation (neglecting rapidly oscillating terms), the interaction term is reduced to the energy conserving processes and the Hamiltonian for the atom–field system becomes

$$H = H_0 + H_{\mathrm{I}} = \frac{\hbar\omega_{\mathrm{ab}}}{2}\sigma_{\mathrm{z}} + \hbar\omega f^\dagger f + \hbar g(f\sigma_+ + \sigma_- f^\dagger), \tag{3.2}$$

with the transition frequency ω_{ab}, the optical frequency ω, the Pauli spin matrix σ_{z}. This is the Jaynes–Cummings Hamiltonian with atom, field and interaction term, with the zero energy level set half-way between the two atomic levels a and b, so that unperturbed the two-level system's energy is $\pm\frac{\hbar\omega_{\mathrm{ab}}}{2}$. The interesting interaction processes are $f\sigma_+$, i.e. one photon gets absorbed and atom correspondingly excited, and $\sigma_- f^\dagger$, the opposite process where one photon is emitted and the atom de-excited. It is the starting point for the description of quantum Rabi oscillations.

In other words, strong interactions between a radiation field (optical mode) and an emitter (electronic mode) can lead to reversible energy exchange (periodic oscillations), which manifests itself in the hybridisation of the coupled-oscillator system and gives rise to new eigen-states in the strong-coupling regime, i.e. mixed states originating from a linear superposition. Indeed, two bosons (oscillators with ω_{ab} and ω) form composite bosons (oscillators with $\omega_\pm$). For exact resonance conditions, the new states represent a half–half hybrid. The detuning-dependent fractions of the constituents (i.e. mode and excitation) are commonly expressed by the Hopfield coefficients [85] or amplitudes of the dressed states in the diagonalised Hamiltonian. To give examples from semiconductor physics, hybridisation of a single exciton with the empty vacuum field can be achieved in special microcavities (see Sect. 4.3), or a quantum-well exciton ensemble with a planar-microcavity mode, or a macroscopic

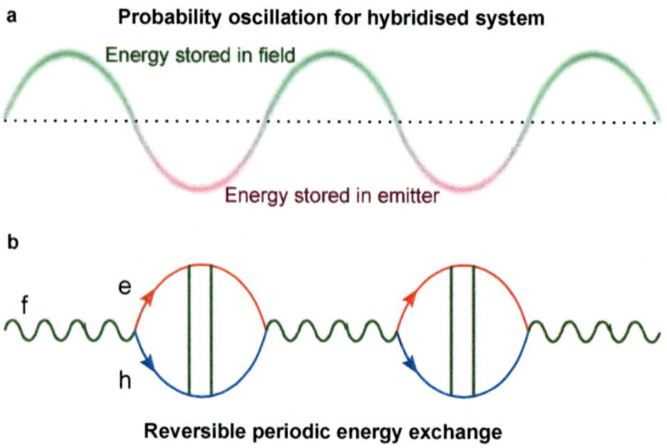

Fig. 3.3 The temporal evolution of an undamped hybrid light–matter system, here representing an exciton–polariton in a semiconductor, is depicted as the probability to find the energy stored in the field or emitter (**a**) and in a diagrammatic representation (**b**). **a** The dotted line represents the average. **b** Its photonic state is indicated as a wavy line and the electronic particles electron and hole in the semiconductor crystal lattice, of which the exciton is comprised, as solid (curved) lines. From left to right: An incident photon creates an electron–hole pair. This dipolar pair, which is bound by Coulomb interactions and exists as an exciton, consecutively recombines under emission of a photon, which can again excite the electronic system and be afterwards re-emitted and so forth. Here, Coulomb interaction is represented by a virtual exchange of photons between the electron and hole indicated by the vertical double line. **b** Freely drawn after [87]. To obtain such hybrid, the dipole oscillator strength of the exciton must be sufficiently large. Note that such coupling regime results in a (periodic) coherent energy exchange and that the initial photon and re-emitted photon are in phase with each other and quantum-mechanically coupled to the matter excitation. Therefore, a polariton can act as a good quantum bit (qubit) within the dephasing time of the coupled system, or can for instance be used in entanglement schemes with single photons

polarisation in bulk with an irradiated electromagnetic wave. A diagrammatic representation of such a hybrid light–matter system, referred to as exciton–polariton, is presented in Fig. 3.3.[3]

[3]This seemingly simple diagram is of very fundamental nature and invites to a deeper consideration of quantum electrodynamics (see double lines connecting the "pathways" of electrons and holes in Fig. 3.3). It deserves more attention than being the content of a footnote, but this would be easily out of the scope for this work, which has a different focus. With photons the fundamental excitations of the electromagnetic vacuum field, the virtual photons are understood as the mediator of force between charged particles (e.g. electrons and positrons, protons, or holes in solid state crystals, the defect electrons). Thus, processes such as charge-carrier scattering processes can be depicted as quantum-mechanical processes involving the exchange of photons (momentum transfer). Note that while hardly anyone knows with certainty what the fermionic (spin-half) electrons "really look like" (the same true for the bosonic photons), based on their behaviour and properties *one is tempted* to make presumptions about these entities. Interestingly, while in classical electrodynamics field lines are drawn from positive to negative charge (indicating force), speaking of springs (in German: 'Quellen') and drainages ('Senken') of the electric field, respectively, for the quantum electrodynamical picture one could literally understand charged particles as such (emitting and absorbing sheer endless numbers of virtual photons), with the density of field lines indicating the density of virtual photons across the path from negative to positive charge. At this point, the attention

Note that in most practical systems and experiments, one is not in the strong but weak-coupling regime and observes or studies behaviour such as linear absorption, irreversible spontaneous emission, ASE/lasing, Purcell effect etc., typically the case for interactions of (e.g. incoherent, broadband, unconfined) light with low-oscillator strength optical dipole transitions (e.g. electron–hole plasma, microscopic polarisation) and strongly-dissipative systems (e.g. leaky cavities, bulk) or systems with high dephasing rate (e.g. with high scattering rates, phonon coupling) (cf. [7, 87, 107]).

Semiconductor Crystal Situation
For the semiconductor crystal interacting with a monochromatic light field, the many-body Hamiltonian taking into account excitations of the mechanical system (lattice) H_{ph}, the electronic system H_{X}, the light field H_{F}, and two interaction terms $H_{\text{F-X}}$ and $H_{\text{X-ph}}$ concerning light and excitons as well as excitons and phonons, respectively, reads:

$$H = \sum \left(H_{\text{ph}} + H_{\text{X}} + H_{\text{F}} + H_{\text{F-X}} + H_{\text{X-ph}}\right). \tag{3.3}$$

For negligible phonon contributions, this Hamiltonian reduces to a similar one as in (3.2). Here, the role of the atom is taken by the exciton, the quantum of matter excitation, with $H_{\text{X}} = E_{\mathbf{Q}} e^{\dagger}_{\mathbf{Q}} e_{\mathbf{Q}}$ (**Q** is the centre-of-mass momentum). The obvious similarity can be easily seen from the time-independent eigen-value equation [87, 108] for the Wannier-like Coulomb-bound electron–hole pair (a composite quasi-particle) in matter:

$$\left(-\frac{\hbar^2 \nabla^2}{2\mu} - \frac{e^2}{4\pi\epsilon\,|r|}\right)\varphi_\alpha(r) = E_\alpha \varphi_\alpha(r), \tag{3.4}$$

whereas μ, $\epsilon = \epsilon_{\text{r}}\epsilon_0$, e and r denote the reduced two-particle system's mass, the permittivity, the electron charge and the electron–hole distance, respectively. E_α and $\varphi_\alpha(r)$ are the exciton's eigen-energies and (envelope) wave-functions, respectively, for the quantum numbers $\alpha = n, l, m$ (main quantum number, angular momentum and magnetic quantum number for atoms, respectively) very similar to the situation for the hydrogen atom. Quantitatively different by the reduced mass and the dielectric screening in the solid, this equation (also referred to as Wannier equation) yields a kinetic free-particle term for the centre-of-mass motion and a series of (bound) hydrogenic states in the Coulomb potential referred to as excitonic Rydberg series, in analogy to the atom Rydberg series. Thus, exciton binding energies are in most semiconductors about 2–3 orders of magnitude smaller than the Rydberg energy 13.6 eV. Of course, the situation is drastically altered in 2D semiconductors due to

---

of the interested reader is also drawn to an effect with practical relevance for nano-(opto-)mechanical systems known as the Casimir effect, one explanation of which involves the pressure from virtual photons in 3D/bulk, or the underpressure in 1D-confined space regarding virtual photons—the other explanation involving van-der-Waals interactions, the explanation of which from a quantum physics point involves vacuum fluctuations (here, an interesting connection to 2D materials, also referred to as vdW materials, can be established).

the influence of the environment (modified screening) and the low-dimensionality of atomically-thin monolayers (strong quantum confinement) (see e.g. [109] for a detailed discussion of 2D excitons, also see Rytova–Keldysh potential for 2D systems [110–112]).

In fact, in the above many-body Hamiltonian (3.3), the bosonic lattice vibrations' Hamiltonian H_{ph} is similar to the one for the bosonic light field, with phonon oscillation frequency and phonon creation/annihilation operators instead of light frequency and photon creation/annihilation operators. For the sake of clarity, the different indices corresponding to the different modes, polarisation, quantum states for the particles involved, i.e. phonons, photons and excitons, respectively, have been skipped in this overview representation. For theoretical details and elaborate descriptions of the individual Hamiltonians and their role in the dynamics of semiconductors, the interested reader is referred to general semiconductor quantum theory described in textbooks (e.g. [113]) and recent monolayer TMDC related theoretical considerations (e.g. [114]).

3.2 Matter Excitations

Optical transitions and the resulting measurable "optical band gap" are very characteristic for absorption and emission properties of semiconductor quantum structures [115, 116]. In the low density regime, they represent excitonic modes hosting correlated Coulomb-bound charge carriers in a dilute gas, whereas at intermediate densities a bath of correlated free charge carriers adds up to the Coulomb-bound excitations of the system (see [117]), where multi-particle scattering processes and excitation-induced dephasing raise. At high densities, the non-equilibrium system is governed by an incoherent electron–hole plasma (see [118]), where it also experiences density-dependent band-gap renormalisation (for a quantum-theoretical description, see [113]). All these fundamental regimes can play an important role in semiconductor device physics. One typical example, where a transition from a low-density to a high-density charge-carrier regime is sought, is that of laser diode operation under the Bernard–Duraffourg condition, i.e. in a non-equilibrium regime of optical transparency at the effective gap wavelength for optical amplification by *stimulated emission*[4] (see [119], and, for the history of laser conditions in semiconductors, [120]). Commonly, "population inversion" is a term used to express

[4]This is a quantum physical effect that refers to the induced emission—the opposite process to (induced) absorption—which relaxes an electronic charge carrier from its excited state to its ground state in a simplified two-level picture under emission of a photon with $\hbar\omega = E_2 - E_1$ with certain probability, provided that a resonant photon is present in the system. This stimulated emission process results in the release of a clone of the existing (irradiated) photon, which has equal energy, phase, polarisation and propagation direction. Thus, the resulting light field is coherent, characterised by a random photon number distribution, and its photons exhibit phase relation over relatively long distances.

the surpassing of a laser threshold condition (see [121] for a definition of a laser threshold).[5] The lower the threshold, the more efficient the laser.

3.2.1 *Excitons as Composite Bosons*

In contrast to higher-density conditions, the picture looks different at the quantum limit of single-particle excitation. Here, the fundamental excitation of matter is characterised by the formation of a hydrogen-like quasi-particle [122] consisting of an excited negatively-charged electron in the conduction band (CB) Coulomb-bound to a positively charged "hole"—a defect electron in the valence band (VB) (Fig. 3.4). These Rydberg-like excitons[6] deliver spectral features very similar to their atomic counterparts, as investigations on cuprous-oxide excitons in bulk crystals with their large binding energies revealed [123, 124]. In a diluted 'gas', the description of these quasi-particles as ideal bosons and individual systems holds true, since the screening effects from other uncorrelated carriers and excitation-induced dephasing can be neglected. In quantum wells, excitons, which are composite bosons, are attractive for investigations of collective properties and quantum phenomena [125, 126], such as Bose–Einstein condensation [127–131]. However, in many semiconductor systems, exciton condensation was not possible due to the typically unfavourable conditions. On the one hand, condensate droplets formation can take place in potential landscapes of real[7] quantum wells, leading to a system without superfluidity, referred to as Bose glass. On the other hand, excitons are prone to disorder and density-dependent dephasing effects which make the formation of a long-range order unlikely in many practical systems even at ultralow temperatures.

---

[5]To achieve lasing conditions, either an optical pump density or an electrical current density exceeds the threshold density, which according to laser rate equations fundamentally depends on the optical losses τ_C and the light–matter interaction strength $W_{if} \propto |M|^2 \propto \mathbf{d}$. Here, W_{if} is the transition rate, which can be expressed through the transition matrix element $M = \langle f | \mathbf{d} \cdot \mathbf{E} | i \rangle$. i and f are initial and final state, respectively. The transition rate is proportional to the emitter's dipole moment $\mathbf{d}$, thus, optical gain as well.

[6]Coulomb-bound electron–hole pairs in solids are quasi-particles with hydrogen-like Rydberg series of energy levels, resembling rather a positronium atom than hydrogen due to the similarity of the masses of electron and hole, and the fact that positronium is also a particle–anti-particle compound, whereas in hydrogen, an electron is bound to the much heavier proton (itself a composite particle built from quarks). The exciton binding energies can vary between weak for Wannier–Mott-like species to strong for Frenkel-like species, with their wave-function delocalised over many lattice sites or strongly localised, respectively. The oscillator strength of the dipole determines the interaction strength with electromagnetic waves. In low-dimensional crystal structures, quantum confinement can raise the binding energies with respect to the bulk crystal.

[7]Real refers to the situation in practical systems achieved by semiconductor growth technology in contrast to ideal systems.

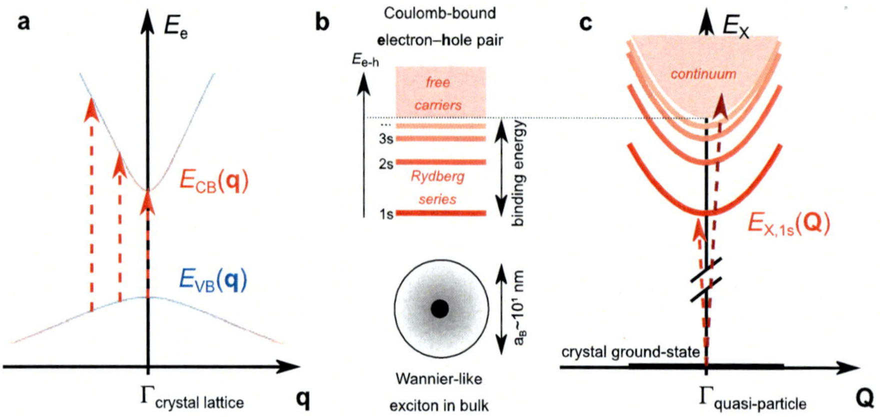

Fig. 3.4 **a** Schematic band structure depiction for a direct-gap semiconductor with electronic transitions from the valence band (VB) to the conduction band (CB). Here, the photo-induced electronic excitation is indicated by arrows (nearly vertical, as the photon momentum is negligible compared to the relevant phase space ranges for crystal electrons). The bands in the energy–momentum-space diagram around the Γ point are oversimplified. q denotes the quasi-momentum for electrons/holes in the lattice. This single-particle excitation picture does not take into account the formation of correlated electron–hole pairs and cannot represent its energetics. **b** Rydberg-series of energy states for the hydrogen-like electronic quasi-particle, which represents a fundamental electronic matter excitation. The formation of Coulomb-bound electron–hole pairs, referred to as excitons, is sketched with centre of mass (dot), envelope function (shaded area) and size (ring), typically given by the Bohr radius for the main quantum number n (states $1s$, $2s$, ...). **c** In contrast to **a**, the quasi-particle excitation picture displays the properties of excitons with centre-of-mass momentum Q that are the dispersions for different resonances ($1s$, $2s$ or higher) and the corresponding effective mass (approximated from the parabolic dispersion around $\Gamma \equiv Q = 0$ point). Resonant (into bound states) and off-resonant (into continuum states, i.e. ionised excitons) optical excitation are indicated by red and dark-red arrows, respectively. By returning to the crystal ground state, the electronic excitation releases its energy to the environment/system either radiatively (photon emission) or nonradiatively (e.g. through Auger-like processes). Note that in the sketches (**a**–**c**) the energies are not correctly scaled and merely serve visualisation purposes

The Long Path to Exciton Condensates

Even after long-lasting intensive research following the predictions of exciton condensation [133, 134] and pioneering work on collective properties of such quasi-particles [125], the desired results were lacking. A high decoherence rate via phonon scattering in appropriate structures posed a major challenge, whereas the Mott transition [135, 136] from an insulator state of matter in the bosonic state to a metallic (electron–hole) plasma in the fermionic regime above a critical particle density [137] set an upper limit to the critical densities for condensation experiments. Many systems such as coupled quantum wells and spatially indirect excitons were considered to circumvent such constraints [130]. Nevertheless, the effective mass of the quasi-particle required experiments to be carried out in the milli-Kelvin range in solids.

To show the role of the effective mass (see Fig. 3.5, in analogy to the 3-axes version in [81]), the equation for the critical temperature for BEC in an ideal 3D system with massive bosons is given here:

$$k_B T_{\text{critical}} = \frac{2\pi\hbar^2}{m_{\text{particle}}} \left(\frac{n_{\text{particle}}}{2.612}\right)^{2/3}, \tag{3.5}$$

with k_B the Boltzmann constant, n_{particle} the particle density and m_{particle} the particle (effective) mass (see e.g. [131]).

Only due to a long stamina and repeated efforts were the experimental signatures obtained with credible results [138–142]. Recently, interlayer excitons in 2D-semiconductors type-II heterostructures have been targeted as a new platform for exciton condensation and superfluidity studies [143, 144] or 2D electronic systems in quantum Hall bilayers [145–147].

On Quantum Emitters and Quantum Droplets

If strongly localised in three directions, excitons can act very nicely as quantum emitters (single-photon emitters) due to the modified energy spectrum in quantum boxes (0D structures, referred to as quantum dots) [12, 16, 19, 20]. Quantum dots can be also found in ultra-short-pulse semiconductor laser chips and in optical elements for such systems [148–150], such as saturable absorbers, and even provide record high output powers if delivered in multi-stacking configuration as heavily-pumped active layers in semiconductor disk lasers [151–153]. In most cases, however, excitons are utilised in 2D systems such as quantum wells, which confine the charge carriers only in one direction (see for instance [87, 154, 155]). Beyond single correlated electron–hole pair entities in quantum-well systems, even higher correlations—quantum droplets of electrons and holes, referred to as "dropletons" in quantum systems—were studied in the literature in non-equilibrium conditions of an excited electron–hole plasma [156].

Coherent Excitons and Exciton–Polaritons

Optical transitions and light propagation are directly related to the polarisability and susceptibility of a medium. In classical and semiclassical linear optics, the concept of the harmonic oscillator with Lorentzian lineshape and oscillator strength ($f \propto N\,|M|^2$, with N the number of contributing resonant dipole oscillators) provides a very effective approach to describe light–matter interactions and particularly macroscopic polarisation waves propagating through matter (see e.g. [87] for a comprehensive description and discussion of semiconductor optics). By employing the implicit polariton equation (also see [88, 157, 158])

$$\frac{c^2\mathbf{k}^2}{\omega^2} = \epsilon_b + \sum_i \frac{f_i}{\omega_{0,i}^2(+2\omega_{0,i}A\mathbf{k}^2) - \omega^2 + \mathrm{i}\omega\gamma_i}, \tag{3.6}$$

based on $\omega = ck$ and $\mathbf{k}^2 = k^2 = \tilde{n}^2(\omega) \cdot k_{\text{vacuum}}^2 = \epsilon(\omega) \cdot (\omega/c)^2$ (with complex functions $\tilde{n}^2(\omega) = \epsilon(\omega)$), the light–matter states' dispersion relation in the crystal can be derived without and with spatial dispersion. The latter is represented in

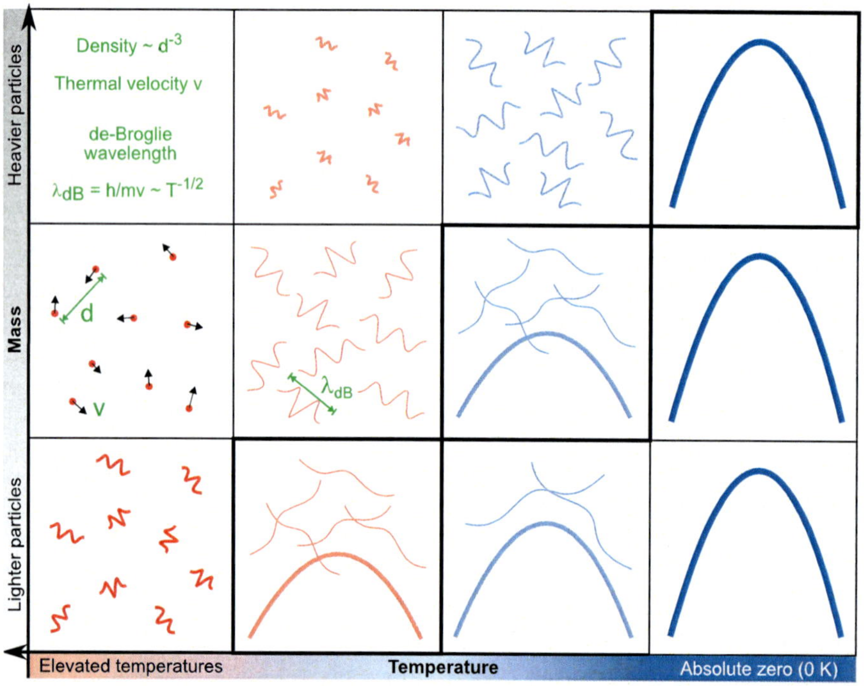

Fig. 3.5 Schematic 2D diagram of the physics behind Bose–Einstein condensation (BEC), showing the impact of the mass on the critical temperature. Boxes illustrate the formation of a condensate from a dilute gas of massive bosons for different temperatures and masses. From left to right, the temperature decreases towards the absolute zero point (0 K). From top to bottom, the particle mass (an effective one for quasi-particles) is decreased. For a given mass configuration (row), the critical temperature for BEC is highlighted by a thick-lined box. At elevated temperatures, the gas of particles obeys the classical Boltzmann statistics. By decreasing the temperature, an increased matter–wave overlap is obtained according to the growing particle de-Broglie wavelength λ_{dB}. Below a critical temperature T_{critical}, a condensate is formed, whereas the ideal gas of bosons is described by the Bose–Einstein distribution function. Note that the majority of particles is in the same quantum degenerate state and described by the same wave-function, rendering the condensed particles indistinguishable. Hence a macroscopically-occupied state with phase-coherent population due to a spontaneous symmetry break is a key signature of BEC. At finite temperatures, a minority of uncondensed particles coexists, expressed by the condensate fraction n_0/n_{total} (ground-state over total particle density). Note the similarity to superfluidity, with normal and superfluid phase coexisting. In contrast, at 0 K, all particles occupy one (ground) state forming a pure Bose condensate. Heavier bosons at given temperature overlap less than the lighter ones due to their shorter λ_{dB}. Thus, by reducing the mass, at given temperature, the critical conditions can be met; or expressed in a more practical way, lighter particles enable condensation studies at elevated temperatures. For bosons with mass of the order of the free electron mass m_0, $T_{\mathrm{critical}} \approx$ mK. This scheme is inspired by the original (1D) chart created by Ketterle and used on his research group's online representation [132], and practically extended to a reduced form of the author's 3D chart shown in [9] (Chap. 1 therein), which even shows a third axis (density n) involving the critical particle density n_{critical}

(3.6) by the term in brackets, whereas $\epsilon \rightarrow \epsilon(\omega, \mathbf{k})$ and related quantities become k-dependent in case of spatial dispersion [87].[8] For an introductory purpose and for clarity, schematic diagrams of these dispersion relations are shown in Fig. 3.6c, d based on the Lorentz oscillator model for the dielectric function (a, b). Note that in (3.6), the sum over different resonances (counting index i) with their specific frequencies and homogeneous linewidths is considered for the dielectric function, which in the case of a single isolated resonance in a given spectral region can be approximated without sum.

Similar to the introductory overview given in [159], a brief summary is analogously provided here: When coupling to a photon mode is impossible—either due to dipole-forbidden transitions or due to the absence of light states at high momentum-space vectors—bare excitons are found in the lattice of the host medium, also known as dark excitons (indeed, there can be different types of dark excitons, see [160] and references therein). In contrast, the optical resonance of this quasi-particle state, which is a consequence of exciton–polariton formation, has its peculiar energy (for simplicity it is often just referred to as the exciton peak): it is on the one hand less energetic than the direct band-gap transition, due to the binding energy of the electronically-neutral bound system, which can be described in the two-particle picture with a free-particle dispersion and excitonic effective mass (cf. Fig. 3.4c). On the other hand, the optical transitions occurring only at very low momenta within the range of light states (within the light cone) exhibit a shifted energy with respect to the bare exciton resonance, since the bright (i.e. optically visible) exciton is usually discussed in the picture of polaritons (see [85, 87, 161–163] and references therein).[9] The shift of the longitudinal branch compared to the commonly discussed transversal exciton strongly depends on the oscillator strength. It can range from 0.08 meV for GaAs [164] over tens of meV to even $\approx$1 eV for transitions in molecular crystals [165].

The polaritons represent the quantisation of the polarisation—i.e. they are the eigen-states (basis vectors) of the light–matter Hamiltonian [113]—and are mixed states of photon modes and the relevant material resonance. One should note that, here, the coupling with the ordinary light cone (photon states) in the material is discussed in contrast to cavity–polaritons that are the product of coupling with a cavity mode. In a typical PL scenario, the excitation takes place in the upper polariton (UP) branch within the light cone [87, 166, 167]. From there, the excitation scatters/relaxes to lower lying states in the upper or lower polariton (LP) branch.

[8] This is not detailed here, as it would be quickly out of the scope of this work, but briefly addressed for clarity. Nevertheless, it shall be noted that while f and γ likely are considered $\mathbf{k}$-independent, the references provided in Chap. 5 of [87] experimentally show a dependency for dipole-allowed transitions. In the sketched polariton dispersions of Fig. 3.6c, d, the literature model only considered a parabolic dispersion $\omega_0(\mathbf{k})$ with prefactor A.

[9] Note that a typical semiconductor PL spectrum with ‘bright’ modes can be dominated by bound-exciton-complexes emission and phonon sidebands as well (see [87]), and the emission at the exciton resonance does not necessarily represent coherent excitons, but rather incoherent excitons (microscopic polarisation) with a significant contribution from electron–hole plasma towards higher densities (cf. [113, 117, 118]).

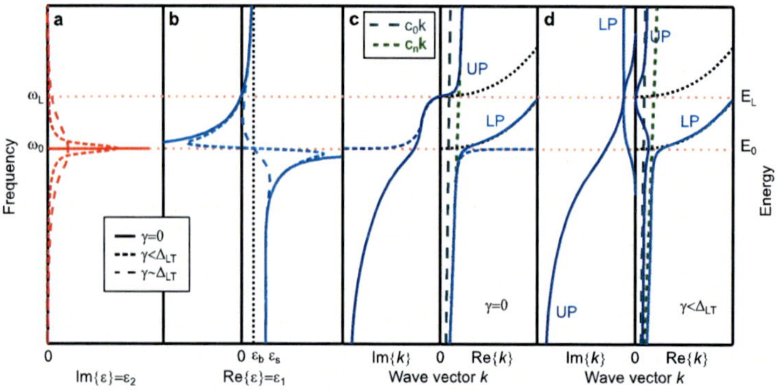

Fig. 3.6 Fundamental exciton–polariton model, freely sketched after [87]. Here, the harmonic oscillator approximation of the macroscopic polarisation in a dielectric medium is the starting point for the description of a coherent exciton's dipole resonance with oscillator strength f according to the material specific dielectric function $\epsilon(\omega)$. This also yields the exciton–polariton dispersion branches. **a** Imaginary part of ϵ in the vicinity of the transversal exciton resonance $\omega_0 = \omega_T$ indicating a peak centered at ω_0 for different damping situations γ. For zero damping, a δ peak would be obtained, for small damping, a Lorentzian line profile results which broadens with increased damping. ω_0 is found in the real part of ϵ in **b** as the frequency of the singularity $\epsilon_1^{\pm}(\omega \to \omega_0) \to \pm\infty$ for the undamped oscillator ($\gamma = 0$). Note that negative ϵ_1 lead to the so-called Reststrahlenbande (i.e. the lack of a propagating mode, cf. photonic stop band), which is smeared out with increasing damping, and reduced with spatial dispersion (k-dependent LP adds a propagating mode with high k in the range of ω_T to ω_L). For small damping (i.e. smaller than the longitudinal–transverse splitting Δ_{LT}), the two opposing branches connect smoothly and ω_0 lies at the crossing of ϵ_1 with the background dielectric constant level ϵ_b. The zero-crossing of ϵ_1 marks the longitudinal frequency ω_L. $\epsilon_s = \epsilon_b + f/\omega_0$ is the static dielectric constant, an approximately constant value well below ω_0. In fact, the low and high frequency values $\epsilon_{b/s}$ are linked to the two resonances by the Lyddane–Sachs–Teller relation $\epsilon_s/\epsilon_b = \omega_L^2/\omega_T^2 > 1$ for $f > 0$. For larger damping ($\gamma \to \Delta_{LT}$), the resonance smears out and no zero crossing of ϵ_1 occurs. This example of an isolated resonance far away from other resonances of the material is well described by the classical Lorentz oscillator model (for details see e.g. [87], Chaps. 4, 5, 13 are particularly inviting in this context). For vanishing damping, $f = (\omega_L^2 - \omega_T^2)\epsilon_b \propto |M|^2$ (the dipole transition matrix element squared). **c** and **d** show the theoretical model for exciton–polaritons with spatial dispersion for two cases, without and with small damping, respectively. Wide/short-dashed nearly-vertical lines sketch the light cone in vacuum/medium ($c_{0/n}k$) and dotted curves the dispersion of the excitonic resonances $\omega_{L/T}$. Due to the interaction of the light field with the polarisation in matter, the coupled system is represented by the lower/upper polariton branches (LP/UP), with clear anticrossing of light and transverse exciton mode for the undamped case. Note that the slope of the light cone changes from $\hbar c/\sqrt{\epsilon_s}$ to $\hbar c/\sqrt{\epsilon_b}$ from LP to UP (i.e. below to above ω_0). (**a**–**d**) Horizontal dotted lines mark the levels of ω_T, to which the LP and UP (real and imaginary part, respectively, short-dashed curves) converge in **c** in the case without spatial dispersion, and ω_L, from where the UP begins in **c** at $k = 0$. For larger damping ($\gamma \to \Delta_{LT}$, not shown here), the interesting and rather common situation leads to a dominant role of the UP within the light cone for the matter resonance with real and imaginary parts of the dispersion relation closely resembling the shape of **b** ϵ_1 and **a** ϵ_2, respectively. This is, however, not surprising, since $\mathrm{Re}/\mathrm{Im}\,\{k\} = \mathrm{Re}/\mathrm{Im}\,\{\tilde{n}\}\,\omega c^{-1}$ (i.e. $\propto$ refractive index n / extinction κ)

Caused by momentum conservation of the parallel part with respect to the matter–air interface (see for instance [87]), only polaritons within the light cone can be emitted to free space if phonon-assisted emission is neglected [167].

From Macroscopic Polarisation to Incoherent Excitons

For systems with a strong oscillator strength, propagating (oscillating) light fields in a medium induce a coherent (oscillating) macroscopic polarisation, i.e. they result in coherent excitons (oscillating polarisation coupled to the electromagnetic wave) understandable as exciton–polaritons. In the literature, this coupling phenomenon for light linked to strong refractive index changes around resonances is also heard of as "slow-light", "dark polaritons" and "stationary light pulses" (mentioned in Chap. 5 of [87], see references therein). However, this macroscopic coherent polarisation wave breaks down after a certain dephasing time into microscopic (local) polarisation, i.e. incoherent excitons. In other cases with weak interaction with matter, light experience retardation proportional to the refractive index $n > 1$ of the medium. In the extreme case, when light states are not allowed in the medium, light simply cannot penetrate into it and will be totally reflected, observed as Reststrahlenbande in the optical spectrum. In fact, a negative permittivity around exciton resonances leads to the same effect as total internal reflection or reflection from a one-dimensional photonic crystal interface (Bragg mirror).

Light states can be easily tailored nowadays using optical resonators [11], and therewith, the absorption, transmission, reflection and radiative-decay behaviour (i.e. energetics, directionality, lifetime) can be drastically modified using weak [168] and strong coupling phenomena [1, 22, 32, 33] (also see [4, 7, 169]).

While the formation of exciton–polaritons can be evidenced by Fourier-space-resolved optical spectroscopy, in which their energy–momentum dispersion exhibits the very light nature of the hybridised quasi-particles represented by the curvature of the dispersion, the build-up of an (incoherent) exciton population on ps-to-ns timescales can be probed effectively using ultrafast optical-pump–THz/IR-probe schemes. With the probe pulses covering the intra-excitonic transitions (see [170, 171]), such pump–probe schemes can serve as a means of verification of formed excitons (composite quasi-particles) and timescale investigations, even if the excitonic species does not couple to the light field. In fact, such method has been also used for the measurement of a dark exciton fraction in light–matter-coupled systems [172].

Ongoing Efforts to Characterise Excitonic Systems

Several achievements in Marburg in the last years opened the pathway to novel studies including the characterisation of indirect excitons in semiconductors via optical-pump–THz-probe spectroscopy [173], the characterisation of charge-transfer excitons in heterostructures by four-wave-mixing and THz spectroscopy [174], the ultrafast gain dynamics in quantum-well chip structures via transient reflectivity measurements [175], and the manipulation and control of quantum-well excitons [176]. The latter experiment combined a four-wave mixing experiment in a unique fashion with a THz pulse that was used to perturb the coherent system in the two-optical-pulses experiment. The aim of such experiment was to monitor the time evolution of a disturbed excitonic polarisation in a multiple quantum-well system. This experiment revealed intra-excitonic Rabi-flopping between 1s and higher-order states in the system in the time domain, whereas the transient four-wave-mixing signal could be systematically reduced as a function of the THz-field strength. However, in the case of exciton ionisation, the signal showed a lack of transfer back into the initial 1*s* state. This allowed to create scenarios of reversible and irreversible intra-excitonic transfer depending on the strength of the transient THz field [176].

In contrast to the experiment by Kaindl et al. which probes exciton population formation, i.e. the generation of incoherent excitons in semiconductors after optical excitation (obtained at times > 100 ps) [170], coherent spectroscopy targets the dynamics of the system in the regime of coherent polarisation, which is the essential situation in microcavities where the photon lifetime of a few ps to tens of ps determines the frame of light–matter coupling phenomena. It is this regime which deserves special attention using THz pulses for manipulation and control of exciton–polaritons, which shall be addressed in one of the author's current projects (DFG RA2841/9-1) using methods such as digital holography [34, 35] and THz pulses (cf. [36, 172]).

3.2.2 *Rich Exciton Physics in 2D Semiconductors*

Detailed group-theory analysis of the possible excitons' branches of TMDCs [177, 178] such as WSe_2 indicated that four different exciton configurations coexist within the light cone at the A-exciton peak as demonstrated in [179]: Among them are two bright excitons ($\Gamma 6$) at K and K', respectively, and two more possible states ($\Gamma 4$ and $\Gamma 3$) arise which represent a coherent superposition of intervalley excitons composed of the spin-forbidden transition across K and K' valley. While the $\Gamma 4$ state is dipole-allowed for z-polarisation corresponding to the out-of-plane dipole orientation and is labeled "grey" exciton, the $\Gamma 3$ state cannot couple to the electromagnetic field and remains completely 'dark' (see Fig. 3.7).

Encouraged by the rich landscape of exciton complexes in encapsulated tungsten diselenide, direct measurements of the radiation pattern for different excitons at cryogenic temperatures were very recently reported in an angle-resolved PL study in [160], revealing the differences between the bright and grey species that were obtained under strong quasi-resonant pulsed excitation of the high-quality 2D stack.

Demonstration of Dispersion Feature Within Light Cone

In addition, the polaritonic nature of the quasi-resonantly excited low-pump-density resonance of high-quality encapsulated WSe_2 was investigated shortly before in [159]. Thereby, a dispersion within the light cone as strong as about 2 meV with nearly-homogeneous-linewidth-broadened PL and reflection contrast modes was demonstrated, corresponding to effective masses between 10^{-3} and $10^{-4}\,m_0$ (see Fig. 3.8).[10] Note that cavity–polaritons exhibit effective masses of the order of

[10] Free electron mass m_0, natural constant.

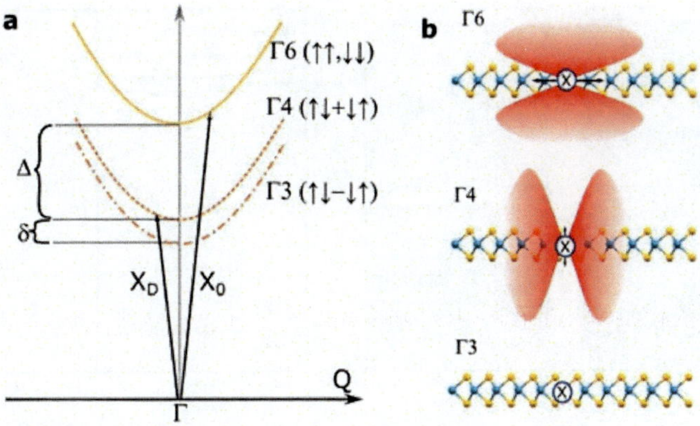

Fig. 3.7 **a** Energy–momentum dispersion diagram for different monolayer WSe_2 exciton configurations and **b** sketch of the possible bright and dark exciton states for the respective monolayer. Schematically, the non-degeneracy of excitonic resonances arising from the crystal symmetry is displayed in **a** with the labels according to group-theory analysis in the literature. Optical excitations are indicated by nearly-vertical arrows. The $\Gamma 6$ branch represents the bright exciton X_0, which is energetically separated from the grey exciton $X_{D,g}$ branch $\Gamma 4$ and the dark exciton X_D branch $\Gamma 3$. The dipole orientation and emission pattern of the radiating exciton species (X) is sketched in **b** by black double-arrows and red lobes, respectively. While the ordinary emitter ($\Gamma 6$) is oriented in the plane and emits out of plane, the grey species ($\Gamma 4$), which is only dipole allowed for z-polarisation is oriented out of the plane and emits into the TMDC plane (x–y plane). The dark species $\Gamma 3$ remains "dark" due to the lack of coupling to the electromagnetic field. Reproduced under the terms of the CC-BY 4.0 Licence (http://creativecommons.org/licenses/by/4.0/). [160] Copyright 2020 The Author(s), published by Springer Nature

$10^{-5}\, m_0$, and typical III/V semiconductor excitons of the order of $10^{-1}\, m_0$. Thus, these 2D-excitonic dispersions imply very light quasi-particles,[11] that are the hybrid exciton–polaritons (light-dressed excitons, macroscopic polarisation). Pump-density-dependent dispersion measurements in the linear excitation regime further indicated a small change in the curvature due to an increased role of an electron–hole plasma fraction in the emission signal [159].

Indeed, the origin and understanding of such a strong dispersion is still subject of ongoing investigations [180], revisiting the concept of renormalisation of longitudinal-mode energies within the light cone and the degeneracy lifting between longitudinal and transverse exciton modes, as historically discussed for III/V quantum-well systems in the view of strong far-field optical dipole coupling

[11] In the study of [159], the formation of polaritons as a result of any cavity effects by a possible Fabry–Pérot structure of the 2D stack on the substrate material was ruled out, because instead of any confined modes only a leaky mode could be found in a simulation for the relevant spectral range. In addition, propagation effects modelled by theory colleagues of the author also did not yield a curved dispersion. Moreover, attributing the effects to resonant coupling of monochromatic intense laser light with the exciton resonance would not explain the dispersion in broadband white-light reflection contrast measurements. In fact, reflection-contrast values of the effective masses as low as few $10^{-4}\, m_0$ represent a zero-density excitation regime and are not affected by any density-related effects.

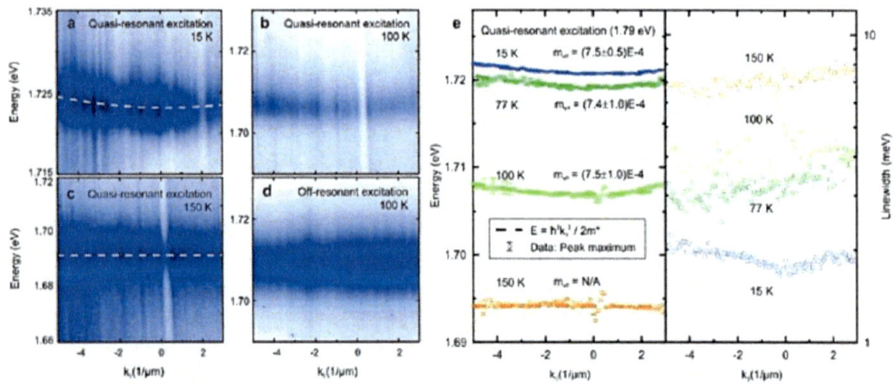

Fig. 3.8 Optical dispersion measurements acquired in a Fourier-space spectroscopy mode with a microscope objective setup (see Fig. 5.17). Energy–$k_{\|}$-resolved PL from hBN-encapsulated WSe_2 at 15 K (**a**), 100 K (**b**) and 150 K (**c**), obtained under quasi-resonant excitation. The dashed line shows a parabolic fit to the data in **a** or a flat line in **c**, which exhibits no measurable dispersion. This behaviour is attributed to the role of thermal effects (that can be faster dephasing times, i.e. larger homogeneous broadening, correspondingly reduced oscillator strength, binding energy and light–matter coupling, loss of macroscopic polarisation, increased fraction of ionised excitons, i.e. e–h plasma) and polaron formation through exciton–optical-phonon interactions. **d** In comparison to **b**, off-resonant excitation results in a loss of measurable dispersion and the data indicates a k-independent constant emission energy, similar to **c**. This behaviour is attributed to an increased fraction of e–h plasma and incoherent excitons (microscopic polarisation). **e** Extracted peak energies in an energy–momentum chart labelled with corresponding effective masses from parabola fits. The right side of **e** shows the respective linewidth values versus $k_{\|}$. Here, the narrowest linewidth values correspond to FWHM of 4–5 meV, given at 15 K. Reproduced with permission. [159] Copyright 2019 Optical Society of America

[181, 182]. In the theory literature, the quantum-mechanically considered long-range exchange interactions of electrons and holes within one valley and across the valleys are understood to give rise to new excitonic eigen-states formed as coherent superpositions with exciton normal mode splitting (see below). Some of these theories (predominantly based on 2D considerations) interestingly suggested a Dirac cone or (nearly-)linear dispersion around zero momentum [183–186] with energy shifts in the few meVs within the light cone, which would excitingly imply light-like or relativistic behaviour, whereas 3D-based approaches [187] still obtain parabolic branches with strongly different effective masses.

Very recently, by the vanishing of a dispersion feature as well as optical helicity with increasing laser–exciton detuning, it was even demonstrated that the formation of a macroscopic polarisation (coherent excitons) becomes highly unlikely for excitation regimes well-detuned above the excitonic resonance, i.e. for off-resonant compared to quasi-resonant pumping [180]. Thus, signatures of uncorrelated electron–hole plasma and microscopic polarisation (incoherent excitons) are obtained for off-resonant pumping, in contrast to exciton–polaritons (coherent excitons) for resonances with nearly-homogeneous linewidth broadening [159] when pumping between the A-exciton 1s and continuum state [180]. Yet, from the obtained data by the author's team (based on the Fourier-space spectroscopy technique) with trending parabolic dispersion relations for 2D excitons in hBN-encapsulated WSe_2 or $MoSe_2$ (the latter with weaker dispersion feature, in agreement with expectations from theory for molybdenum-based TMDCs), a linear or nearly-linear behaviour around $k_{\parallel} = 0$ can neither be clearly verified nor ruled out.

In fact, given these examples, one has to abandon the common view that an excitonic quasi-particle dispersion cannot be obtained optically. That view existed due to the traditional knowledge for conventional semiconductors that the exciton mass is too heavy (1–0.1 m_0) to feature angle-dependent energy shifts larger than a few μeV within the light cone—typically impossible to measure reliably with exciton linewidth of the order of meV.

Fine-Structure's Pseudo-Spin and Angle-Resolved Valley Polarisation

Additional examinations of the fine-structure's pseudo-spin texture for this specific 2D stack (i.e. the hBN-encapsulated WSe_2 monolayer) showed considerable differences for continuous-wave quasi-resonant excitation detunings and off-resonant pumping cases [180]. A systematic decrease of the measurable dispersion feature in Fourier-space resolved PL spectra was evidenced due to contributions of incoherent excitons and emission from plasma when moving the excitation energy step-wise above the electronic band gap. A step-wise increase of the electron–hole plasma fraction and faster decoherence of the coherently mixed exciton states[12] was further indicated by the step-wise reduction of the PL helicity (degree of circular polarisation) with its peculiar far-field pattern changes for the excitonic resonance. By having revealed that there is a complex locking between valley-pseudo-spin and centre-of-mass momentum present [180], further investigations of angle-resolved polarisation anisotropy in exciton-complexes-rich 2D semiconductors are encouraged, such as time-resolved Fourier-space mapping in order to shed light on valley decoherence mechanisms as a function of the centre-of-mass momentum.

While several theoretical predictions [183–187] previously claimed that the neutral exciton of TMDCs splits into a transversal and longitudinal exciton branch (whereas the longitudinal one, which is the upper branch, exhibits an extraordinarily strong dispersion in the meV range within the light cone), current understanding from the studies by the author and co-workers let one recall that long-range exchange interactions (already proposed for quantum-well exciton–polaritons in III/V semiconductors [181, 182]) and strong far-field optical dipole coupling may indeed describe two sides of the same coin.

[12]Dipole resonance states that are coherently superposed due to long-range exchange interaction between charge carriers within one valley and with those of opposite valleys, with the hybridised states resembling the Rabi oscillations [183] between the Rabi-split states of strong optical dipole–dipole coupling.

3.3 Strong Exciton–Photon Coupling and Polariton Bose–Einstein Condensation

Strong quantum coupling of excitons and photons in semiconductors results in coherent superpositions of the two components, or quasi-particles called exciton–polaritons [1, 188]. In recent years, quantum-well exciton–polaritons in optical microresonator structures (cf. Fig. 3.9a, b), also referred to as cavity–polaritons [189–194], attracted considerable interest beyond the quantum optics community [8, 15, 42, 51, 81, 102]. Quantum wells—the more wells (emitters in resonance), the more robust the strong coupling—placed at anti-node positions of a microcavity mode's standing wave electric field deliver the matter component, which gets dressed with the cavity field in the strong coupling regime [1, 38, 169, 190, 195]. The appropriate description of the hybridised quantum system, i.e. the eigen-states of the coupled-oscillator system, is based on composite quasi-particles, the polaritons, with their appealing bosonic character and ultralight effective masses [2, 3, 7, 42]. Thus, the hybrid modes are part light (bosonic) and part matter excitation (composite boson made of two fermions), with the Rabi frequency characterizing the interaction strength between the two components. A typical in-plane-momentum-dependent diagram for such hybrid system is shown with Hopfield coefficients as well as lower and upper polariton dispersions with Rabi splitting in the case of zero detuning at $k = 0$ in Fig. 3.9c.

Favourable Properties of Polaritons
The effective mass of polaritons is heavily influenced by the photon dispersion in the microcavity, which is quadratic close to the zero transverse momentum point (the effective mass concept applies), and is typically four to five orders of magnitude smaller than the mass of a bare electron. The appearance of such mixed modes is interesting from both the point of view of fundamental and applied research. Strong interparticle interactions result from the exciton component, whereas the extremely low effective mass and the high mobility originate from the photonic component. Thus, polaritons with their much more delocalised wave-function are less prone to trapping, dephasing and non-radiative recombination due to inhomogeneities in semiconductor structures than quantum-well excitons.

BEC-Like Phase Transition for Exciton–Polaritons
The field of Bose–Einstein-like condensation of exciton–polaritons opened in 2006, when the first fully convincing report of BEC-like condensation of exciton–polaritons was published which included the demonstration of a spontaneous symmetry breaking with extended spatial correlations (as a signature of off-diagonal long-range order) [37]. However, previous experiments dating back to 2002 clearly showed condensation effects for polaritons in microcavities [38], later amended by various studies including spatial coherence studies [39, 198, 199]. It was demonstrated that exciton–polaritons can exhibit behaviour analogous to Bose–Einstein condensation [37, 39, 40], which is a phase transition to an exquisite quantum state of bosonic particles (see polariton-related [41–45, 59, 200–203] and BEC-related [128, 131]).

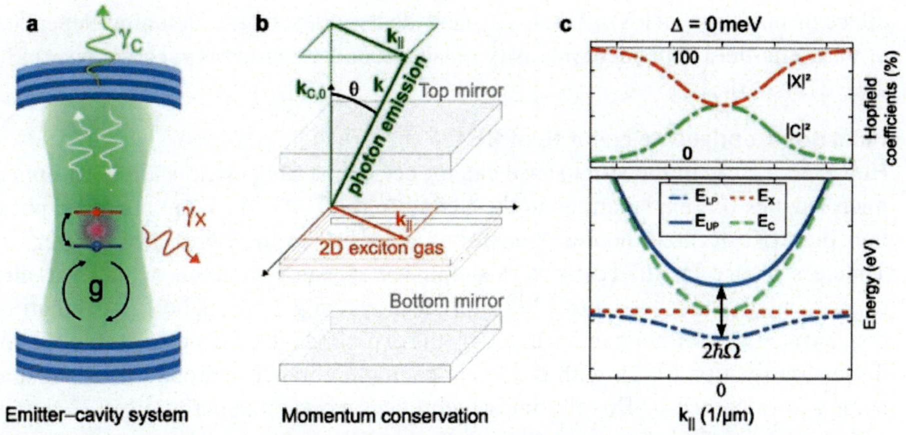

Fig. 3.9 **a** Emitter–cavity system with coupling strength g between resonant light field and quantum emitter. **b** Planar Fabry–Pérot resonator with multiple quantum wells as active region, displaying momentum conservation for emitted photons from an in-plane 2D exciton gas with centre-of-mass momentum $k_{\|}$. In the wells, excitons can only couple to cavity photons of total photon momentum **k** when their in-plane momentum projection $k_{\|}$ matches. Correspondingly, the photon momentum of the angle (θ) dependent emission can be decomposed into an in-plane component $k_{\|}$ and an out-of-plane component $|\mathbf{k}_{C,0}| = 2\pi/\lambda_{C,0}$ (with resonance wavelength in vacuum at zero incidence angle $\lambda_{C,0}$). **c** For the hybrid cavity–polaritons, the Hopfield coefficients can be given for any detuning and momentum configuration, here displayed for resonance conditions at zero momentum (top) with the corresponding polariton dispersions (bottom) in a planar quantum-well microresonator system. The characteristic normal mode splitting between upper and lower polariton branch (LP/UP) of $2\hbar\Omega$ (the Rabi splitting) is displayed at resonance conditions for the cavity C and exciton mode X. **a–c** Drawn after the author's schematics in [196], based on [197]

First Realisation of Bose–Einstein Condensation

Originally, condensation was predicted from statistical properties of a system of bosons at low temperatures by S. Bose and A. Einstein, in the mid-twenties of the 20th century. The first true Bose–Einstein condensate (BEC) of dilute atomic gases was observed in 1995 (by three groups working on this subject) [204–206], opening a new chapter in the field of atomic and molecular physics and quantum optics. The historic evolution of condensate experiments and the path towards polariton quasi-BEC is highlighted in [81].

The Super Flow of Condensation Studies

Condensation of polaritons was a major breakthrough in this field from both fundamental and applied point of view. In the past two decades, polaritons in microcavities evolved to a unique, easily-accessible and versatile testbed for condensation studies in solids, owing to the remarkable properties of exciton–polaritons (see aforementioned literature on these hybrid quasi-particles, also cf. [196, 207] for an overview of the fundamentals of cavity–polaritons). Condensates of polaritons were achieved away from ideal thermal equilibrium, in a gas of particles with only picoseconds lifetime [42, 199, 208–210], and exhibited interesting properties such as superfluidity (up to room temperature) [60, 66], a spin-Meissner effect (spin analogue of the Meissner effect in magnetic fields) [211, 212], and density-dependent, detuning-dependent or magnetic-field dependent intensity noise (second-order temporal coherence) [38, 209, 213–215].

Polariton Condensate Not a Real BEC
However, a BEC in the strict sense cannot occur in a 2D system such as in optical microcavities (cf. explanations in the literature [5, 7, 42, 45]), but a similar phase transition (Kosterlitz–Thouless transition) according to the Berezinski–Kosterlitz–Thouless theory facilitated by vortex–anti-vortex pair formation, which was first experimentally evidenced in 2011 [216]. The interacting nature of bosons in polariton condensates was further demonstrated by the experimental evidence of a Bogoliubov dispersion relation [201], with the linear excitation branches around the final state named in reference to Bogoliubov's theoretical work on superfluidity [217]. The experimental optical characterisation of condensates and special features of polariton systems—ranging from typical cavity–polariton condensates to those of the kind at their extremes—are summarised for instance in [46].

Towards Room-Temperature Operation and Electrical Injection
The ability to achieve polariton condensation at room temperature [218] and in organic materials [63, 64] opened a way to practical applications. It came into reach due to technological achievements in the production of high-quality optical microcavities (cf. [4, 7, 169]). This motivated room-temperature polariton-lasing (condensation) studies using large-bandgap semiconductors such as GaN [219, 220] and ZnO [221, 222], and the hunt for high-binding energy materials such as the currently highly attractive TMDCs (2D semiconductors) [223–227] or polymers [66, 228]. In recent years, even electrically-driven polariton condensates, also referred to as polariton lasers, have emerged [48–50].

Plethora of Polariton Systems and Features
Although polaritons are already well studied in many regards, still intriguing features come up in late studies involving the dynamics of exciton–polaritons [34, 172, 215, 229] and the behaviour of condensed gases in fluids [59, 61, 62, 66, 230, 231], relaxation oscillations in the formation of polariton condensates [232] and interactions at the quantum limit [79].

Within the last decade, various examples of potential landscapes (e.g. for quantum simulators [8]) were explored with eyes towards zero-state, π-state, d-wave or Dirac-point condensates [233–235], gap solitons [236] and a flat band in lattices [237, 238], as well as condensates in engineered polariton band structures using patterned structures [239–241], which can even give rise to topological edge states [241].

Many more exciting examples and important findings are not covered in this brief overview, whereas even the scope of [9] could not provide a complete summary of all activities in these directions. An overview of numerous examples of potential landscapes, polariton devices and alternative material systems for polariton studies is delivered in [242], of optical condensation studies in [46] and an overview of the concept of polariton lasers is found in [58].

3.3.1 Cavity–Polaritons Exposed to External Fields

When placing strongly-coupled light–matter systems in external electric or magnetic fields, numerous interesting effects can be studied and the response of polaritons in the linear and nonlinear regime (i.e. dynamical condensates of polaritons) explored. A topical introduction into this subject is provided for instance in [243]. To start with a practical example, one simple effect arising from the exposure to electric bias in polariton diodes on the coupling regime can for instance allow one to switch on ultrafast timescales between a polariton lasing and conventional lasing situation [244], exploiting the quantum-confined Stark effect (QCSE) on the quantum-well excitons. In the following, a few examples from the domain of magneto-optical and transient-field studies are briefly summarised.

Polaritons in Magnetic Fields

Exposing polaritons to external magnetic fields (typically in Faraday configuration) offers an important playground for the study of polariton interactions and degeneracy-lifted spinor condensates. On the one hand, with increasing magnetic flux, threshold conditions are modified due to diamagnetic shifts [245], exciton Bohr-radius changes [246], and altered polariton–polariton interactions [211, 247] when the spin-opposite quasi-particles align oppositely in the external field (see [48, 49]). On the other hand, the degeneracy lifting of spinor condensates due to Zeeman splitting offers an important tool to identify the matter component of polariton modes, even when quantisation spoils the typical dispersion, as shown by the works of the author and co-workers (see [48, 248]).

In fact, the electronic g-factor and diamagnetic coefficient become effective ones for hybrid light–matter modes in relation to the Hopfield coefficients (the understanding of which was developed through [212, 248–250]), that are the excitonic/photonic fractions of the polariton mode. In addition, the spin-Meissner effect increasingly suppresses (elliptically-polarised) superfluidity towards a critical magnetic field until above the critical field spin degeneracy is lifted and the (circularly-polarised) superfluidity reappears for Zeeman-split spinor-polariton condensates [211]. This was indicated in the circular-polarisation resolved electroluminescence from polariton lasers at finite magnetic flux by Schneider and Rahimi-Iman et al. [48], and further discussed in a later experimental demonstration of the non-equilibrium spin-Meissner effect with optically-pumped spinor condensates by Fischer et al. [212], which similarly discussed modified polariton–polariton interactions.

Another experiment with spinor condensates (Fig. 3.10) revealed that one of the two coexisting circularly-polarised condensates quickly approached a second-order temporal autocorrelation function of $G^{(2)}(0) = 1$ above condensation threshold for 5 T (for the lower-energetic condensate). This improved degree of coherence for one of the condensates corresponded to changes in the populations of the spin-polarised Zeeman-split sublevels, whereas the counterpart (spin-opposite condensate) still exhibited noticeable intensity fluctuations [214]—indicating modified polariton–polariton scattering rates and effective uncoupling of the spin-opposite baths of polaritons. Chernenko et al. could also show with polarisation-sensitive pho-

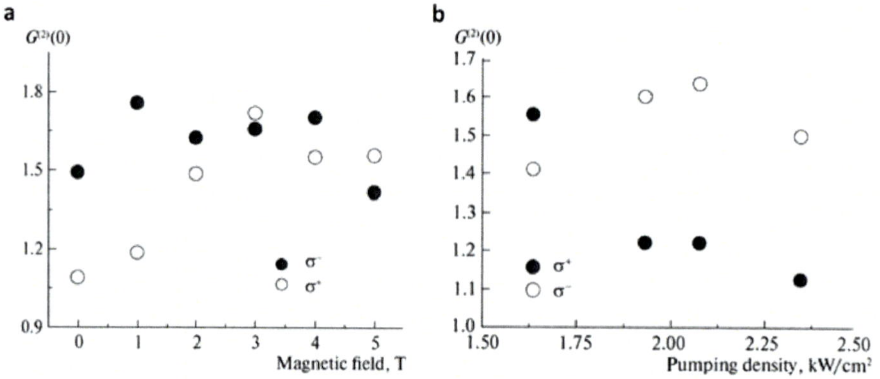

Fig. 3.10 Temporal photon statistics measured for oppositely circularly-polarised spinor condensate emission. Second-order temporal autocorrelation functions were acquired as a function of an external magnetic field B (see [214]). **a** At fixed pump density, spin-split polariton modes featured a magnetic-field dependent degree of circular polarisation above the condensate threshold, which had a direct impact on the photon correlations measured for coexisting spinor condensates with opposing polarisations. **b** At 5 T, an increase in pump density above the threshold resulted in a considerable increase in coherence for the spinor condensate with lower-energetic mode in the external magnetic field, reflected by a decrease in polarisation-resolved $G^{(2)}(0)$ towards unity. In contrast, the oppositely polarised condensate exhibited an elevated level of intensity fluctuations. Note that the use of symbols in **a** and **b** differs. Reused with permission [214] Copyright 2016 Springer Nature

ton statistics measurements that with increasing field, spinor condensates of opposing polarisations approached each others $G^{(2)}(0)$, which exhibited noticeable intensity fluctuations, whereas at 0 T, one of the two circularly-polarised condensate signals featured a high degree of coherence with near unity value (signal from the lower-energetic condensate in magnetic field series) [214]. This result indicated that the symmetry break through condensation gives preference to one polariton polarisation intrinsically without external field.[13]

Ultimately, the understanding of conditions for good second-order coherence and low intensity noise are expected to pave the way for coherent exciton–polariton devices, such as described in the literature [54, 57].

Polaritons and Terahertz Radiation

Furthermore, the relationship of polaritons with terahertz radiation in terms of the Rabi-oscillation frequency being of the order of terahertz waves in common polariton systems has also triggered research activities which promise the realisation of bosonic lasers, e.g. proposed bosonic cascade lasers [77], and novel polariton-based

[13]In fact, the low-density (below-threshold) lower-polariton mode remained unpolarised over the whole flux range of such experiment, whereas in stark contrast the helicity for the dynamical BEC was initially high at 0 T, where no energy-degeneracy lifting through an external magnetic field occurred, and vanished towards 5 T, where the intensity noise (measured $G^{(2)}(0)$) became comparable. In the absence of the external field, polariton-condensate formation was accompanied by an abrupt change in the degree of circular polarisation from 0 below to 80% above the bosonic stimulated polariton–polariton scattering threshold.

THz emitters [74], with important physics still not studied well enough to exploit many-particle effects in novel device concepts and quantum information schemes. Nevertheless, many terahertz-emitting polariton systems have been proposed so far [71–73, 75, 76].

Promises of Light–Matter Coupled Systems
Exploring this domain of light–matter interaction thus has still the potential to unravel new applications of quantum gases and microcavity systems, such as the development of polariton qubits [78], ultrafast optical switches based on coherent control [34], electronic bias [244] or operation bistability [251], manipulation [35] and cooling of quantum gases [252], and generation of coherent light in nonlinear regimes [71].

Polariton States Exposed to Terahertz Pulses
Within the past decade, a couple of independent efforts have been made to combine THz radiation and polariton systems in experiments. For instance, Tomaino et al. used terahertz radiation to depopulate the upper/higher-energy polariton branch (UP/HEP) by a THz-induced transition to the excitonic $2p$ state as was predicted by theory [253], while Ménard et al. combined a THz-probe experiment with a polariton condensation study, "revealing the dark side of a bright exciton–polariton condensate"—in other words, showing the pronounced existence of an uncondensed high-energy fraction of "hot" excitons, which do not couple to the light field and thus coexist with a condensate [172]. Such uncondensed fraction typically contributes to intensity noise and is considered parasitic for coherent condensates (see [54, 209, 254, 255]). In addition, a single THz-pulse protocol was proposed and demonstrated as a polarisation "reset" in a transient reflection experiment on a microcavity, which was disturbed by the THz pulse at certain times in order to probe light–matter interaction time scales [36].

Hunt for Terahertz-Generating Polaritons
On the other hand, a few experimentalist groups worldwide tried to obtain THz emission as proposed by expert theoreticians from polariton condensates [71–73, 75], demonstrating a strong interest in the field of THz-related polariton physics. On such a path, Rojan et al. found that in CdTe microcavities the Rabi oscillation resonates with the phonon–polariton, rendering the Rabi oscillation as a potentially active source of THz radiation, whereas the LP–UP transition itself is else dipole forbidden [76]. Nevertheless, the wider lack of success with polariton-based THz-generation schemes can be attributed to different technical aspects such as THz-detection sensitivity, design parameters, and condensate sizes, as summarised in [243].

Filling the Knowledge Gaps on Terahertz Interactions
In order to gain a better understanding of how THz radiation and polaritonic gases interact with each other, still further studies are needed both in the linear regime, such as performed by Pietka et al. on doubly-dressed bosons [256], and, more importantly, in the nonlinear regime. It can be expected that practical generation of THz waves from polariton systems will strongly benefit from such endeavours by the polariton

community. This had motivated a systematic investigation of THz-induced effects in various configurations involving ultrafast spectroscopy experiments and microcavity polaritons, which are outlined in one of the author's projects (DFG RA2841/9-1).

References

1. C. Weisbuch, M. Nishioka, A. Ishikawa, Y. Arakawa, Observation of the coupled exciton-photon mode splitting in a semiconductor quantum microcavity. Phys. Rev. Lett. **69**, 3314 (1992)
2. Y. Yamamoto, A. Imamoglu, *Mesoscopic Quantum Optics* (Wiley, Hoboken, 1999)
3. Y. Yamamoto, F. Tassone, H. Cao, *Semiconductor Cavity Quantum Electrodynamics* (Springer, Berlin, 2000)
4. K. Vahala, *Optical Microcavities* (World Scientific, Singapore, 2004)
5. B. Deveaud, *The Physics of Semiconductor Microcavities* (Wiley-VCH Verlag, Weinheim, 2007)
6. D. Sanvitto, V. Timofeev, *Exciton Polaritons in Microcavities: New Frontiers* (Springer, Berlin, 2012)
7. A.V. Kavokin, J.J. Baumberg, G. Malpuech, F.P. Laussy, *Microcavities*, vol. 1 (Oxford University Press, Oxford, 2017)
8. N.Y. Kim, Y. Yamamoto, Exciton-polariton quantum simulators, in *Quantum Science and Technology* (Springer International Publishing, Berlin, 2017), pp. 91–121
9. A. Rahimi-Iman, *Polariton Physics: From Dynamic Bose–Einstein Condensates in Strongly-Coupled Light–Matter Systems to Polariton Lasers*. Springer Series in Optical Sciences, vol. 229 (Springer International Publishing, Cham, 2020)
10. H. Yokoyama, Physics and device applications of optical microcavities. Science **256**, 66–70 (1992)
11. K.J. Vahala, Optical microcavities. Nature **424**, 839–846 (2003)
12. S. Reitzenstein, A. Forchel, Quantum dot micropillars. J. Phys. D: Appl. Phys. **43**(3), 033001 (2010)
13. E. Yablonovitch, Inhibited spontaneous emission in solid-state physics and electronics. Phys. Rev. Lett. **58**, 2059–2062 (1987)
14. T.C.H. Liew, I.A. Shelykh, G. Malpuech, Polaritonic devices. Physica E **43**(9), 1543–1568 (2011)
15. D. Sanvitto, S. Kéna-Cohen, The road towards polaritonic device. Nat. Mater. **15**(10), 1061–1073 (2016)
16. J. M. Gérard, D. Barrier, J.Y. Marzin, R. Kuszelewicz, L. Manin, E. Costard, V. Thierry Mieg, T. Rivera, Quantum boxes as active probes for photonic microstructures: the pillar microcavity case. Appl. Phys. Lett. **69**, 449 (1996)
17. Z. Yuan, B.E. Kardynal, R.M. Stevenson, A.J. Shields, C.J. Lobo, K. Cooper, N.S. Beattie, D.A. Ritchie, M. Pepper, Electrically driven single-photon source. Science **295**(5552), 102–105 (2002)
18. C. Santori, D. Fattal, J. Vučković, G.S. Solomon, Y. Yamamoto, Indistinguishable photons from a single-photon device. Nature **419**, 594–597 (2002)
19. P. Michler (ed.), *Single Quantum Dots: Fundamentals, Applications and New Concepts* (Springer, Berlin, 2003)
20. P. Michler (ed.), *Single Semiconductor Quantum Dots* (Springer, Berlin, 2009)
21. J.-M. Gérard, Solid-state cavity-quantum electrodynamics with self-assembled quantum dots, in *Single Quantum Dots: Fundamentals, Applications, and New Concepts* (Springer, Berlin, 2003), pp. 269–314

22. E. Peter, P. Senellart, D. Martrou, A. Lemaître, J. Hours, J.M. Gérard, J. Bloch, Exciton-photon strong-coupling regime for a single quantum dot embedded in a microcavity. Phys. Rev. Lett. **95**, 067401 (2005)
23. D. Englund, D. Fattal, E. Waks, G. Solomon, B. Zhang, T. Nakaoka, Y. Arakawa, Y. Yamamoto, J. Vučković, Controlling the spontaneous emission rate of single quantum dots in a two-dimensional photonic crystal. Phys. Rev. Lett. **95**, 013904 (2005)
24. S. Noda, Seeking the ultimate nanolaser. Science **314**, 260 (2006)
25. N. Na, Y. Yamamoto, Massive parallel generation of indistinguishable single photons via the polaritonic superfluid to Mott-insulator quantum phase transition. New J. Phys. **12**, 123001 (2010)
26. S. Strauf, F. Jahnke, Single quantum dot nanolaser. Laser Photonics Rev. **5**, 607–633 (2011)
27. X. Lin, X. Dai, C. Pu, Y. Deng, Y. Niu, L. Tong, W. Fang, Y. Jin, X. Peng, Electrically-driven single-photon sources based on colloidal quantum dots with near-optimal antibunching at room temperature. Nat. Commun. **8**, 1132 (2017). https://doi.org/10.1038/s41467-017-01379-6
28. J.L. O'Brien, A. Furusawa, J. Vučković, Photonic quantum technologies. Nat. Photonics **3**, 687–695 (2009)
29. A. Boretti, L. Rosa, A. Mackie, S. Castelletto, Electrically driven quantum light sources. Adv. Opt. Mater. **3**, 1012–1033 (2015)
30. I. Aharonovich, D. Englund, M. Toth, Solid-state single-photon emitters. Nat. Photonics **10**, 631–641 (2016)
31. S. Rodt, S. Reitzenstein, T. Heindel, Deterministically fabricated solid-state quantum-light sources. J. Phys. Condens. Matter **32**, 153003 (2020)
32. J.P. Reithmaier, G. Sek, A. Löffler, C. Hofmann, S. Kuhn, S. Reitzenstein, L.V. Keldysh, V.D. Kulakovskii, T.L. Reinecke, A. Forchel, Strong coupling in a single quantum dot-semiconductor microcavity system. Nature **432**, 197–200 (2004)
33. T. Yoshie, A. Scherer, J. Hendrickson, G. Khitrova, H.M. Gibbs, G. Rupper, C. Ell, O.B. Shchekin, D.G. Deppe, Vacuum Rabi splitting with a single quantum dot in a photonic crystal nanocavity. Nature **432**, 200–203 (2004)
34. L. Dominici, D. Colas, S. Donati, J.P. Restrepo Cuartas, M. De Giorgi, D. Ballarini, G. Guirales, J.C. López Carreño, A. Bramati, G. Gigli, E. Del Valle, F.P. Laussy, D. Sanvitto, Ultrafast control and Rabi oscillations of polaritons. Phys. Rev. Lett. **113**, 226401 (2014)
35. D. Colas, L. Dominici, S. Donati, A.A. Pervishko, T.C. Liew, I.A. Shelykh, D. Ballarini, M. De Giorgi, A. Bramati, G. Gigli, E. Del Valle, F.P. Laussy, A.V. Kavokin, D. Sanvitto, Polarization shaping of Poincaré beams by polariton oscillations. Light Sci. Appl. **4**, e350 (2015)
36. A.D. Jameson, J.L. Tomaino, Y.-S. Lee, G. Khitrova, H.M. Gibbs, C.N. Böttge, A.C. Klettke, M. Kira, S.W. Koch, Direct measurement of light-matter energy exchange inside a microcavity. Optica **1**, 276 (2014)
37. J. Kasprzak, M. Richard, S. Kundermann, A. Baas, P. Jeambrun, J.M.J. Keeling, F.M. Marchetti, M.H. Szymańska, R. André, J.L. Staehli, V. Savona, P.B. Littlewood, B. Deveaud, L.S. Dang, Bose-Einstein condensation of exciton polaritons. Nature **443**, 409–414 (2006)
38. H. Deng, G. Weihs, C. Santori, J. Bloch, Y. Yamamoto, Condensation of semiconductor microcavity exciton polaritons. Science **298**(5591), 199–202 (2002)
39. H. Deng, G.S. Solomon, R. Hey, K.H. Ploog, Y. Yamamoto, Spatial coherence of a polariton condensate. Phys. Rev. Lett. **99**, 126403 (2007)
40. R. Balili, V. Hartwell, D. Snoke, L. Pfeiffer, K. West, Bose-Einstein condensation of microcavity polaritons in a trap. Science **316**, 1007–1010 (2007)
41. D. Snoke, Polariton condensates: a feature rather than a bug. Nat. Phys. **4**, 673 (2008)
42. H. Deng, H. Haug, Y. Yamamoto, Exciton-polariton Bose-Einstein condensation. Rev. Mod. Phys. **82**, 1489 (2010)
43. V.B. Timoffeev, Bose condensation of exciton polaritons in microcavities. Fizika i Tekhnika Poluprovodnikov **46**(7), 865–883 (2012). (Engl. translation in Semiconductors **46**(7), S. 843) (Springer, 2012)

44. B. Deveaud-Plédran, On the condensation of polaritons. J. Opt. Soc. Am. B **29**, A138–A145 (2012)
45. A. Rahimi-Iman, *On the Condensation of Polaritons*. Springer Series in Optical Sciences, vol. 229 (Springer International Publishing, Cham, 2020), pp. 65–85
46. A. Rahimi-Iman, *Optically-Excited Polariton Condensates*. Springer Series in Optical Sciences, vol. 229 (Springer International Publishing, Cham, 2020), pp. 195–240
47. A.V. Kavokin, Exciton-polaritons in microcavities: recent discoveries and perspectives. Phys. Status Solidi B **247**(8), 1898–1906 (2010)
48. C. Schneider, A. Rahimi-Iman, N.Y. Kim, J. Fischer, I.G. Savenko, M. Amthor, M. Lermer, A. Wolf, L. Worschech, V.D. Kulakovskii, I.A. Shelykh, M. Kamp, S. Reitzenstein, A. Forchel, Y. Yamamoto, S. Höfling, An electrically pumped polariton laser. Nature **497**, 348–352 (2013)
49. P. Bhattacharya, B. Xiao, A. Das, S. Bhowmick, J. Heo, Solid state electrically injected exciton-polariton laser. Phys. Rev. Lett. **110**, 206403 (2013)
50. P. Bhattacharya, T. Frost, S. Deshpande, M.Z. Baten, A. Hazari, A. Das, Room temperature electrically injected polariton laser. Phys. Rev. Lett. **112**, 236802 (2014)
51. A. Imamoglu, R.J. Ram, S. Pau, Y. Yamamoto, Nonequilibrium condensates and lasers without inversion: exciton-polariton lasers. Phys. Rev. A **53**, 4250 (1996)
52. A. Kavokin, G. Malpuech, F.P. Laussy, Polariton laser and polariton superfluidity in microcavities. Phys. Lett. A **306**, 187–199 (2003)
53. D. Bajoni, Polariton lasers. Hybrid light-matter lasers without inversion. J. Phys. D: Appl. Phys. **45**(31), 313001 (2012)
54. S. Kim, B. Zhang, Z. Wang, J. Fischer, S. Brodbeck, M. Kamp, C. Schneider, S. Höfling, H. Deng, Coherent polariton laser. Phys. Rev. X **6**, 011026 (2016)
55. S.A. Moskalenko, I.M. Tiginyanu, Exciton-polariton laser. Low Temp. Phys. **42**, 330–339 (2016)
56. A. Kavokin, T.C.H. Liew, C. Schneider, S. Höfling, Bosonic lasers: the state of the art (review article). Low Temp. Phys. **42**, 323–329 (2016)
57. M.D. Fraser, Coherent exciton-polariton devices. Semicond. Sci. Technol. **32**, 093003 (2017)
58. A. Rahimi-Iman, *The Concept of Polariton Lasing*. Springer Series in Optical Sciences, vol. 229 (Springer International Publishing, Cham, 2020), pp. 87–118
59. A. Amo, D. Sanvitto, F.P. Laussy, D. Ballarini, E.d. Valle, M.D. Martín, A. Lemaître, J. Bloch, D.N. Krizhanovskii, M.S. Skolnick, C. Tejedor, L. Viña, Collective fluid dynamics of a polariton condensate in a semiconductor microcavity. Nature **457**, 291–295 (2009)
60. A. Amo, J. Lefrere, S. Pigeon, C. Adrados, C. Ciuti, I. Carusotto, R. Houdré, E. Giacobino, A. Bramati, Superfluidity of polaritons in semiconductor microcavities. Nat. Phys. **5**, 805–810 (2009)
61. A. Amo, S. Pigeon, D. Sanvitto, V.G. Sala, R. Hivet, I. Carusotto, F. Pisanello, G. Lemé-nager, R. Houdré, E. Giacobino, C. Ciuti, A. Bramati, Polariton superfluids reveal quantum hydrodynamic solitons. Science **332**, 1167–1170 (2011)
62. L. Dominici, G. Dagvadorj, J.M. Fellows, D. Ballarini, M. De Giorgi, F.M. Marchetti, B. Piccirillo, L. Marrucci, A. Bramati, G. Gigli, M.H. Szymañska, D. Sanvitto, Vortex and half-vortex dynamics in a nonlinear spinor quantum fluid. Sci. Adv. **1**, e1500807 (2015)
63. J.D. Plumhof, T. Stöferle, L. Mai, U. Scherf, R.F. Mahrt, Room-temperature Bose–Einstein condensation of cavity exciton-polaritons in a polymer. Nat. Mater. **13**, 247 (2014). Published online Dec 2013
64. K.S. Daskalakis, S.A. Maier, R. Murray, S. Kéna-Cohen, Nonlinear interactions in an organic polariton condensate. Nat. Mater. **13**(3), 271–278 (2014)
65. K.S. Daskalakis, S.A. Maier, S. Kéna-Cohen, Spatial coherence and stability in a disordered organic polariton condensate. Phys. Rev. Lett. **115**, 035301 (2015)
66. G. Lerario, A. Fieramosca, F. Barachati, D. Ballarini, K.S. Daskalakis, L. Dominici, M.D. Giorgi, S.A. Maier, G. Gigli, S. Kéna-Cohen, D. Sanvitto, Room-temperature superfluidity in a polariton condensate. Nat. Phys. **13**, 837 (2017)
67. I.A. Shelykh, R. Johne, D.D. Solnyshkov, G. Malpuech, Optically and electrically controlled polariton spin transistor. Phys. Rev. B **82**, 153303 (2010)

68. T.C.H. Liew, A.V. Kavokin, I.A. Shelykh, Optical circuits based on polariton neurons in semiconductor microcavities. Phys. Rev. Lett. **101**, 016402 (2008)
69. T. Espinosa-Ortega, T.C.H. Liew, Complete architecture of integrated photonic circuits based on AND and NOT logic gates of exciton polaritons in semiconductor microcavities. Phys. Rev. B **87**, 195305 (2013)
70. D.D. Solnyshkov, O. Bleu, G. Malpuech, All optical controlled-NOT gate based on an exciton-polariton circuit. Superlattices Microstruct. **83**, 466–475 (2015)
71. K.V. Kavokin, M.A. Kaliteevski, R.A. Abram, A.V. Kavokin, S. Sharkova, I.A. Shelykh, Stimulated emission of terahertz radiation by exciton-polariton lasers. Appl. Phys. Lett. **97**, 201111–3 (2010)
72. I.G. Savenko, I.A. Shelykh, M.A. Kaliteevski, Nonlinear terahertz emission in semiconductor microcavities. Phys. Rev. Lett. **107**, 027401 (2011)
73. A.V. Kavokin, I.A. Shelykh, T. Taylor, M.M. Glazov, Vertical cavity surface emitting terahertz laser. Phys. Rev. Lett. **108**, 197401 (2012)
74. S. Huppert, O. Lafont, E. Baudin, J. Tignon, R. Ferreira, Terahertz emission from multiple-microcavity exciton-polariton lasers. Phys. Rev. B **90**, 241302(R) (2014)
75. K. Kristinsson, O. Kyriienko, I.A. Shelykh, Terahertz laser based on dipolaritons. Phys. Rev. A **89**, 023836 (2014)
76. K. Rojan, Y. Léger, G. Morigi, M. Richard, A. Minguzzi, Enhanced second-order nonlinearity for THz generation by resonant interaction of exciton-polariton Rabi oscillations with optical phonons. Phys. Rev. Lett. **119**, 127401 (2017)
77. T.C.H. Liew, M.M. Glazov, K.V. Kavokin, I.A. Shelykh, M.A. Kaliteevski, A.V. Kavokin, Proposal for a bosonic cascade laser. Phys. Rev. Lett. **110**, 047402 (2013)
78. S.S. Demirchyan, I.Y. Chestnov, A.P. Alodjants, M.M. Glazov, A.V. Kavokin, Qubits based on polariton Rabi oscillators. Phys. Rev. Lett. **112**, 196403 (2014)
79. Á. Cuevas, J.C.L. Carreño, B. Silva, M. De Giorgi, D.G. Suárez-Forero, C.S. Muñoz, A. Fieramosca, F. Cardano, L. Marrucci, V. Tasco, G. Biasiol, E. Del Valle, L. Dominici, D. Ballarini, G. Gigli, P. Mataloni, F.P. Laussy, F. Sciarrino, D. Sanvitto, First observation of the quantized exciton-polariton field and effect of interactions on a single polariton. Sci. Adv. **4**, eaao6814 (2018)
80. M. Ramezani, A. Halpin, A.I. Fernez-Domuez, J. Feist, S.R.-K. Rodriguez, F.J. Garcia-Vidal, J.G. Rivas, Plasmon-exciton-polariton lasing. Optica **4**, 31–37 (2017)
81. A. Rahimi-Iman, *Towards Polariton Condensates and Devices*. Springer Series in Optical Sciences, vol. 229 (Springer International Publishing, Cham, 2020), pp. 1–32
82. A. Rahimi-Iman, *Spectroscopy Techniques for Polariton Research*. Springer Series in Optical Sciences, vol. 229 (Springer International Publishing, Cham, 2020), pp. 167–193
83. A. Benz, S. Campione, S. Liu, I. Monta, J.F. Klem, A. Allerman, J.R. Wendt, M.B. Sinclair, F. Capolino, I. Brener, Strong coupling in the sub-wavelength limit using metamaterial nanocavities. Nat. Commun. **4**, 2882 (2013)
84. S.I. Pekar, The theory of electromagnetic waves in a crystal in which excitons are produced. Sov. Phys. JETP **6**, 785 (1958)
85. J.J. Hopfield, Theory of the contribution of excitons to the complex dielectric constant of crystals. Phys. Rev. **112**, 1555 (1958)
86. C.H. Henry, J.J. Hopfield, Raman scattering by polaritons. Phys. Rev. Lett. **15**, 964–966 (1965)
87. C.F. Klingshirn, *Semiconductor Optics* (Springer, Berlin, 2012)
88. K. Huang, Lattice vibrations and optical waves in ionic crystals. Nature **167**, 779–780 (1951)
89. D.W. Ward, J.D. Beers, T. Feurer, E.R. Statz, N.S. Stoyanov, K.A. Nelson, Coherent control of phonon-polaritons in a terahertz resonator fabricated with femtosecond laser machining. Opt. Lett. **29**, 2671–2673 (2004)
90. S. Dai, Q. Ma, T. Andersen, A.S. Mcleod, Z. Fei, M.K. Liu, M. Wagner, K. Watanabe, T. Taniguchi, M. Thiemens, F. Keilmann, P. Jarillo-Herrero, M.M. Fogler, D.N. Basov, Subdiffractional focusing and guiding of polaritonic rays in a natural hyperbolic material. Nat. Commun. **6**, 6963 (2015)

91. J.D. Caldwell, L. Lucas, G. Vincenzo, V. Igor, L. Reinecke Thomas, A. Maier Stefan, J. Glembocki Orest, Low-loss, infrared and terahertz nanophotonics using surface phonon polaritons. Nanophotonics **4**(1), 44 (2015)
92. J.-S. Wu, D.N. Basov, M.M. Fogler, Topological insulators are tunable waveguides for hyperbolic polaritons. Phys. Rev. B **92**, 205430 (2015)
93. M. Sidler, P. Back, O. Cotlet, A. Srivastava, T. Fink, M. Kroner, E. Demler, A. Imamoglu, Fermi polaron-polaritons in charge-tunable atomically thin semiconductors. Nat. Phys. **13**, 255 (2016)
94. N. Wu, J. Feist, F.J. Garcia-Vidal, When polarons meet polaritons: exciton-vibration interactions in organic molecules strongly coupled to confined light fields. Phys. Rev. B **94**, 195409 (2016)
95. M. Irie, S. Kawata, M. Ohtsu (eds.), *Near-Field Optics and Surface Plasmon Polaritons* (Springer, Berlin, 2001)
96. V.W. Brar, M.S. Jang, M. Sherrott, S. Kim, J.J. Lopez, L.B. Kim, M. Choi, H. Atwater, Hybrid surface-phonon-plasmon polariton modes in graphene/monolayer h-bn heterostructures. Nano Lett. **14**, 3876–3880 (2014)
97. V.A. Kosobukin, Plasmon-excitonic polaritons in superlattices. Phys. Solid State **59**, 999–1007 (2017)
98. D.N. Basov, M.M. Fogler, F.J. García De Abajo, Polaritons in van der Waals materials. Science **354**, aag1992 (2016)
99. A. Frisk Kockum, A. Miranowicz, S. De Liberato, S. Savasta, F. Nori, Ultrastrong coupling between light and matter. Nat. Rev. Phys. **1**, 19–40 (2019)
100. P. Forn-D, L. Lamata, E. Rico, J. Kono, E. Solano, Ultrastrong coupling regimes of light-matter interaction. Rev. Mod. Phys. **91**, 025005 (2019)
101. F. Wall, O. Mey, L.M. Schneider, A. Rahimi-Iman, Continuously-tunable light-matter coupling in optical microcavities with 2D semiconductors. Sci. Rep. **10**, 8303 (2020). https://doi.org/10.1038/s41598-020-64909-1
102. M. Hertzog, M. Wang, J. Mony, K. Börjesson, Strong light-matter interactions: a new direction within chemistry. Chem. Soc. Rev. **48**, 937–961 (2019)
103. P. Qing, J. Gong, X. Lin, N. Yao, W. Shen, A. Rahimi-Iman, W. Fang, L. Tong, A simple approach to fiber-based tunable microcavity with high coupling efficiency. Appl. Phys. Lett. **114**, 021106 (2019). https://doi.org/10.1063/1.5083011
104. S.R.K. Rodriguez, Classical and quantum distinctions between weak and strong coupling. Eur. J. Phys. **37**, 25802 (2016)
105. R. El-Ganainy, K.G. Makris, M. Khajavikhan, Z.H. Musslimani, S. Rotter, D.N. Christodoulides, Non-Hermitian physics and PT symmetry. Nat. Phys. **14**(1), 11–19 (2018)
106. W.D. Heiss, The physics of exceptional points. J. Phys. A: Math. Theor. **45**, 444016 (2012)
107. M. Orszag, *Quantum Optics: Including Noise Reduction, Trapped Ions, Quantum Trajectories, and Decoherence*, 3rd edn. (Springer International Publishing, Berlin, 2016)
108. G.H. Wannier, The structure of electronic excitation levels in insulating crystals. Phys. Rev. **52**, 191 (1937)
109. G. Wang, A. Chernikov, M.M. Glazov, T.F. Heinz, X. Marie, T. Amand, B. Urbaszek, Colloquium: excitons in atomically thin transition metal dichalcogenides. Rev. Mod. Phys. **90**, 21001 (2018)
110. N.S. Rytova, Screened potential of a point charge in a thin film. MSU Phys. Astron. **3**(30) (1967). arXiv:1806.00976
111. L.V. Keldysh, Coulomb interaction in thin semiconductor and semimetal films. Sov. JETP Lett. **29**, 658 (1979)
112. P. Cudazzo, I.V. Tokatly, A. Rubio, Dielectric screening in two-dimensional insulators: implications for excitonic and impurity states in graphane. Phys. Rev. B **84**, 085406 (2011)
113. H. Haug, S.W. Koch, *Quantum Theory of the Optical and Electronic Properties of Semiconductors*, 5th edn. (World Scientific Publishing Co., Singapore, 2009)
114. S. Brem, A. Ekman, D. Christiansen, F. Katsch, M. Selig, C. Robert, X. Marie, B. Urbaszek, A. Knorr, E. Malic, Phonon-assisted photoluminescence from indirect excitons in monolayers of transition-metal dichalcogenides. Nano Lett. **20**(4), 2849–2856 (2020)

115. D.A.B. Miller, D.S. Chemla, T.C. Damen, A.C. Gossard, W. Wiegmann, T.H. Wood, C.A. Burrus, Electric field dependence of optical absorption near the band gap of quantum-well structures. Phys. Rev. B **32**, 1043–1060 (1985)
116. R.J. Elliott, Intensity of optical absorption by excitons. Phys. Rev. **108**, 1384–1389 (1957)
117. M. Kira, S. Koch, Many-body correlations and excitonic effects in semiconductor spectroscopy. Prog. Quantum Electron. **30**(5), 155–296 (2006)
118. S. Chatterjee, C. Ell, S. Mosor, G. Khitrova, H.M. Gibbs, W. Hoyer, M. Kira, S.W. Koch, J.P. Prineas, H. Stolz, Excitonic photoluminescence in semiconductor quantum wells: plasma versus excitons. Phys. Rev. Lett. **92**(6), 067402 (2004)
119. M.G.A. Bernard, G. Duraffourg, Laser conditions in semiconductors. Phys. Status Solidi B **1**(7), 699–703 (1961)
120. M. Bernard, The history of laser conditions in semiconductors. Semicond. Sci. Technol. **27**, 090201 (2012)
121. G. Björk, A. Karlsson, Y. Yamamoto, Definition of a laser threshold. Phys. Rev. A **50**, 1675–1680 (1994)
122. S. Schmitt-Rink, D.S. Chemla, D.A.B. Miller, Linear and nonlinear optical properties of semiconductor quantum wells. Adv. Phys. **38**, 89–188 (1989)
123. T. Kazimierczuk, D. Fröhlich, S. Scheel, H. Stolz, M. Bayer, Giant Rydberg excitons in the copper oxide Cu_2O. Nature **514**, 343–347 (2014)
124. J. Thewes, J. Heckötter, T. Kazimierczuk, M. Aßmann, D. Fröhlich, M. Bayer, M. Semina, M. Glazov, Observation of high angular momentum excitons in cuprous oxide. Phys. Rev. Lett. **115**, 027402 (2015)
125. L.V. Keldysh, A.N. Kozlov, Collective properties of excitons in semiconductors. Sov. Phys. JETP **27**, 521–528 (1968)
126. D. Snoke, Spontaneous Bose coherence of excitons and polaritons. Science **298**(5597), 1368–1372 (2002)
127. E. Hanamura, H. Haug, Condensation effects of excitons. Phys. Rep. **33**(4), 209 (1977)
128. A. Griffin, D.W. Snoke, S. Stringari, *Bose-Einstein Condensation* (Cambridge University Press, Cambridge, 1995)
129. P.B. Littlewood, X. Zhu, Possibilities for exciton condensation in semiconductor quantum-well structures. Phys. Scr. **1996**(T68), 56–67 (1996)
130. S.A. Moskalenko, D.W. Snoke, *Bose-Einstein Condensation of Excitons and Biexcitons: And Coherent Nonlinear Optics with Excitons* (Cambridge University Press, Cambridge, 2000)
131. L.P. Pitaevskii, S. Stringari, *Bose-Einstein Condensation* (Oxford University Press, Oxford, 2003)
132. W. Ketterle, What is B.E.C.? (webpage 2014: www.cua.mit.edu/ketterle_group/intro/whatbec/whtisbec.html) (2014)
133. S.A. Moskalenko, Reversible optico-hydrodynamic phenomena in a nonideal exciton gas. Sov. Phys. Solid State **4**(1), 199–204 (1962)
134. J.M. Blatt, K.W. Boer, W. Brandt, Bose-Einstein condensation of excitons. Phys. Rev. **126**(5), 1691 (1962)
135. N.F. Mott, Conduction in polar crystals. II. The conduction band and ultra-violet absorption of alkali-halide crystals. Trans. Faraday Soc. **34**, 500 (1938)
136. N.F. Mott, *Metal Insulator Transitions*, 2nd edn. (CRC Press, Boca Raton, 1990)
137. L. Kappei, J. Szczytko, F. Morier-Genoud, B. Deveaud, Direct observation of the Mott transition in an optically excited semiconductor quantum well. Phys. Rev. Lett. **94**, 147403 (2005)
138. L.V. Butov, C.W. Lai, A.L. Ivanov, A.C. Gossard, D.S. Chemla, Towards Bose-Einstein condensation of excitons in potential traps. Nature **417**, 47–52 (2002)
139. L.V. Butov, A.C. Gossard, D.S. Chemla, Macroscopically ordered state in an exciton system. Nature **418**, 751–754 (2002)
140. K. Yoshioka, E. Chae, M. Kuwata-Gonokami, Transition to a Bose-Einstein condensate and relaxation explosion of excitons at sub-Kelvin temperatures. Nat. Commun. **2**, 328 (2011)
141. A.A. High, J.R. Leonard, A.T. Hammack, M.M. Fogler, L.V. Butov, A.V. Kavokin, K.L. Campman, A.C. Gossard, Spontaneous coherence in a cold exciton gas. Nature **483**, 584–588 (2012)

142. A.A. High, J.R. Leonard, M. Remeika, L.V. Butov, M. Hanson, A.C. Gossard, Condensation of excitons in a trap. Nano Lett. **12**, 2605–2609 (2012)
143. M.M. Fogler, L.V. Butov, K.S. Novoselov, High-temperature superfluidity with indirect excitons in van der Waals heterostructures. Nat. Commun. **5**, 4555 (2014)
144. Z. Wang, D.A. Rhodes, K. Watanabe, T. Taniguchi, J.C. Hone, J. Shan, K.F. Mak, Evidence of high-temperature exciton condensation in two-dimensional atomic double layers. Nature **574**, 76–80 (2019)
145. J.P. Eisenstein, A.H. MacDonald, Bose-Einstein condensation of excitons in bilayer electron systems. Nature **432**, 691–694 (2004)
146. X. Liu, K. Watanabe, T. Taniguchi, B.I. Halperin, P. Kim, Quantum Hall drag of exciton condensate in graphene. Nat. Phys. **13**, 746–750 (2017)
147. J.P. Eisenstein, L.N. Pfeiffer, K.W. West, Precursors to exciton condensation in quantum Hall bilayers. Phys. Rev. Lett **123**, 066802 (2019)
148. E.U. Rafailov, M.A. Cataluna, W. Sibbett, Mode-locked quantum-dot lasers. Nat. Photonics **1**, 395–401 (2007)
149. M. Gaafar, D.A. Nakdali, C. Möller, K.A. Fedorova, M. Wichmann, M.K. Shakfa, F. Zhang, A. Rahimi-Iman, E.U. Rafailov, M. Koch, Self-mode-locked quantum-dot vertical-external-cavity surface-emitting laser. Opt. Lett. **39**, 4623–4626 (2014)
150. M.A. Gaafar, A. Rahimi-Iman, K.A. Fedorova, W. Stolz, E.U. Rafailov, M. Koch, Mode-locked semiconductor disk lasers. Adv. Opt. Photonics **8**, 370–400 (2016)
151. D. Al Nakdali, M.K. Shakfa, M. Gaafar, M. Butkus, K.A. Fedorova, M. Zulonas, M. Wichmann, F. Zhang, B. Heinen, A. Rahimi-Iman, W. Stolz, E.U. Rafailov, M. Koch, High-power quantum-dot vertical-external-cavity surface-emitting laser exceeding 8 W. IEEE Photonics Technol. Lett. **26**, 1561–1564 (2014)
152. D. Al Nakdali, M. Gaafar, M.K. Shakfa, F. Zhang, M. Vaupel, K.A. Fedorova, A. Rahimi-Iman, E.U. Rafailov, M. Koch, High-power operation of quantum-dot semiconductor disk laser at 1180 nm. IEEE Photonics Technol. Lett. **27**, 1128–1131 (2015)
153. A. Rahimi-Iman, Recent advances in VECSELs. J. Opt. **18**, 093003 (2016)
154. C. Weisbuch, B. Vinter, *Quantum Semiconductor Structures: Fundamentals and Applications* (Academic, Cambridge, 1991)
155. J.H. Davies, *The Physics of Low-dimensional Semiconductors: An Introduction* (Cambridge University Press, Cambridge, 1998)
156. A.E. Almand-Hunter, H. Li, S.T. Cundiff, M. Mootz, M. Kira, S.W. Koch, Quantum droplets of electrons and holes. Nature **506**(7489), 471–475 (2014)
157. K.B. Tolpygo, Physical properties of a rock salt lattice made up of deformable ions. Sov. Phys. JETP **20**(6), 497–509 (1950). in Russian
158. S.I. Pekar, Theory of electromagnetic waves in a crystal with excitons. J. Phys. Chem. Solids **5**, 11 (1958)
159. L.M. Schneider, S.S. Esdaille, D.A. Rhodes, K. Barmak, J.C. Hone, A. Rahimi-Iman, Shedding light on exciton's nature in monolayer quantum material by optical dispersion measurements. Opt. Express **27**, 37131–37149 (2019). https://doi.org/10.1364/OE.27.037131
160. L.M. Schneider, S.S. Esdaille, D.A. Rhodes, K. Barmak, J.C. Hone, A. Rahimi-Iman, Direct measurement of the radiative pattern of bright and dark excitons and exciton complexes in encapsulated tungsten diselenide. Sci. Rep. **10**, 8091 (2020). https://doi.org/10.1038/s41598-020-64838-z
161. J.J. Hopfield, D.G. Thomas, Theoretical and experimental effects of spatial dispersion on the optical properties of crystals. Phys. Rev. **132**, 563–572 (1963)
162. S. Suga, K. Cho, P. Hiesinger, T. Koda, Luminescence of exciton-polaritons in semiconductors. J. Lumin. **12–13**, 109–117 (1976)
163. V.M. Agranovich, V. Ginzburg, *Chrystal Optics with Spatial Dispersion, and Excitons*, 2nd edn. (Springer, Berlin, 1984)
164. W.J. Rappel, L.F. Feiner, M.F.H. Schuurmans, Exciton-polariton picture of the free-exciton lifetime in GaAs. Phys. Rev. B **38**, 7874–7876 (1988)

165. H. Kuzmany, Light scattering experiments and nonlinear behaviour of the acoustoelectric oscillator. Phys. Status Solidi A **34**, 347–353 (1976)
166. P. Wiesner, U. Heim, Dynamics of exciton-polariton recombination in CdS. Phys. Rev. B **11**, 3071–3077 (1975)
167. R. Hauschild, H. Lange, H. Priller, C. Klingshirn, R. Kling, A. Waag, H.J. Fan, M. Zacharias, H. Kalt, Stimulated emission from ZnO nanorods. Phys. Status Solidi B **243**, 853–857 (2006)
168. E.M. Purcell, Spontaneous emission probabilities at radio frequencies. Phys. Rev. **69**, 681 (1946)
169. A. Rahimi-Iman, *Optical Microcavities for Polariton Studies*. Springer Series in Optical Sciences, vol. 229 (Springer International Publishing, Cham, 2020), pp. 119–137
170. R.A. Kaindl, M.A. Carnahan, D. Hägele, R. Lövenich, D.S. Chemla, Ultrafast terahertz probes of transient conducting and insulating phases in an electron-hole gas. Nature **423**, 734–738 (2003)
171. C. Poellmann, P. Steinleitner, U. Leierseder, P. Nagler, G. Plechinger, M. Porer, R. Bratschitsch, C. Schüller, T. Korn, R. Huber, Resonant internal quantum transitions and femtosecond radiative decay of excitons in monolayer WSe_2. Nat. Mater. **14**, 889–893 (2015)
172. J.M. Ménard, C. Poellmann, M. Porer, U. Leierseder, E. Galopin, A. Lemaître, A. Amo, J. Bloch, R. Huber, Revealing the dark side of a bright exciton-polariton condensate. Nat. Commun. **5**, 4648 (2014)
173. M. Stein, C. Lammers, P. Springer, P.-H. Richter, S.W. Koch, M. Koch, M. Kira, Density-dependent exciton dynamics and L-valley anisotropy in germanium. Phys. Rev. B **95**, 155207 (2017)
174. M. Stein, C. Lammers, P.-H. Richter, C. Fuchs, W. Stolz, M. Koch, O. Vänskä, M.J. Weseloh, M. Kira, S.W. Koch, Dynamics of charge-transfer excitons in type-II semiconductor heterostructures. Phys. Rev. B **97**, 125306 (2018)
175. C. Lammers, M. Stein, C. Berger, C. Möller, C. Fuchs, A. Ruiz Perez, A. Rahimi-Iman, J. Hader, J.V. Moloney, W. Stolz, S.W. Koch, M. Koch, Gain spectroscopy of a type-II VECSEL chip. Appl. Phys. Lett. **109**, 232107 (2016)
176. M.J. Drexler, R. Woscholski, S. Lippert, W. Stolz, A. Rahimi-Iman, M. Koch, Disturbing the coherent dynamics of an excitonic polarization with strong terahertz fields. Phys. Rev. B **90**, 195304 (2014)
177. Y. Song, H. Dery, Transport theory of monolayer transition-metal dichalcogenides through symmetry. Phys. Rev. Lett. **111**, 026601 (2013)
178. H. Dery, Y. Song, Polarization analysis of excitons in monolayer and bilayer transition-metal dichalcogenides. Phys. Rev. B **92**, 125431 (2015)
179. C. Robert, T. Amand, F. Cadiz, D. Lagarde, E. Courtade, M. Manca, T. Taniguchi, K. Watanabe, B. Urbaszek, X. Marie, Fine structure and lifetime of dark excitons in transition metal dichalcogenide monolayers. Phys. Rev. B **96**, 755423 (2017)
180. L.M. Schneider, S.S. Esdaille, D.A. Rhodes, K. Barmak, J.C. Hone, A. Rahimi-Iman, Optical dispersion of valley-hybridised coherent excitons with momentum-dependent valley polarisation in monolayer semiconductor. 2D Mater. **8**, 015009 (2020)
181. E. Hanamura, Rapid radiative decay and enhanced optical nonlinearity of excitons in a quantum well. Phys. Rev. B **38**, 1228–1234 (1988)
182. L.C. Andreani, F. Bassani, Exchange interaction and polariton effects in quantum-well excitons. Phys. Rev. B **41**, 7536–7544 (1990)
183. H. Yu, G.-B. Liu, P. Gong, X. Xu, W. Yao, Dirac cones and Dirac saddle points of bright excitons in monolayer transition metal dichalcogenides. Nat. Commun. **5**, 3876 (2014)
184. F. Wu, F. Qu, A.H. MacDonald, Exciton band structure of monolayer MoS_2. Phys. Rev. B **91**, 075310 (2015)
185. D.Y. Qiu, T. Cao, S.G. Louie, Nonanalyticity, valley quantum phases, and lightlike exciton dispersion in monolayer transition metal dichalcogenides: theory and first-principles calculations. Phys. Rev. Lett. **115**, 176801 (2015)
186. M.M. Glazov, T. Amand, X. Marie, D. Lagarde, L. Bouet, B. Urbaszek, Exciton fine structure and spin decoherence in monolayers of transition metal dichalcogenides. Phys. Rev. B **89**, 201302 (2014)

187. T. Deilmann, K.S. Thygesen, Finite-momentum exciton landscape in mono- and bilayer transition metal dichalcogenides. 2D Mater. **6**, 035003 (2019)
188. M.S. Skolnick, T.A. Fisher, D.M. Whittaker, Strong coupling phenomena in quantum microcavity structures. Semicond. Sci. Technol. **13**, 645–669 (1998)
189. R. Houdré, C. Weisbuch, R.P. Stanley, U. Oesterle, P. Pellandini, M. Ilegems, Measurement of cavity-polariton dispersion curve from angle-resolved photoluminescence experiments. Phys. Rev. Lett. **73**, 2043–2046 (1994)
190. R. Houdré, R.P. Stanley, U. Oesterle, M. Ilegems, C. Weisbuch, Room-temperature cavity polaritons in a semiconductor microcavity. Phys. Rev. B **49**, 16761 (1994)
191. R. Houdré, J.L. Gibernon, P. Pellandini, R.P. Stanley, U. Oesterle, C. Weisbuch, J. O'Gorman, B. Roycroft, M. Ilegems, Saturation of the strong-coupling regime in a semiconductor microcavity: free-carrier bleaching of cavity polaritons. Phys. Rev. B **52**, 7810 (1995)
192. R. André, F. Boeuf, D. Heger, L.S. Dang, R. Romestain, J. Bleuse, M. Müller, Cavity-polariton effects in II-VI microcavities. Acta Phys. Pol. A **96**(5), 511–524 (1999)
193. C. Sturm, D. Tanese, H. Nguyen, H. Flayac, E. Galopin, A. Lemaitre, A. Amo, G. Malpuech, J. Bloch, All-optical phase modulation in a cavity-polariton Mach-Zehnder interferometer. Nat. Commun. **5**, 3278 (2014)
194. R.T. Juggins, J. Keeling, M.H. Szymańska, Coherently driven microcavity-polaritons and the question of superfluidity. Nat. Commun. **9**, 4062 (2018)
195. J. Bloch, T. Freixanet, J.Y. Marzin, V. Thierry-Mieg, R. Planel, Giant Rabi splitting in a microcavity containing distributed quantum wells. Appl. Phys. Lett. **73**, 1694–1696 (1998)
196. A. Rahimi-Iman, *Fundamentals of Polariton Physics*. Springer Series in Optical Sciences, vol. 229 (Springer International Publishing, Cham, 2020), pp. 33–64
197. A. Rahimi-Iman, Winkelaufgelöste Spektroskopie an Exziton-Polaritonen in optisch und elektrisch angeregten III/V Quantenfilm-Mikroresonatoren. Engineering diploma thesis, Julius-Maximilians-Universität Würzburg, Lehrstuhl für Technische Physik, 2010
198. H. Deng, G. Weihs, D. Snoke, J. Bloch, Y. Yamamoto, Polariton lasing vs. photon lasing in a semiconductor microcavity. Proc. Natl. Acad. Sci. **100**(26), 15318–15323 (2003)
199. H. Deng, D. Press, S. Götzinger, G.S. Solomon, R. Hey, K.H. Ploog, Y. Yamamoto, Quantum degenerate exciton-polaritons in thermal equilibrium. Phys. Rev. Lett. **97**, 146402 (2006)
200. V. Savona, D. Sarchi, Bose-Einstein condensation of microcavity polaritons. Phys. Status Solidi B **242**, 2290–2301 (2005)
201. S. Utsunomiya, L. Tian, G. Roumpos, C.W. Lai, N. Kumada, T. Fujisawa, M. Kuwata-Gonokami, A. Löffler, S. Höfling, A. Forchel, Y. Yamamoto, Observation of Bogoliubov excitations in exciton-polariton condensates. Nat. Phys. **4**, 700–705 (2008)
202. J. Keeling, N.G. Berloff, Exciton-polariton condensation. Contemp. Phys. **52**, 131–151 (2011)
203. T. Byrnes, N.Y. Kim, Y. Yamamoto, Exciton–polariton condensates. Nat. Phys. **10**, 803–813 (2014)
204. K.B. Davis, M.O. Mewes, M.R. Andrews, N.J. van Druten, D.S. Durfee, D.M. Kurn, W. Ketterle, Bose-Einstein condensation in a gas of sodium atoms. Phys. Rev. Lett. **75**, 3969 (1995)
205. M.H. Anderson, J.R. Ensher, M.R. Matthews, C.E. Wieman, E.A. Cornell, Observation of Bose-Einstein condensation in a dilute atomic vapor. Science **269**, 198–201 (1995)
206. C.C. Bradley, C.A. Sackett, J.J. Tollett, R.G. Hulet, Evidence of Bose-Einstein condensation in an atomic gas with attractive interactions. Phys. Rev. Lett. **75**, 1687 (1995)
207. A. Kavokin, G. Malpuech, *Cavity Polaritons* (Academic, Cambridge, 2003)
208. J. Kasprzak, M. Richard, A. Baas, B. Deveaud, R. Andre, J.-P. Poizat, L.S. Dang, Second-order time correlations within a polariton Bose-Einstein condensate in a CdTe microcavity. Phys. Rev. Lett. **100**, 067402 (2008)
209. A. Rahimi-Iman, A.V. Chernenko, J. Fischer, S. Brodbeck, M. Amthor, C. Schneider, A. Forchel, S. Höfling, S. Reitzenstein, M. Kamp, Coherence signatures and density-dependent interaction in a dynamical exciton-polariton condensate. Phys. Rev. B **86**, 155308 (2012)
210. J.-S. Tempel, F. Veit, M. Assmann, L.E. Kreilkamp, A. Rahimi-Iman, A. Löffler, S. Höfling, S. Reitzenstein, L. Worschech, A. Forchel, M. Bayer, Characterization of two-threshold behavior of the emission from a GaAs microcavity. Phys. Rev. B **85**, 075318 (2012)

211. Y.G. Rubo, A. Kavokin, I. Shelykh, Suppression of superfluidity of exciton-polaritons by magnetic field. Phys. Lett. A **358**, 227–230 (2006)
212. J. Fischer, S. Brodbeck, A.V. Chernenko, I. Lederer, A. Rahimi-Iman, M. Amthor, V.D. Kulakovskii, L. Worschech, M. Kamp, M. Durnev, C. Schneider, A.V. Kavokin, S. Höfling, Anomalies of a nonequilibrium spinor polariton condensate in a magnetic field. Phys. Rev. Lett. **112**, 093902 (2014)
213. M. Assmann, J.-S. Tempel, F. Veit, M. Bayer, A. Rahimi-Iman, A. Löffler, S. Höfling, S. Reitzenstein, L. Worschech, A. Forchel, From polariton condensates to highly photonic quantum degenerate states of bosonic matter. Proc. Natl. Acad. Sci. **108**, 1804–1809 (2011)
214. A.V. Chernenko, A. Rahimi-lman, J. Fischer, M. Amthor, C. Schneider, S. Reitzenstein, A. Forchel, S. Hoefling, Polariton condensate coherence in planar microcavities in a magnetic field. Semiconductors **50**, 1609–1613 (2016). https://doi.org/10.1134/S1063782616120058
215. E. Estrecho, T. Gao, N. Bobrovska, M.D. Fraser, M. Steger, L. Pfeiffer, K. West, T.C.H. Liew, M. Matuszewski, D.W. Snoke, A.G. Truscott, E.A. Ostrovskaya, Single-shot condensation of exciton polaritons and the hole burning effect. Nat. Commun. **9**, 2944 (2018)
216. G. Roumpos, M.D. Fraser, A. Löffler, S. Höfling, A. Forchel, Y. Yamamoto, Single vortex-antivortex pair in an exciton-polariton condensate. Nat. Phys. **7**, 129–133 (2011)
217. N.N. Bogoliubov, On the theory of superfluidity. J. Phys. USSR **11**, 23 (1947)
218. J.J. Baumberg, A.V. Kavokin, S. Christopoulos, A.J.D. Grundy, R. Butté, G. Christmann, D.D. Solnyshkov, G. Malpuech, G. Baldassarri Höger von Högersthal, E. Feltin, J.-F. Carlin, N. Grandjean, Spontaneous polarization buildup in a room-temperature polariton laser. Phys. Rev. Lett. **101**, 136409 (2008)
219. G. Malpuech, A. Di Carlo, A. Kavokin, J.J. Baumberg, M. Zamfirescu, P. Lugli, Room-temperature polariton lasers based on GaN microcavities. Appl. Phys. Lett. **81**, 412–414 (2002)
220. G. Christmann, R. Butté, E. Feltin, J.-F. Carlin, N. Grandjean, Room temperature polariton lasing in a GaN/AlGaN multiple quantum well microcavity. Appl. Phys. Lett. **93**, 051102 (2008)
221. H. Franke, C. Sturm, R. Schmidt-Grund, G. Wagner, M. Grundmann, Ballistic propagation of exciton-polariton condensates in a ZnO-based microcavity. New J. Phys. **14**, 013037 (2012)
222. F. Li, L. Orosz, O. Kamoun, S. Bouchoule, C. Brimont, P. Disseix, T. Guillet, X. Lafosse, M. Leroux, J. Leymarie, M. Mexis, M. Mihailovic, G. Patriarche, F. Réveret, D. Solnyshkov, J. Zuniga-Perez, G. Malpuech, From excitonic to photonic polariton condensate in a ZnO-based microcavity. Phys. Rev. Lett. **110**, 196406 (2013)
223. X. Liu, T. Galfsky, Z. Sun, F. Xia, E.C. Lin, Y.H. Lee, S. Kéna-Cohen, V.M. Menon, Strong light-matter coupling in two-dimensional atomic crystals. Nat. Photonics **9**, 30–34 (2014). https://doi.org/10.1038/nphoton.2014.304
224. S. Dufferwiel, S. Schwarz, F. Withers, A.A. Trichet, F. Li, M. Sich, O. Del Pozo-Zamudio, C. Clark, A. Nalitov, D.D. Solnyshkov, G. Malpuech, K.S. Novoselov, J.M. Smith, M.S. Skolnick, D.N. Krizhanovskii, A.I. Tartakovskii, Exciton-polaritons in van der Waals heterostructures embedded in tunable microcavities. Nat. Commun. **6**, 8579 (2015)
225. L.C. Flatten, Z. He, D.M. Coles, A.A. Trichet, A.W. Powell, R.A. Taylor, J.H. Warner, J.M. Smith, Room-temperature exciton-polaritons with two-dimensional WS_2. Sci. Rep. **6**, 33134 (2016)
226. N. Lundt, S. Klembt, E. Cherotchenko, S. Betzold, O. Iff, A.V. Nalitov, M. Klaas, C.P. Dietrich, A.V. Kavokin, S. Höfling, C. Schneider, Room-temperature Tamm-plasmon exciton-polaritons with a WSe_2 monolayer. Nat. Commun. **7**, 13328 (2016)
227. S. Wang, S. Li, T. Chervy, A. Shalabney, S. Azzini, E. Orgiu, J.A. Hutchison, C. Genet, P. Samorì, T.W. Ebbesen, Coherent coupling of WS_2 monolayers with metallic photonic nanostructures at room temperature. Nano Lett. **16**(7), 4368–74 (2016)
228. F. Scafirimuto, D. Urbonas, U. Scherf, R.F. Mahrt, T. Stöferle, Room-temperature exciton-polariton condensation in a tunable zero-dimensional microcavity. ACS Photonics **5**, 85 (2018)
229. L.A. Smirnov, D.A. Smirnova, E.A. Ostrovskaya, Y.S. Kivshar, Dynamics and stability of dark solitons in exciton-polariton condensates. Phys. Rev. B **89**, 235310 (2014)

230. E. Cancellieri, F.M. Marchetti, M.H. Szymańska, D. Sanvitto, C. Tejedor, Frictionless flow in a binary polariton superfluid. Phys. Rev. Lett. **108**, 065301 (2012)
231. L. Dominici, M. Petrov, M. Matuszewski, D. Ballarini, M. De Giorgi, D. Colas, E. Cancellieri, B. Silva Fernández, A. Bramati, G. Gigli, A. Kavokin, F. Laussy, D. Sanvitto, Real-space collapse of a polariton condensate. Nat. Commun. **6**, 8993 (2015)
232. M. De Giorgi, D. Ballarini, P. Cazzato, G. Deligeorgis, S.I. Tsintzos, Z. Hatzopoulos, P.G. Savvidis, G. Gigli, F.P. Laussy, D. Sanvitto, Relaxation oscillations in the formation of a polariton condensate. Phys. Rev. Lett. **112**, 113602 (2014)
233. C.W. Lai, N.Y. Kim, S. Utsunomiya, G. Roumpos, H. Deng, M.D. Fraser, T. Byrnes, P. Recher, N. Kumada, T. Fujisawa, Y. Yamamoto, Coherent zero-state and π-state in an exciton-polariton condensate array. Nature **450**, 529–532 (2007)
234. N.Y. Kim, K. Kusudo, C. Wu, N. Masumoto, A. Löffler, S. Höfling, N. Kumada, L. Worschech, A. Forchel, Y. Yamamoto, Dynamical d-wave condensation of exciton-polaritons in a two-dimensional square-lattice potential. Nat. Phys. **7**, 681–686 (2011)
235. N.Y. Kim, K. Kusudo, A. Löffler, S. Höfling, A. Forchel, Y. Yamamoto, Exciton-polariton condensates near the Dirac point in a triangular lattice. New J. Phys. **15**, 035032 (2013)
236. E.A. Cerda-Méndez, D. Sarkar, D.N. Krizhanovskii, S.S. Gavrilov, K. Biermann, M.S. Skolnick, P.V. Santos, Exciton-polariton gap solitons in two-dimensional lattices. Phys. Rev. Lett. **111**, 146401 (2013)
237. T. Jacqmin, I. Carusotto, I. Sagnes, M. Abbarchi, D.D. Solnyshkov, G. Malpuech, E. Galopin, A. Lemae, J. Bloch, A. Amo, Direct observation of Dirac cones and a flatband in a honeycomb lattice for polaritons. Phys. Rev. Lett **112**, 116402 (2014)
238. M. Sun, I.G. Savenko, S. Flach, Y.G. Rubo, Excitation of localized condensates in the flat band of the exciton-polariton Lieb lattice. Phy. Rev. B **98**, 161204 (2018)
239. K. Winkler, J. Fischer, A. Schade, M. Amthor, R. Dall, J. Geßler, M. Emmerling, E.A. Ostrovskaya, M. Kamp, C. Schneider, S. Höfling, A polariton condensate in a photonic crystal potential landscape. New J. Phys. **17**, 023001 (2015)
240. A. Amo, J. Bloch, Exciton-polaritons in lattices: a non-linear photonic simulator. C. R. Phys. **17**, 934–945 (2016)
241. S. Klembt, T.H. Harder, O.A. Egorov, K. Winkler, R. Ge, M.A. Bandres, M. Emmerling, L. Worschech, T.C.H. Liew, M. Segev, C. Schneider, S. Höfling, Exciton-polariton topological insulator. Nature **562**, 552–556 (2018)
242. A. Rahimi-Iman, *Technological Realization of Polariton Systems*. Springer Series in Optical Sciences, vol. 229 (Springer International Publishing, Cham, 2020), pp. 139–166
243. A. Rahimi-Iman, *Polaritons in External Fields*. Springer Series in Optical Sciences, vol. 229 (Springer International Publishing, Cham, 2020), pp. 241–262
244. M. Amthor, S. Weissenseel, J. Fischer, M. Kamp, C. Schneider, S. Höfling, Electro-optical switching between polariton and cavity lasing in an InGaAs quantum well microcavity. Opt. Express **22**, 31146–31153 (2014)
245. A. Armitage, T.A. Fisher, M.S. Skolnick, D.M. Whittaker, P. Kinsler, J.S. Roberts, Exciton polaritons in semiconductor quantum microcavities in a high magnetic field. Phys. Rev. B **55**, 16395 (1997)
246. A.V. Kavokin, J.J. Baumberg, G. Malpuech, F.P. Laussy, *Microcavities* (Oxford University Press, 2007)
247. F. Tassone, Y. Yamamoto, Exciton-exciton scattering dynamics in a semiconductor microcavity and stimulated scattering into polaritons. Phys. Rev. B **59**, 10830 (1999)
248. A. Rahimi-Iman, C. Schneider, J. Fischer, S. Holzinger, M. Amthor, S. Höfling, S. Reitzenstein, L. Worschech, M. Kamp, A. Forchel, Zeeman splitting and diamagnetic shift of spatially confined quantum-well exciton polaritons in an external magnetic field. Phys. Rev. B **84**, 165325 (2011)
249. T.A. Fisher, A.M. Afshar, M.S. Skolnick, D.M. Whittaker, J.S. Roberts, Vacuum Rabi coupling enhancement and Zeeman splitting in semiconductor quantum microcavity structures in a high magnetic field. Phys. Rev. B **53**, R10469 (1996)

250. A. Rahimi-Iman, C. Schneider, J. Fischer, S. Holzinger, M. Amthor, L. Worschech, S. Reitzenstein, S. Höfling, A. Forchel, M. Kamp, Magnetic-field interaction of spatially confined quantum-well exciton-polaritons. J. Phys. Conf. Ser. **456**, 012033 (2013)
251. M. Amthor, T.C.H. Liew, C. Metzger, S. Brodbeck, L. Worschech, M. Kamp, I.A. Shelykh, A.V. Kavokin, C. Schneider, S. Hoefling, Optical bistability in electrically driven polariton condensates. Phys. Rev. B **91**, 081404 (2015)
252. S. Klembt, E. Durupt, S. Datta, T. Klein, A. Baas, Y. Léger, C. Kruse, D. Hommel, A. Minguzzi, M. Richard, Exciton-polariton gas as a nonequilibrium coolant. Phys. Rev. Lett. **114**, 186403 (2015)
253. J.L. Tomaino, A.D. Jameson, Y.-S. Lee, G. Khitrova, H.M. Gibbs, A.C. Klettke, M. Kira, S.W. Koch, Terahertz excitation of a coherent Λ-type three-level system of exciton-polariton modes in a quantum-well microcavity. Phys. Rev. Lett. **108**, 267402 (2012)
254. P. Schwendimann, A. Quattropani, Statistics of the polariton condensate. Phys. Rev. B **77**, 085317 (2008)
255. T.D. Doan, H.T. Cao, D.B.T. Thoai, H. Haug, Coherence of condensed microcavity polaritons calculated within Boltzmann-Master equations. Phys. Rev. B **78**, 205306 (2008)
256. B. Piętka, N. Bobrovska, D. Stephan, M. Teich, M. Kr, S. Winnerl, A. Pashkin, R. Mirek, K. Lekenta, F. Morier-Genoud, H. Schneider, B. Deveaud, M. Helm, M. Matuszewski, J. Szczytko, Doubly dressed bosons: exciton polaritons in a strong terahertz field. Phys. Rev. Lett. **119**, 077403 (2017)

Chapter 4
In the Field of Quantum Technologies

Abstract Quantum technologies are an emerging field one can think of in two categories. The first generation, which used principles of quantum physics to deliver technologies of the last century, and the second generation, which harnesses quantum phenomena to achieve novel device functionalities in the current century. In general, various sophisticated tools for the control and manipulation of quantum states are required and under development, fundamental concepts based on nonclassical behaviour or properties are used, or new capabilities are enabled which would have been out of reach for classical devices. Indeed, the role of semiconductor photonics as well as quantum structures is pronounced. In this chapter, two aspects of work in the field of quantum technologies are summarised. The first part will deal with coherent light sources, ranging from semiconductor lasers to novel sources of coherent light. The second part provides an introduction into quantum optics based on quantum light generation and tailored light–matter coupling. Serving as a starting point, here given examples shall motivate further reading in the dedicated literature on these subjects.

4.1 Into the Quantum Realm

Entering the nonclassical world, quantum technologies have revealed themselves as an emerging topic. In fact, (previously discussed) quantum structures have been indispensable when addressing the needs in the domain of quantum technologies of the second generation. In many ways, future technologies such as quantum information processing (e.g. quantum computation and communication) will heavily rely on various sub-components such as quantum networks, quantum channels, quantum nodes, quantum memories, quantum light, quantum repeaters, quantum sensors, quantum radars/LIDARs etc. The discussion of these novel technologies and their subsystems—even in a superficial manner—clearly appears out of the scope of this work. However, as the wider field of quantum optics and related subjects has attracted considerable attention and increasingly becomes important in postgraduate courses, seminars and lectures, the interested reader is referred to the expert literature (see for instance [1–5] and references therein).

A. Rahimi-Iman, *Semiconductor Photonics of Nanomaterials and Quantum Structures*, Springer Series in Solid-State Sciences 196,
https://doi.org/10.1007/978-3-030-69352-7_4

Quantum Technology Basics
These quantum technologies of the second generation were preceded by quantum technologies of the first generation, such as semiconductor light source technologies (i.e. LEDs and lasers) or micro-/nano-electronics (i.e. circuitry with classical gate structures), which have remained attractive due to their maturity and good integrability in existing technologies and have been subject of further development. Numerous textbooks discuss fundamentals and advances in this important field, such as [2, 6–9], and their underlying theoretical framework [10] and semiconductor optics [11–14].

Quantum Computing on the Verge
Likely, quantum computing is one of the most prominent subjects, as the news coverage on the latest achievements by major information-technology companies worldwide shows, such as on quantum supremacy [15]. So far, different complex techniques are at play, such as those employing superconducting qubits (such as in [16]). Trapping of ions for quantum control and logic operations is also a possible path, while the use of photons and solid-state quantum/photonic structures becomes increasingly popular [17–21].

Quantum Light for Quantum Applications
In recent years, major leaps have been evidenced in the field of quantum light source development [22–26], which is important for future optical quantum computation [20, 21] and communication schemes. This has set free a quantum space race in which labs around the world try to bridge farther and farther distances using quantum teleportation for photons (see for instance the news feature from 2012, [27]). A record distance for 'spooky action' was for instance in recent years achieved via satellite-based transmission over more than thousand kilometers [28].

Entanglement for Communications
Typically, heralded photons from entangled photon pairs or nonclassical states of light are utilised to enable secure quantum channels for quantum key distribution. Future long-haul transmission schemes may rely on devices sending out polarisation-entangled photons (see for instance [29]) bidirectionally from a quantum repeater station, whereas optical microcavity devices with embedded quantum emitters at the ends of the transmission line provide indistinguishable single photons on demand, or may be used as quantum storage. Also, microcavities have been proposed for memory-based quantum repeaters [30]. Moreover, quantum measurements (Bell state measurements) could swap the entanglement between individual photons from different quantum repeaters. Thereby, two parties (famously named) Alice and Bob can exchange flying quantum bits (qubits) with the chance to discover any eavesdropping event (by person Eve) via the evaluation of the error rate—as the no-cloning theorem renders the mere copying and replacing of photons in the quantum channel technically impossible, apart from measuring the quantum state of the photon without manipulating it.

Advances in Quantum Light Generation
The demand for the fundamental equipment for such schemes is rapidly growing. However, most commercial products are based on conventional heavily attenuated

pulsed lasers, which contain with very high probability no photons per pulse and sometimes one single photon, whereas the probability to obtain a second photon is non-zero due to the underlying Poissonian photon statistics for the coherent state. Quantum emitters in microcavities (making use of controlled spontaneous emission by tailored low-mode-volume light fields [31–33]), on the other hand, can provide single photons with record-high repetition rates [34] and become more and more practical platforms for bright sources of indistinguishable single photons due to recent technological improvements (see above references on quantum light sources and [35–41], to name but a few).

Concerning the generation of entangled photons (e.g. for quantum communication or ghost imaging applications), typically spontaneous parametric down conversion (SPDC) is used. This technique requires high-repetition rate laser pulses (for ultrafast communication schemes) and femtoseconds pulse durations (for high pulse peak powers) to make the low-probability conversion effect as efficient and frequent as possible, linking first-generation and second-generation quantum technologies. Recently, also other sources of entangled photons based on quantum dots have been investigated (see for instance [42]).

4.2 Coherent Light Sources

The development of semiconductor lasers[1] heavily relied on concepts of the early quantum mechanics world and fundamental theory of light–matter interactions. Among the members of the laser family, such as solid-state lasers (see for instance [8]), semiconductor lasers (e.g. [43, 44]) are widely used in our modern world for various applications and in different frequency ranges. They became increasingly reliable, mass-producible and, therefore, cheap owing to the well-matured field of semiconductor technology.

Coherence in Advantageous Light

Another reason to pay attention to coherent light sources is motivated by the very nature of their light output: A directional beam of phase-coherent, polarised and typically monochromatic light can be conveniently and precisely used to deliver high intensities, ultrashort pulses or spectrally well-defined irradiances to experiments and applications. Moreover, the coherent state of light is a very prominent example of a light state, which represents the vacuum state displaced by a coherent amplitude (certain mean number of photons).[2] The photon statistics for an ideal coherent state is totally random[3] and centred around the expectation value of the photon number (N,

---

[1]Laser is the acronym for *Light Amplification by Stimulated Emission of Radiation*.

[2]See vacuum state $|0\rangle$ displaced by the displacement operator $D(\alpha)$ representing the coherent state $|\alpha\rangle = D(\alpha)\,|0\rangle$, e.g., in [5].

[3]The photon number of the coherent state is Poisson distributed. Correspondingly, the second-order temporal autocorrelation function is $g^{(2)}(0) = 1$.

in the *Glauber state* $|\alpha\rangle$).[4] A fundamental and theoretical overview can be obtained for instance through [5].

Their temporal and spatial coherence properties render coherent light sources indispensable tools in modern optics and light-based techniques. Numerous excellent descriptions of laser theory and technology exist in the vast pool of literature (see for instance [6]), and it is not the aim of this section to deliver a comprehensive summary of this field. Instead, a bridge to semiconductor lasers shall be offered, which allows the summary of specific investigations in this domain.

4.2.1 Semiconductor Lasers: From Efficient Nanolasers to Powerful External–Cavity Lasers

Miniaturisation and diversification of employable gain materials have played important roles in the achievement of compact devices, higher efficiencies and wavelength versatility. In addition, semiconductor lasers can be well integrated in electronic devices. In fact, numerous scientists have been awarded with Nobel prices for their foundational works that enabled new lasers and laser technologies. The class of semiconductor lasers can be further separated into two main types of lasers, which are known as edge emitters and surface emitters, according to the optical output direction obtained from the respective device with regard to the semiconductor chip. Among the vertical emitters, one typically finds laser diodes in the form of VCSELs,[5] and their siblings with external cavity referred to as VECSELs,[6] which are predominantly optically-pumped semiconductor disk lasers (SDLs). A review on the recent advances in VECSELs is given in [49], and the interested reader is further referred to [50–52].

Semiconductor Laser Development

While major aims in the semiconductor laser community can be defined as achieving ultra-low-threshold or even threshold-less lasers [53–56], or ecologically–friendly ("green") diode lasers [57], or the ultimate nanolaser [58], others target ultrashort pulse generation or high output powers. One key element for very efficient devices are high-quality microresonators (see [59]). Another key element for optimised light–matter interactions are low-dimensional quantum structures, which can be incorporated into low-mode-volume optical cavities, such as in the case of quantum-dot micropillar [60] or quantum-dot microdisk [61] lasers, to name but a few.

In contrast to the ultra-efficiency schemes, ensemble-quantum-dot systems in VECSELs combined with appropriate thermal management can even show record high output powers from a single laser chip with optically-pumped quantum-dots

---

[4]The quantum mechanical description of photon correlations and optical coherence was first provided by Glauber [45–47], after which the coherent state was named, and Sudarshan [48].

[5]Acronym for *v*ertical-*c*avity *s*urface-*e*mitting *l*aser.

[6]Acronym for *v*ertical-*e*xternal-*c*avity *s*urface-*e*mitting *l*aser.

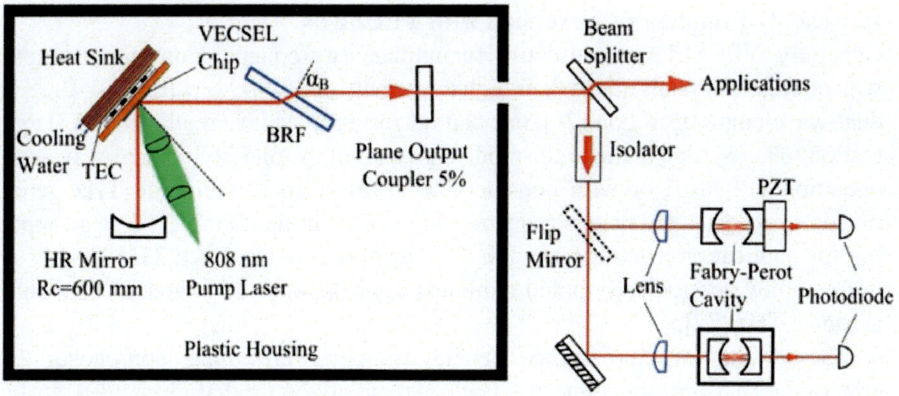

Fig. 4.1 Scheme of a passively-stabilised single-frequency VECSEL with confocal scanning Fabry-Pérot (FP) interferometer and frequency discriminator for emission linewidth analysis. Investigations of the narrow linewidth with a reference FP cavity as a function of the sampling time showed the influence of different noise components such as thermal and mechanical ones (see [65]). Here, the cavity design was optimised for record-high output from a single semiconductor disk laser chip up to 23.6 W in the single-frequency regime. The angled cavity with active mirror (chip) as folding mirror of the optical resonator and birefringent filter as polarisation and wavelength selective element allowed for maximum gain for a single longitudinal mode. Different passive methods were employed to promote mechanical, acoustic and thermal stability, which can be further amended by active stabilisation for a temporally stable single-frequency output. Reproduced with permission. [65] Copyright 2014 Optical Society of America

gain region [62, 63], whereas their optically-pumped quantum-well counterparts even exceeded the 100-W level due to their naturally higher density of emitters [64].

Single-Frequency Semiconductor Disk Lasers

Using similar chip structures with optimised cavity configuration and passive stability measures (Fig. 4.1), record-high single-frequency output from a VECSEL in excess of 23 W was demonstrated at 1 μm emission wavelength [65]. Since the quasi-monochromatic output of a single-frequency laser exhibits very low noise and, therefore, a very narrow linewidth (and, additionally, a very good beam profile), these laser sources are considered as key components for applications in a wide range of areas, such as metrology, optical free-space telecommunication, spectroscopy, and laser cooling.

Intracavity frequency doubling to the green could be for instance used to deliver a narrow-linewidth laser mode for interaction with corresponding ions used in atomic clocks. Such a clock is a quantum-technological device which provides very accurate time information. Single-frequency VECSELs at 852 nm for Cesium atomic clocks had been previously reported [66]. In astronomy, excitation of artificial stars, referred to as guide stars, in earth's atmosphere can also benefit from well-matched emission from a single-frequency VECSEL. For instance, yellow light from a frequency-doubled 1180-nm VECSEL output can be tuned to match the sodium line in the atmosphere [67, 68].

Intracavity Frequency Conversion with VECSELs
Generally, VECSELs are attractive for intracavity frequency conversion schemes with nonlinear crystals due to their high intracavity optical powers [69–73]. Based on dual-wavelength operation [74], one can for instance achieve multi-mode THz generation [69, 75] (due to the multi-mode character of the underlying dual-wavelength operation [75, 76]) or with considerable efforts also single-mode THz generation in a special two-single-frequency VECSEL configuration [70] (e.g. appealing for application as strong-enough THz local oscillator). Such THz sources can even employ aperiodically-poled nonlinear crystals for widely tunable THz output around 1 THz [77].

These schemes can further benefit from power-scaling approaches [78] employing two serially-connected [79] or parallely-connected chips [80]. In fact, a two-chip TECSEL[7] was recently reported by Guoyu et al. and characterised in comparison to a single-chip counterpart [81]. While the long-term output stability of TECSELs over hours of operation time is good [81], the noise from multi-mode operation and mechanical vibrations on various time scales [82], mode competition [76] and antiphase intensity fluctuations due to typically-shared spatial and spectral gain regions [83] have to be taken into account for further improvements and control of the stability.

Mode-Locked VECSELs
In addition to the aforementioned operation modes, various schemes for mode-locking can be realised with VECSELs [84], which can be used for applications such as multi-photon microscopy [85] or the generation of frequency combs[8] that can be used for spectroscopy [87]. Ultrashort pulsing down to the 100-fs level has been reported in recent years [88, 89]. VECSELs have recently also demonstrated their potential use as ultrafast pump source for quantum emitters with high single-photon flux [34] (see Fig. 4.2) which are highly desired for high-repetition rate quantum cryptography schemes.

The Self-Mode-Locking Phenomenon
One recent breakthrough in the field of ultrashort-pulse generation from VECSELs has been heavily discussed and investigated in recent years, namely the self-starting SESAM[9]-free pulse formation, "magic" mode-locking or "self-mode-locking" phenomenon in these devices [90–98]. Although not an undisputed phenomenon among the mode-locked SDL community due to the early claims with incomplete characterisation in support of a mode-locking claim, this remarkable self-mode-locking effect has been successfully obtained for different VECSEL platforms, which are based on quantum-well [95] or quantum-dot [94] gain media. Their pulsed light reaches from the IR over the near-IR (most examples were demonstrated around 1 μm emission

[7]Acronym for ***t***erahertz ***e***xternal-***c***avity ***s***urface-***e***mitting ***l***aser.

[8]Frequency combs were first demonstrated by W. Hänsch and co-workers and offer a unique optical metrology tool for ultra-precise measurements of frequencies with the help of mode-locked lasers. For background information on this subject, see [86].

[9]Acronym for ***se***miconductor ***s***aturable ***a***bsorber ***m***irror.

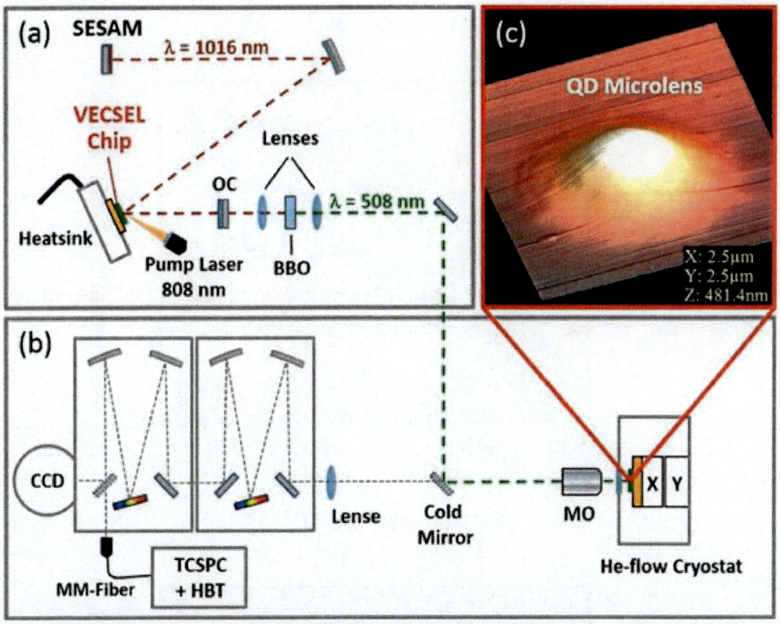

Fig. 4.2 Experimental setup schematically displaying an ultrafast mode-locked VECSEL (**a**) employed as a pump source for a single-photon source (**b**), from which a record-high single-photon flux was obtained [34] (also see Fig. 4.7). **a** A z-shaped VECSEL resonator with semiconductor saturable-absorber mirror (SESAM) delivers a 500-MHz pulse train of ps-pulses in the IR region, which are frequency-doubled externally in a nonlinear crystal (BBO) to a wavelength of 508 nm. These green laser pulses are directed to a typical micro-PL experiment (**b**) with a cryostat containing a quantum-dot–microlens (an AFM image of the produced 2 μm-diameter microlens device is displayed in **c** and a Hanbury–Brown-and-Twiss (HBT) setup behind a double monochromator for wavelength-sensitive second-order temporal autocorrelation function measurements. OC: Output coupler. MM: Multi-mode. MO: Microscope objective. TCSPC: Time-correlated single-photon counting. Reproduced with permission. [34] Copyright 2015 AIP Publishing

wavelength) to the visible (see [99] for a red-emitting VECSEL exhibiting self-mode-locking under stable single-pulse operation at a repetition rate of 3.5 GHz). Only due to more advanced characterisation in support of a mode-locked operation claim, this field could be established as a promising research direction (for an overview of self-mode-locked VECSELs, see [100]).

Nonlinear Optics in the VECSEL Chip

To understand whether the mechanism behind pulse formation through self-mode-locking is governed by a Kerr-lensing effect, as has been one major work hypothesis in this field, or another effect or, likely, an interplay of different complementary effects, preliminary studies aimed at the characterisation of the effective nonlinear refractive index in "black-box" VECSEL chips [101–105].

Among these studies, the works by Kriso et al. not only investigated nonlinear refraction, but also nonlinear absorption. More importantly, these optical nonlinearities were systematically studied as a function of the microcavity resonance, as well as of the probe wavelength in the spectral vicinity of the laser's output mode [101], and as a function of the optical irradiance [105]. By optical irradiance (pump

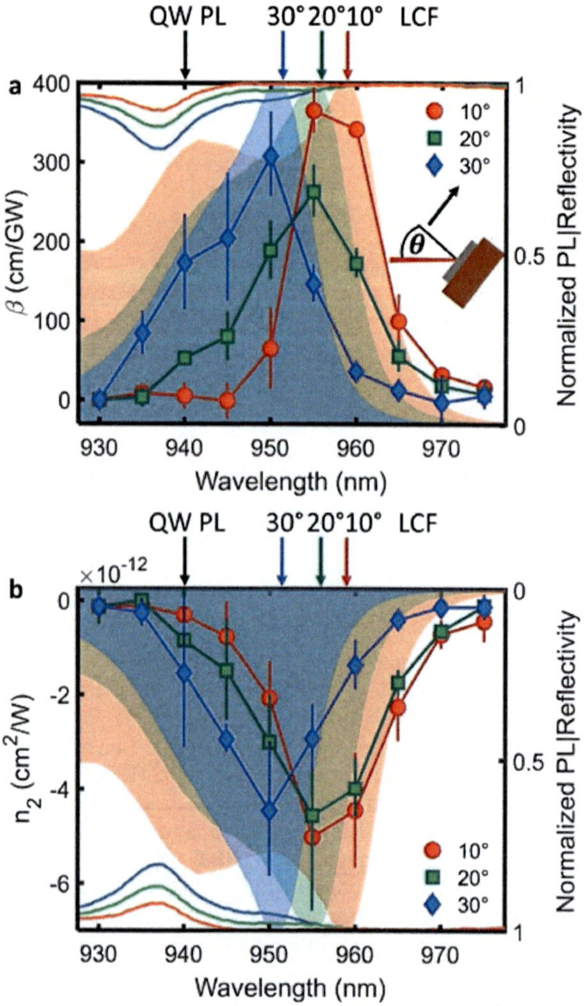

Fig. 4.3 Measurement data from VECSEL-chip Z-scan experiments revealing the wavelength- and angle-dependent nonlinear absorption (**a**) and nonlinear refraction (**b**): Both nonlinear coefficients are displayed for three incidence angles, that are 10, 20 and 30° with regard to the left axis in **a**, **b** as a function of the wavelength. Arrows on top of the diagram indicate the PL from the quantum-well (QW) gain medium and the different angle-dependent longitudinal-confinement factor (LCF) peaks. Shaded plots show the corresponding angle-dependent surface PL from the chip for comparison. Similarly, reflectivity spectra for the three angles are displayed as line plots. Both the PL and reflectivity are normalised to 1 (right axes). Inset: sketch of the probe geometry with respect to the VECSEL chip. Details and explanations about the experiment are provided in [101]. Reproduced with permission. [101] Copyright 2019 Optical Society of America

light), nonresonantly-excited charge-carrier densities were introduced—in the relevant range for operation, below and above the laser's threshold density. Thereby, an angle-dependent, wavelength-dependent negative nonlinear refraction profile as well as nonlinear absorption profile was revealed (Fig. 4.3) which is strongly shaped by the sub-microcavity's longitudinal confinement factor[10] but little affected by optical pumping. Furthermore, in these experiments, a typical third-order nonlinearity was determined through variations of the optical probe power [101, 105]. It is worth noting that the same technique in transmission geometry had been recently used to characterise nonlinear refraction and absorption of a single-crystalline perovskite material [106].

Type-II Heterostructure Gain Chips

In the past decade, also the concept of type-II heterostructures in the active region of VECSEL chips—previously introduced in and known from the domain of edge emitters—had been explored [107, 108]. Recent demonstrations showed that with an appropriate design of W-shaped quantum-well structures with two separate confinement layers for electrons around a central confinement layer for holes can indeed deliver sufficient optical gain at 1.2 μm to extract output powers in excess of 4 W from such SDL chip structure under optical pumping [107]. At low output powers in the hundreds of mW, even a very good beam profile can be obtained [109].

In such quantum-well configuration with electron double-well (electronic confinement in a molecule-like situation), the wave-function for electrons can deliver an up to approximately 60% overlap integral (wave-function overlap) due to the high probability to have electrons present spatially on top of the hole's wave-function maximum, resulting in remarkable optical transition probabilities for such tailored band alignments. Moreover, the dynamics in such a type-II (W-type) VECSEL chip with regard to gain build-up were studied using ultrafast optical pump–probe spectroscopy in reflection geometry. This was performed in direct comparison to a type-I chip designed for the same wavelength region, showing a clear delay in gain availability for the W-type chip under pulsed excitation [110].

Laser Pulse Applications

Pulsed lasers (not only the mode-locked ones, but also Q-switched devices) are attractive for numerous applications, such as optical spectroscopy (not only for time-resolved experiments), optical sensing, optical tweezers, two-photon polymerisation in nano-size 3D printing schemes, fluorescence or super-resolution or multi-photon microscopy, and also quantum imaging or quantum light generation (see also the briefly discussed subjects related to entangled photon pair or single-photon generation), to name but a few.

One of the uses in spectroscopic material characterisation employs high-energy pulses focused onto samples to create tiny plasma plumes, which contain evaporated sample material and irradiate characteristic element lines during the cooling process after the ionised matter enters the electronic relaxation processes towards its initial state. This technique referred to as laser-induced plasma spectroscopy (LIPS) or breakdown spectroscopy (LIBS) can be for instance used to detect different materials in samples contactless, real-time and nearly-nondestructively [111–114], not only in vacuum or air, but also under water, in a microscope system with strong spatial selectivity or robot-mounted and remotely controlled on another planet. Some attempts

---

[10]Denoting spectral and spatial resonance with respect to the gain medium, here quantum wells.

by the author's team to analyse plastics, to detect microplastics or to specify debris in tissue samples led to exciting insights into the possibilities involving LIPS, although preliminary works have not reached an advanced stage to exploit this method systematically. Nevertheless, plasma parameters were studied for molybdenum samples to improve the measurement capabilities collaboratively with partners [115]. The molybdenum-I lines were corrected against self-absorption and the corrected values of the temperature and the electron density were obtained.

4.2.2 Novel Coherent Light Sources

The field of semiconductor laser research is so vast, that these mentioned achievements in fact appear like a small drop in a large ocean. The selection of examples and achievements is strongly related to the work of the author, and works by a plethora of other research groups and institutions worldwide would be hardly adequately reflected in this section if the goal was to provide a topical overview.

Lasers can be widely used in numerous domains, such as for industrial manufacturing, health care, scientific diagnostics and fabrication tools, information processing, security applications and many more, and no attempt has been made to summarise them accurately in this section. Instead, here, another aspect of coherent light source engineering shall be briefly highlighted, that is the development of a platform not based on stimulated emission of light but stimulated scattering of bosonic polaritons into a macroscopically-occupied final state (an overview of the polariton condensation subject is for instance provided in [116]).

Polariton Laser as Energy-Efficient Coherent Light Source

Towards novel sources of coherent light, the polariton laser had been proposed [122, 123] and realised in multiple forms, both optically [124–134] and electrically pumped [117, 118, 135, 136] (cf. Fig. 4.4). Recently, even coherent polariton lasers were demonstrated based on adjusted ground-state–reservoir interactions (e.g. in single-mode devices) [133], as promoted by the discrete emission modes in micropillars [137]. The prospects of coherent polariton devices and their applications are for instance further discussed in [138, 139]. More about the topic of polariton lasers, the difference to conventional lasers and the identification of lasing in the strong-coupling regime is summarised in [120].

Spinor Polariton Condensates

To understand their properties better, studies on optically pumped polariton condensates (such as in [144–146]) in magnetic fields were performed [142, 143] in continuation of previous works [118, 145, 147, 202]. This was pursued for instance to shed light on the suppression of a Zeeman splitting at a critical magnetic field due to the spin analogue of the Meissner effect—the "spin Meissner effect" [140] (cf. Fig. 4.5)—for spinor condensates. Moreover, this was also pursued to analyse their polarisation-resolved second-order coherence ($g^{(2)}_{\sigma\pm}(0)$) as a function of the excitation density and magnetic flux. In fact, polaritons in external fields offer many possibilities to manipulate and control the hybrid quasi-particles formed in optical microcavities (see for instance [148]).

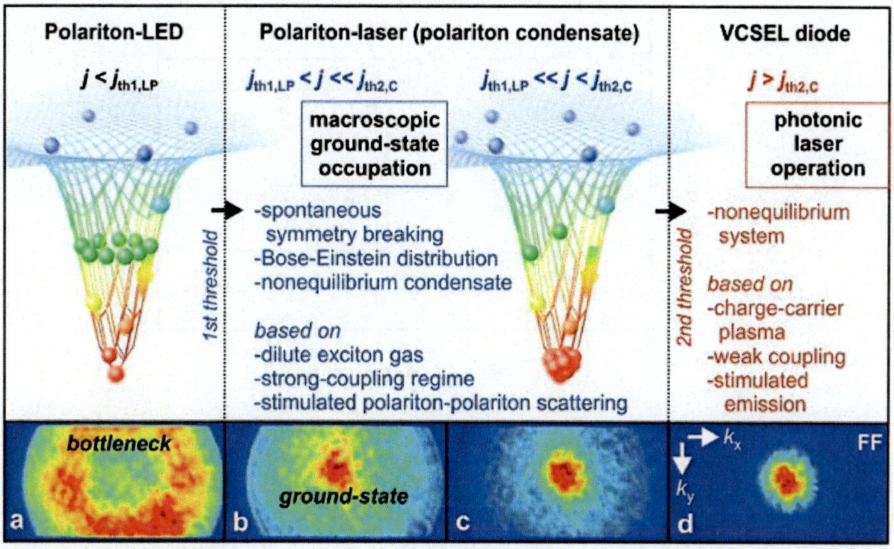

Fig. 4.4 Three regimes of operation and overview of the momentum-space signatures for a polariton-laser diode with two-threshold behaviour (DC-operated quantum-well microcavity device reported in [117]). From left to right, the current density j (particle density n) is increased to surpass the first and the second threshold of the device, attributed to polariton lasing (polariton condensation) and photon lasing (conventional VCSEL operation), respectively [117, 118] (further information about polariton lasers can be found for instance in [119, 120]). Top: Here, sketches of the energy–momentum dispersion relation in $k_{\|}$ for the lower polariton (LP) branch indicate the occupation distribution below and above the condensation threshold density. A brief description of the two distinct nonlinear regimes for such device is given in analogy to [116]. Bottom: Fourier-space (far-field: FF) projection of the electrically-driven microcavity's emission profile at corresponding current densities, showing the relaxation bottleneck (**a**) in the linear polariton regime, as well as the phase transitions towards a condensate (**b**, **c**) and photonic laser (**d**) with strongly narrowed far-field signature. Adapted from [117]. Courtesy of the author

4.3 Quantum Optics

Within the field of quantum optics, cavity quantum electrodynamics (cQED) research (see for instance [56, 149]) has grown to an important subject. It allows the exploration of fundamental light–matter coupling regimes and the development of novel coherent (e.g. single-atom or polariton laser) as well as nonclassical light sources (e.g. single-photon, squeezed-light or entangled photon-pair sources). Two important regimes are that of weak coupling, where optical modes alter the emission properties due to the Purcell effect [150], and that of strong coupling, for which the involved modes hybridise to form new eigen-states and experience a periodic coherent energy exchange (cf. [151]). Such system is characterised by Rabi oscillations [152] in the time domain and a Rabi splitting in the frequency domain, first demonstrated for quantum-well microcavities in 1992 [153].

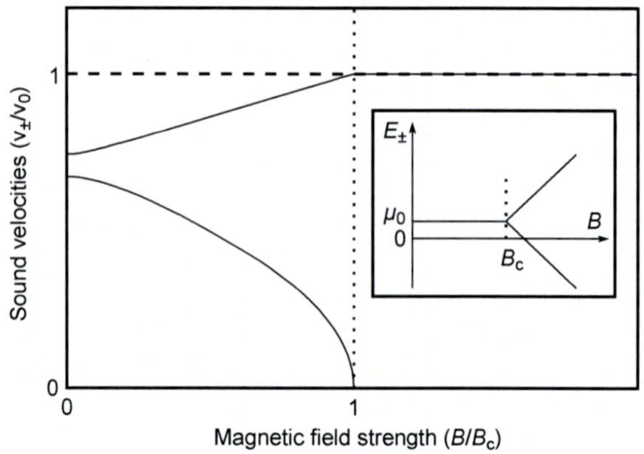

Fig. 4.5 Diagram of the magnetic-field dependent superfluidic behaviour of a polariton condensate, freely sketched after [140]. For interacting massive bosons, Bogoliubov modes describe the energy spectrum of excitations of the Bose condensed quantum fluid (experimentally demonstrated for polaritons in [141]), with quasi-linear k dependence around the final state's zero momentum. According to theoretical considerations by Rubo et al., the long-wavelength sound velocities $v_\pm$ for the spinor polariton condensate with $E_\pm$ in an external magnetic field deviate increasingly for small fields towards reaching a critical field strength B_c [140], at which superfluidity is suppressed. Correspondingly, the Zeeman splitting at $k = 0$ (sketched as inset) exhibits a quenching up to B_c in order to keep the chemical potential μ_0 of the condensate constant. This effect has been referred to as spin analogue of the Meissner effect, which is phenomenologically understood as a redistribution of polaritons between two Zeeman levels, caused by polariton–polariton interactions within the condensate that compensate the effect of the field B at $k = 0$. It is represented by an elliptically polarised superfluid below and circularly polarised one above B_c (see [140]). Above B_c, the Zeeman splitting occurs $\propto B_{\mathrm{eff}} = B - B_c$ (effective field). The non-equilibrium spin Meissner effect (with quasi-equilibrium model assuming thermal equilibrium within each spinor population of opposite polarisation) had been studied both experimentally and theoretically and delivered an additional tool to distinguish polariton condensation from conventional weak-coupling lasing from polariton microcavities [142]. This phenomenon motivated the investigation of photon statistics for spinor condensates [143] and further studies on the superfluidic system's response perturbed by transient electromagnetic fields

Weak and Strong Light–Matter Coupling

Weak coupling simply specifies the scenario, in which decay rates exceed the coupling strength and, thus, reversible exchange of energy is inhibited through the loss of excitation before one cycle of a Rabi oscillation can be completed. In this Purcell regime, the mere presence of a vacuum field (empty optical mode) is enough for the resonant emitter to couple to it and to release its energy radiatively. An optical cavity with no photon inside, but a defined resonance, can indeed also couple strongly with a single quantum emitter, provided that the coupling strength is high enough and spatial and spectral resonance are established.

Such strong-coupling experiments in the quantum regime between a single two-level matter system and a confined state of the vacuum field have been achieved with quantum dots embedded in high-quality photonic microresonators [154–156], resulting in a vacuum Rabi splitting of the "anticrossing" (coupled-oscillator) modes.

In those cases, where an ensemble of emitters couples to the light field in an optical microcavity, one typically refers to the system as semiclassical. Nonetheless, in both cases, hybridisation leads to exciton–polariton formation.

A basic introduction to cavity–polaritons is for instance provided in [121], whereas more detailed theoretical analysis of the coupling regimes is given for example in [157–159], and a rich background on (bulk, 2D and confined) exciton–polaritons in general can be acquired through [14].

Quantum Emitters in Photonic Quantum Boxes

In the field of cQED, these single quantum emitters provided for example by epitaxially-grown quantum dots (artificial atoms in the form of nanoislands) are investigated with regard to coupling regimes and photon statistics (cf. Fig. 4.6). They can strongly couple to the light field, as well as deliver photon antibunching (a signature for nonclassical light) from the hybridised eigen-modes in optical micropillars [161]. Similarly, a one-atom laser, which also operated in the strong-

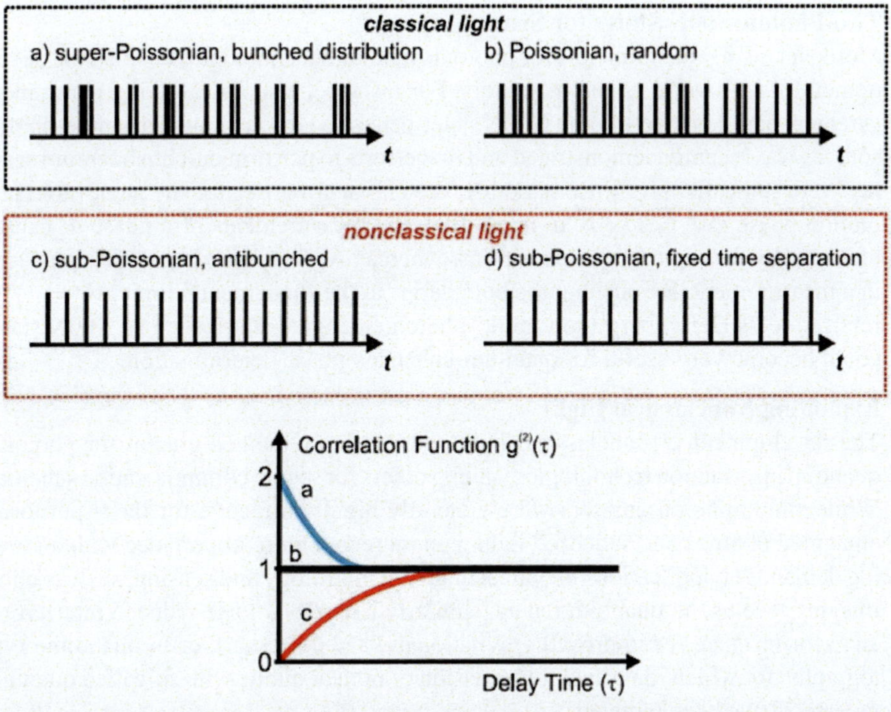

Fig. 4.6 Sketch of the temporal statistics for thermal, coherent (top) and nonclassical light (centre), as well as a schematic second-order temporal autocorrelation function diagram around zero temporal delay (bottom) for these three light sources, with **a** super-Poissonian (bunching), **b** Poissonian (random distribution), and **c** sub-Poissonian statistics (antibunching), respectively. Lines in the schematic pulse trains indicate individual photons. The ideal case of a deterministic single-photon emitter (**d**) corresponds to a Fock state with $N = 1$ and an output with temporally equally-separated pulses of single photons (right pulse train in the red box). Reused with permission. [160] Copyright 2009 Arash Rahimi-Iman

coupling regime, showed sub-Poissonian photon statistics [162]. Interesting experiments with quantum-dot micropillar systems had furthermore addressed photon–photon coupling mediated by the exciton spin state [163] or the climbing of the Jaynes–Cummings ladder [164], to name but a few examples in this section.

4.3.1 Tailored Light–Matter Interactions for Quantum Light Generation

In the past decades, quantum-dot-based photonic structures have been widely utilised for the development of quantum-light sources, which are particularly characterised by their nonclassical light output.[11] To achieve the Purcell effect (resulting in directional output, i.e. a 1D quantum emitter, and a fast and highly likely recombination), (various) optical microcavities have for many years remained highly attractive for quantum emitter studies [26, 165].

Two-Photon N00N States for Sensing
Moreover, using quantum dots for photonic quantum technologies enabled the development of real-world quantum sensors. For instance, a super-resolving phase measurement based on two-photon N00N states generated by quantum-dot single-photon sources was recently demonstrated and first efforts to perform on-chip quantum sensing were indicated [166]. Remarkably, the classical barrier for the achievable resolution phase $\Delta\Phi = 1/\sqrt{N}$ in interferometric measurements of a phase Φ can be undercut by entangled light conditions, whereas N is the number of photons used in a measurement. By pushing the boundaries to the fundamental limit $\Delta\Phi = 1/N$, referred to as Heisenberg limit, multi-photon entangled states such as N00N states could become very useful for quantum-enhanced phase determination.

Exploiting Nonclassical Light
The development of nonclassical light sources has remained crucial for photonic quantum information technologies, among others for secure communication schemes. While single-photon emitters were generally highly attractive for these purposes, entangled photons and squeezed light gain more and more importance with a growing demand for applications in optical quantum metrology and sensing, such as ghost imaging (see use in quantum radars, similarly LIDARs), noise-reduced interferometry (for instance for improved gravitational wave detection), or in quantum cryptography, for which long-haul transmission of optical qubits with so-called quantum repeaters have been targeted.

[11] Single-photon sources exhibit light output being in a Fock state with photon number $N = 1$, which can be conveniently probed by the second-order temporal autocorrelation function $g^{(2)}$, which features a zero probability at (delay) time zero ($g^{(2)}(\tau = 0)$) of obtaining more than one photon per pulse/time period. This is typically characterised by the famous antibunching phenomenon, which accompanies systems that exhibit a sub-Poissonian probability distribution, but it doesn't necessarily mean the same (see [5]).

Efficient Single-Photon Sources

In addition to conventional concepts incorporating epitaxially-grown quantum dots, scientists recently demonstrated electrically-pumped room-temperature single-photon sources based on different quantum emitter platforms (cf. low- and high-temperature current-driven platforms such as [31, 33, 167], respectively). In fact, different approaches lead to efficient sources, but for practical applications compactness, scalability, room-temperature operation and the possibility to electrically excite and control the emission have been particularly desirable [22].

For instance, single neutral nitrogen-vacancy centers in a novel diamond diode structure were used to achieve a stable, room-temperature, electrically-driven single-photon source [168]. Similarly, the development of a room-temperature device was pursued with the help of organic molecules, which feature strong exciton binding energies on the order of 1 eV. From such material platform, stable single-molecule emission was achieved by incorporating carefully chosen molecular emitters in a solid-state matrix in the form of a specially designed organic light-emitting diode structure [169]. Alternatively, single defect systems in silicon carbide were exploited to obtain bright single-photon emitting diodes at room temperature [170].

However, to bring the demonstrated antibunching behaviour that characterises single-photon emission closer to an ideal result, solution-processed devices based on CdSe/CdS core/shell quantum dots were recently reported, from which high-purity single photons with $g^{(2)}(0) < 0.05$ was obtained under electrical pumping at room temperature [167]. Without carefully tailoring hole-transport and electron-transport layers in such devices, in which isolated quantum dots were embedded in a PMMA layer, a high quantum yield would be impossible from electrical injection of opposite charge-carriers from two sides of the sample.

While many practical platforms for single-photon generation have evolved in recent years, even defects in the host lattice of atomically-thin 2D materials have emerged as promising room-temperature quantum emitters and moved into the focus of single-photon source research based on novel material systems (see [171–176]).

Advances in the Field of Quantum Emitters

Still, quantum emitters based on conventional III/V semiconductors are heavily developed towards higher device fidelity, pulse repetition rate and photon indistinguishability. Numerous studies were devoted to the generation, guiding and detection of triggered single photons, for instance from resonantly-excited quantum dots in a photonic circuit [38, 39].

On the one hand, the advantages of nonclassical light emission from advanced devices such as quantum-dot–microlense structures have led to the deterministic fabrication of bright quantum-dot-based single-photon sources. On the other hand, on-chip waveguide structures defined by means of in-situ electron-beam lithography (EBL) have promised better yield and faster implementation of such structures in quantum cryptography schemes with semiconductor quantum emitters. The recently demonstrated integration of pre-selected quantum dots into photonic devices such as microlenses and waveguide systems with nm-accuracy has offered a flexible device design and high process yield [37, 177, 178]. Similarly, molecule-based quantum

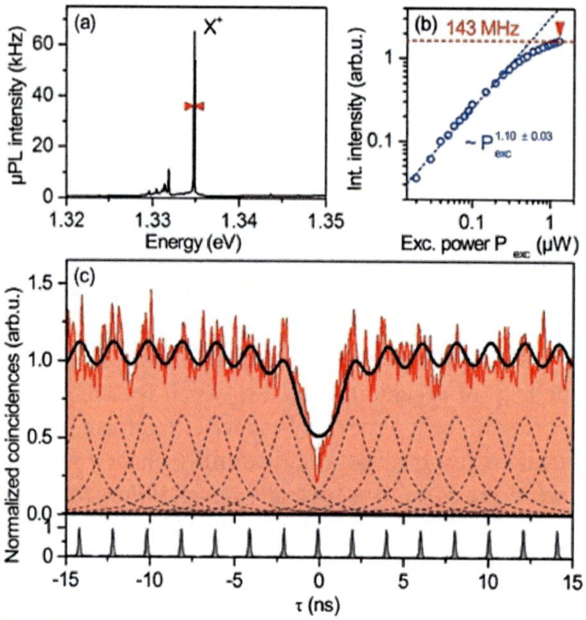

Fig. 4.7 High-repetition-rate deterministic single-photon output from a VECSEL-pumped quantum-dot–microlens structure, as illustrated in Fig. 4.2. **a** Micro-PL spectrum of the quantum emitter, here a positively charged exciton of a III/V quantum dot grown in Stranski–Krastanov epitaxy mode. **b** Pump-dependent integrated intensity from the pulsed single-photon source. For the highest flux acquired through the first lens of the setup, a maximum effective repetition rate of 143 MHz is obtained, which is lower than the 494 MHz of the optical pulse train from the VECSEL owing to the quantum-dot lifetimes. **c** Temporal photon statistics measured at quantum-dot saturation (marked by red label in **b**) following spectral filtering (indicated by red arrows in **a**) with normalised photon counts as a function of the delay time, reaching a minimum of 0.22 at $\tau = 0$. A clear antibunching with second-order temporal autocorrelation function value $g^{(2)}(0) < 0.03$ well below 0.5 according to the model (solid black line) is obtained. Further details on the experiment and analysis are provided in [34]. The laser pulse train is displayed at the bottom. Dashed curves in **c** model individual equidistant photon pulses represented by Lorentzian profiles (FWHM of 2.3 ns, repetition rate 494 MHz, constant counts area per pulse for $\tau \neq 0$). Reproduced with permission. [34] Copyright 2015 AIP Publishing

emitters can be nowadays incorporated into flexibly 3D laser-written polymer (photoresist) matrices forming optical host structures for guidance or efficient out-coupling [179, 180].

As briefly summarised in [181], such deterministic quantum-dot microlenses delivered a platform for the exploration of the limits of photon indistinguishability by delay-time and temperature dependent Hong-Ou–Mandel experiments [182]. Furthermore, they paved the way towards entanglement swapping [183], for the coherent control of the quantum-dot's biexciton–exciton cascade [184], and for the efficient generation of polarisation-entangled photon pairs [42].

Prospects of Quantum Channels with Record-High Single-Photon Flux
In addition, in terms of overcoming the limitations of optical pulsed excitation with common Titanium-Sapphire oscillators running at a repetition rate of 80 MHz, the combination of ultrafast mode-locked VECSELs with quantum-dot–microlens single-photon sources by the author and co-workers has resulted in a record-high single-photon flux with effective duty cycle of currently as high as 143 MHz [34] (Fig. 4.7). To combine the self-built laser source and the emitter in one quantum device in Berlin, the picosecond-short pulses from the (breadboard-mounted and thereby mobile) mode-locked VECSEL had to be frequency-doubled externally in a nonlinear BBO crystal (by second-harmonic generation) to enable optical pumping[12] of the microlens-embedded quantum emitter (see Fig. 4.2), which was addressed in a μ-PL setup for photon statistics measurements (Fig. 4.7c). Thereby, the combination of quantum technologies of the first and of the second generation in an overarching quantum-technological system has become a natural success story which will be further exploited in different device schemes.

4.3.2 Strong Light–Matter Coupling for Polariton Research

Strong light–matter coupling as described above, although mainly for quantum-well-based systems, is at the verge of being utilised. Many proposals and experiments regarding application-oriented polariton systems have emerged in recent years (cf. [185, 186]).

Different approaches were suggested in the literature to employ polariton systems for optoelectronic or quantum optical applications, such as switches based on polaritons for use in all-optical/integrated photonic circuits [187, 188], spin switches [189], optically-imprinted polaritonic logic circuits [190], and transistors based on polariton spin [191]. For some, condensates of polaritons with their long-range spatial coherence had become attractive candidates for a polariton transistor based on ultrafast coherent switches [192], whereas others discussed a two-fluid polariton switch [193]. As part of the evolution of polaritonics as a new field for light-based devices, resonant-tunneling diodes [194], polariton transistors [195], routers [196], and interferometers [197] were recently introduced, as summarised for instance in [185].

The subject of polariton lasing has been raised in the previous section and remains an exciting playground for device development and the delivery of condensates on demand (also see [119, 120]).

[12] Here, strongly off-resonant optical pumping with few-ps pulses at 507 nm with 500 MHz repetition rate was achieved due to device limitations. However, resonant p-shell excitation and repetition rates better adjusted to the quantum-dot exciton (or trion, here X^+, see Fig. 4.7a) lifetime would be more desired for future high-repetition rate single-photon sources, in order to provide for the best possible photon indistinguishability and optimised single-photon duty cycle, respectively.

Manipulation and Control of Coherent States

In this context, also the ultrafast manipulation and control of the polariton state with external fields gained importance, as recent works on optically modifying states on the polariton's Bloch sphere [198] or polarisation scrambling [199] show. External fields were previously known as a successful tool to alter a polariton configuration reversibly, such as by magnetic fields [118, 142, 143, 147, 200–202], even to control strong coupling in the single-quantum-emitter regime [203]. Similarly, electric fields can be utilised, allowing one to spectrally tune individual quantum dots within a quantum strong-coupling regime [204] or to switch between a polaritonic and photonic regime of microcavity emission [137].

Advances in Polariton Research

Recently, the topic of 2D materials in optical microcavities gained strong momentum after pioneering works [205–208] as a logical consequence of first insights into monolayer exciton properties (also see the summary in Chap. 2, and [209]). 2D semiconductors with their strong excitonic binding energy and correspondingly high oscillator strength in the ideal quantum-well-like 2D crystal were predestined for strong light–matter coupling, which was previously already considered for graphene islands ([210]). The monolayers of TMDCs show a clear advantage with respect to high temperature operation and have become very promising testbeds for room-temperature polaritonics and BEC studies. Similarly, few-layer materials have come into focus, such as homobilayers (targeted with configurations such as in [211]), or heterobilayers with interlayer and moiré features (some of which indicating a pronounced oscillator strength as reflection-contrast measurements in [212] show), leading to a unique revival of polariton research. Recently, even exotic "topolaritons"—quasiparticles with topologically nontrivial states (polaritonic bands with so-called chiral edge modes) and unidirectional polariton flow—were proposed in the literature [213] which could possibly be achieved with monolayer TMDCs when the system is properly prepared.

The ability to use valley physics—already discussed with regard to "*valleytronics*" with 2D materials [214, 215]—in combination with polarisation-selective optical cavities, such as chiral microcavities based on polarisation-sensitive mirrors and structures (addressed by the author's team with preliminary design and fabrication attempts for chiral polaritons within a Master thesis project 2017/2018) remains very attractive and requires advanced (nano-)fabrication capabilities. In fact, helicity-favouring microcavities were previously investigated for monolithically-grown quantum-well and quantum-dot microcavity systems [216–218]. Valley-polarised polaritons based on optical pumping schemes had been demonstrated shorty after the first TMDC polaritons [219–222], whereas the natural next step would involve the employment of chiral optical microstructure approaches which could open up new possibilities related to polarisation- as well as spin-sensitive electro-optical devices, i.e. optical valleytronics.

Tunable Light–Matter Interactions Around an Exceptional Point

One major advantage which is linked to the high oscillator strength of monolayer semiconductors, other than the room-temperature operability, lies with the reduced dependence on high-quality microresonators. Moderately high Q-factors in the range of a few hundred up to 1000 can be enough to obtain strong coupling. This allows one to employ various tunable-cavity concepts (cf. [206, 207, 223]). With optical cavities such as those proposed in [224], the material independent continuous adjustability of cavity resonances could be utilised for polariton-based chemistry allowing a precise adjustment of the polariton energy levels [225]. Furthermore, they could also become useful for an efficient study of chemical reactions influenced by different light–matter coupling situations.

One motivation to employ such open tunable microcavities is given by the investigation of the transition between weak and strong coupling by smoothly tuning the coupling strength [224] (see Fig. 4.8). Therefor two concepts have been discussed and simulated that could allow one to probe this interesting transition situation, since the two coupling regimes are linked by an exceptional point (EP). According to the literature, such point, at which only one complex solution exists for the coupled-oscillator system, brings up exotic phenomena [226–228], including chiral behaviour [229] and topological energy transfer between the two coupled modes [230]. Moreover, EPs in optical microcavity systems are understood to enhance sensing [231] due to the sharp transition between the coupling regimes.

Prospects of Open Tunable Microcavity Systems

While typically monolithic cavities are used for polariton research, tunable cavities such as in [206, 207, 223] not only allow to flexibly set the cavity length,[13] they also provide experimental access to the intracavity space which can be used for coupling experiments with nanosheets (see proposed tunable-coupling experiment in [224]), colloidal quantum emitters, molecules, dispersed particles or particles in a polymer matrix (cf. [223]), or even for coupling between the resonator-light field and two different emitter systems inside the cavity (e.g. [232]). Thus, tunable open microresonators qualify for numerous experiments, including opto-mechanical coupling experiments such as in [233, 234] (which were also intended for 2D materials through works by the author and his students), and can be conveniently used in connection with 2D, 1D and 0D excitonic systems.

Polariton Research on Solid Foundations

The enrichment of knowledge throughout recent work further facilitated the study of a variety of polariton systems under optical and electrical excitation, e.g. Esaki-diodes [235], 1D polariton gases in microcavity wires [236], electroluminescence from polariton traps [237], as well as the study of condensates in textured landscapes [238, 239]. Further examples about the technological realisation of various polariton systems are found in [240].

The acquired background in polariton physics also enabled the author of this work to investigate together with cooperation partners on the nature of spinor condensates [142, 143], and to use correlation experiments [143, 145] and magnetic fields [142, 143, 147, 202] to alter the exciton–polariton system's dynamics and features. In

[13]In experiments, tuning of the cavity length (i.e. mirror position) is practically done via piezoelectric actuators, which can provide positioning control on the nm scale.

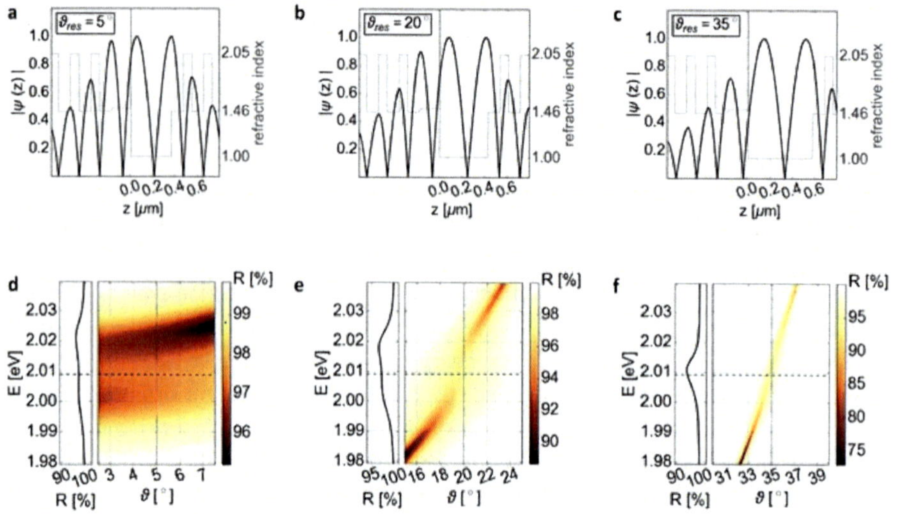

Fig. 4.8 Modelling data for tunable exciton–photon coupling in a planar open Fabry–Pérot-type optical microcavity with fixed 2D semiconductor sheet on a PMMA buffer on one dielectric end mirror. The effective cavity length at a given PMMA thickness changes as a function of the incidence angle, which alters the strength of the resonator standing wave's electric field at the position of the here employed TMDC monolayer. The structure is sketched in [224]. Here, the air gap for the second cavity mode increases from about **a** 361 nm over **b** 426 nm to **c** 578 nm. The refractive index modulation of the empty DBR–DBR cavity with air spacer is displayed for clarity (grey dotted line, right axis). The simulated field profile (absolute amplitude as a function of distance from the monolayer in z) is shown at resonance conditions ($E_C(\theta_{res}) - E_X = 0$), correspondingly marked in the calculated angle-dependent reflectivity spectra (**d–f**) by a vertical dotted line based on neglected absorption of WS_2, whereas the exciton's energy E_X (determined by the peak position within the imaginary part of the complex refractive index $\tilde{n}$) is indicated by the horizontal dashed line. Cross-sectional spectra for θ_{res} are displayed at the sides of each false-colour reflectivity profile, showing (**d**) strong and (**e**) reduced strong coupling, as well as (**f**) weak coupling with vanished mode anti-crossing. As a supplement for comparison, the situation for the first three cavity modes can be understood with the help of the Supporting Information Figs. SI.4 and 5 of [224]; in addition, calculated spectra for transmission and absorption corresponding to (**d–f**) as well as reflectivity spectra for p-polarised light are provided in Figs. SI.8 and 9, respectively. Reproduced under the terms of the CC-BY Creative Commons Attribution 4.0 International Licence (http://creativecommons.org/licenses/by/4.0/). [224] Copyright 2020 The Author(s), published by Springer Nature

this context, the $G^{(2)}$ second-order temporal autocorrelation function (i.e. photon statistics) of pulsed condensates in external magnetic fields was studied and the different degrees of coherence for circularly-polarised Zeeman-split spinor condensates unraveled [143]. The interested reader is referred to the literature on polaritons in external fields for further details [148].

References

1. J.L. O'Brien, A. Furusawa, J. Vučković, Photonic quantum technologies. Nat. Photonics **3**, 687–695 (2009)
2. M. Razeghi, *Technology of Quantum Devices* (Springer, New York, 2010)
3. W.P. Schleich, K.S. Ranade, C. Anton, M. Arndt, M. Aspelmeyer, M. Bayer, G. Berg, T. Calarco, H. Fuchs, E. Giacobino, M. Grassl, P. Hänggi, W.M. Heckl, I.-V. Hertel, S. Huelga, F. Jelezko, B. Keimer, J.P. Kotthaus, G. Leuchs, N. Lütkenhaus, U. Maurer, T. Pfau, M.B. Plenio, E.M. Rasel, O. Renn, C. Silberhorn, J. Schiedmayer, D. Schmitt-Landsiedel, K. Schönhammer, A. Ustinov, P. Walther, H. Weinfurter, E. Welzl, R. Wiesendanger, S. Wolf, A. Zeilinger, P. Zoller, Quantum technology: from research to application. Appl. Phys. B **122**, 130 (2016)
4. A. Acín, I. Bloch, H. Buhrman, T. Calarco, C. Eichler, J. Eisert, D. Esteve, N. Gisin, S.J. Glaser, F. Jelezko, S. Kuhr, M. Lewenstein, M.F. Riedel, P.O. Schmidt, R. Thew, A. Wallraff, I. Walmsley, F.K. Wilhelm, The quantum technologies roadmap: a European community view. New J. Phys. **20**, 080201 (2018)
5. M. Orszag, *Quantum Optics: Including Noise Reduction, Trapped Ions, Quantum Trajectories, and Decoherence*, 3rd edn. (Springer International Publishing, Berlin, 2016)
6. F.K. Kneubühl, M.W. Sigrist, *Laser* (Teubner, Stuttgart, 2005)
7. E.F. Schubert, *Light-Emitting Diodes*, 2nd edn. (Cambridge University Press, Cambridge, 2006)
8. W. Koechner, *Solid-State Laser Engineering* (Springer, New York, 2006)
9. M. Grundmann, *The Physics of Semiconductors: An Introduction Including Nanophysics and Applications*, 3rd edn. (Springer International Publishing, Berlin, 2016)
10. H. Haug, S.W. Koch, *Quantum Theory of the Optical and Electronic Properties of Semiconductors*, 5th edn. (World Scientific Publishing Co., Singapore, 2009)
11. C. Weisbuch, B. Vinter, *Quantum Semiconductor Structures: Fundamentals and Applications* (Academic, Cambridge, 1991)
12. N. Peyghambarian, S.W. Koch, A. Mysyrowicz, *Introduction to Semiconductor Optics* (Prentice Hall, Upper Saddle River, 1993)
13. J.H. Davies, *The Physics of Low-dimensional Semiconductors: An Introduction* (Cambridge University Press, Cambridge, 1998)
14. C.F. Klingshirn, *Semiconductor Optics* (Springer, Berlin, 2012)
15. F. Arute, K. Arya, R. Babbush, D. Bacon, J.C. Bardin, R. Barends, R. Biswas, S. Boixo, F.G.S.L. Brandao, D.A. Buell, B. Burkett, Y. Chen, Z. Chen, B. Chiaro, R. Collins, W. Courtney, A. Dunsworth, E. Farhi, B. Foxen, A. Fowler, C. Gidney, M. Giustina, R. Graff, K. Guerin, S. Habegger, M.P. Harrigan, M.J. Hartmann, A. Ho, M. Hoffmann, T. Huang, T.S. Humble, S.V. Isakov, E. Jeffrey, Z. Jiang, D. Kafri, K. Kechedzhi, J. Kelly, P.V. Klimov, S. Knysh, A. Korotkov, F. Kostritsa, D. Landhuis, M. Lindmark, E. Lucero, D. Lyakh, S. Mandrà, J.R. McClean, M. McEwen, A. Megrant, X. Mi, K. Michielsen, M. Mohseni, J. Mutus, O. Naaman, M. Neeley, C. Neill, M.Y. Niu, E. Ostby, A. Petukhov, J.C. Platt, C. Quintana, E.G. Rieffel, P. Roushan, N.C. Rubin, D. Sank, K.J. Satzinger, V. Smelyanskiy, K.J. Sung, M.D. Trevithick, A. Vainsencher, B. Villalonga, T. White, Z.J. Yao, P. Yeh, A. Zalcman, H. Neven, J.M. Martinis, Quantum supremacy using a programmable superconducting processor. Nature **574**, 505–510 (2019)
16. J. Kelly, R. Barends, A.G. Fowler, A. Megrant, E. Jeffrey, T.C. White, D. Sank, J.Y. Mutus, B. Campbell, Y. Chen, Z. Chen, B. Chiaro, A. Dunsworth, I.-C. Hoi, C. Neill, P.J.J. O'Malley, C. Quintana, P. Roushan, A. Vainsencher, J. Wenner, A.N. Cleland, J.M. Martinis, State preservation by repetitive error detection in a superconducting quantum circuit. Nature **519**, 66–69 (2015)
17. A. Crespi, R. Osellame, R. Ramponi, V. Giovannetti, R. Fazio, L. Sansoni, F.D. Nicola, F. Sciarrino, P. Mataloni, Anderson localization of entangled photons in an integrated quantum walk. Nat. Photonics **7**, 322–328 (2013)

18. A. Crespi, R. Osellame, R. Ramponi, D.J. Brod, E.F. Galvão, N. Spagnolo, C. Vitelli, E. Maiorino, P. Mataloni, F. Sciarrino, Integrated multimode interferometers with arbitrary designs for photonic boson sampling. Nat. Photonics **7**, 545–549 (2013)
19. N. Spagnolo, C. Vitelli, M. Bentivegna, D.J. Brod, A. Crespi, F. Flamini, S. Giacomini, G. Milani, R. Ramponi, P. Mataloni, R. Osellame, E.F. Galvão, F. Sciarrino. Experimental validation of photonic boson sampling. Nat. Photonics **8**, 615–620 (2014)
20. J. Loredo, M. Broome, P. Hilaire, O. Gazzano, I. Sagnes, A. Lemaitre, M. Almeida, P. Senellart, A. White, Boson sampling with single-photon fock states from a bright solid-state source. Phys. Rev. Lett. **118** (2017)
21. C. Antón, J.C. Loredo, G. Coppola, H. Ollivier, N. Viggianiello, A. Harouri, N. Somaschi, A. Crespi, I. Sagnes, A. Lemaître, L. Lanco, R. Osellame, F. Sciarrino, P. Senellart, Interfacing scalable photonic platforms: solid-state based multi-photon interference in a reconfigurable glass chip. Optica **6**, 1471 (2019)
22. A. Boretti, L. Rosa, A. Mackie, S. Castelletto, Electrically Driven Quantum Light Sources. Adv. Opt. Mater. **3**, 1012–1033 (2015)
23. N. Somaschi, V. Giesz, L. De Santis, J.C. Loredo, M.P. Almeida, G. Hornecker, S.L. Portalupi, T. Grange, C. Antón, J. Demory, C. Gómez, I. Sagnes, N.D. Lanzillotti-Kimura, A. Lemaître, A. Auffeves, A.G. White, L. Lanco, P. Senellart, Near-optimal single-photon sources in the solid state. Nat. Photonics **10**, 340–345 (2016)
24. T. Grange, N. Somaschi, C. Antón, L. De Santis, G. Coppola, V. Giesz, A. LemaîEtre, I. Sagnes, A. Auffèves, P. Senellart, Reducing phonon-induced decoherence in solid-state single-photon sources with cavity quantum electrodynamics. Phys. Rev. Lett. **118**, 253602 (2017)
25. P. Senellart, G. Solomon, A. White, High-performance semiconductor quantum-dot single-photon sources. Nat. Nanotechnol. **12**, 1026–1039 (2017)
26. S. Rodt, S. Reitzenstein, T. Heindel, Deterministically fabricated solid-state quantum-light sources. J. Phys. Condens. Matter **32**, 153003 (2020)
27. Z. Merali, Data teleportation: the quantum space race. Nat. News Featur. **492**, 22 (2012)
28. J. Yin, Y. Cao, Y.-H. Li, S.-K. Liao, L. Zhang, J.-G. Ren, W.-Q. Cai, W.-Y. Liu, B. Li, H. Dai, G.-B. Li, Q.-M. Lu, Y.-H. Gong, Y. Xu, S.-L. Li, F.-Z. Li, Y.-Y. Yin, Z.-Q. Jiang, M. Li, J.-J. Jia, G. Ren, D. He, Y.-L. Zhou, X.-X. Zhang, N. Wang, X. Chang, Z.-C. Zhu, N.-L. Liu, Y.-A. Chen, C.-Y. Lu, R. Shu, C.-Z. Peng, J.-Y. Wang, J.-W. Pan, Satellite-based entanglement distribution over 1200 kilometers. Science **356**, 1140–1144 (2017)
29. S. Atzeni, A.S. Rab, G. Corrielli, E. Polino, M. Valeri, P. Mataloni, N. Spagnolo, A. Crespi, F. Sciarrino, R. Osellame, Integrated sources of entangled photons at the telecom wavelength in femtosecond-laser-written circuits. Optica **5**, 311 (2018)
30. H.J. Kimble, The quantum internet. Nature **453**, 1023–1030 (2008)
31. D.J.P. Ellis, A.J. Bennett, S.J. Dewhurst, C.A. Nicoll, D.A. Ritchie, A.J. Shields, Cavity-enhanced radiative emission rate in a single-photon-emitting diode operating at 0.5GHz. New J. Phys. **10**, 043035 (2008)
32. J. Claudon, J. Bleuse, N.S. Malik, M. Bazin, P. Jaffrennou, N. Gregersen, C. Sauvan, P. Lalanne, J.-M. Gérard, A highly efficient single-photon source based on a quantum dot in a photonic nanowire. Nat. Photonics **4**, 174–177 (2010)
33. T. Heindel, C. Schneider, M. Lermer, S.H. Kwon, T. Braun, S. Reitzenstein, S. Höfling, M. Kamp, A. Forchel, Electrically driven quantum dot-micropillar single photon source with 34% overall efficiency. Appl. Phys. Lett. **96**, 011107 (2010)
34. A. Schlehahn, M. Gaafar, M. Vaupel, M. Gschrey, P. Schnauber, J.-H. Schulze, S. Rodt, A. Strittmatter, W. Stolz, A. Rahimi-Iman, T. Heindel, M. Koch, S. Reitzenstein, Single-photon emission at a rate of 143 MHz from a deterministic quantum-dot microlens triggered by a mode-locked vertical-external-cavity surface-emitting laser. Appl. Phys. Lett. **107**, 041105 (2015). https://doi.org/10.1063/1.4927429
35. O. Gazzano, S.M. de Vasconcellos, C. Arnold, A. Nowak, E. Galopin, I. Sagnes, L. Lanco, A. Lemaître, P. Senellart, Bright solid-state sources of indistinguishable single photons. Nat. Commun. **4**, 1425 (2013)

36. Y.-M. He, Y. He, Y.-J. Wei, D. Wu, M. Atatüre, C. Schneider, S. Höfling, M. Kamp, C.-Y. Lu, J.-W. Pan, On-demand semiconductor single-photon source with near-unity indistinguishability. Nat. Nanotechnol. **8**, 213–217 (2013)
37. M. Gschrey, A. Thoma, P. Schnauber, M. Seifried, R. Schmidt, B. Wohlfeil, L. Krüger, J.H. Schulze, T. Heindel, S. Burger, F. Schmidt, A. Strittmatter, S. Rodt, S. Reitzenstein, Highly indistinguishable photons from deterministic quantum-dot microlenses utilizing three-dimensional in situ electron-beam lithography. Nat. Commun. **6**, 7662 (2015)
38. U. Rengstl, M. Schwartz, T. Herzog, F. Hargart, M. Paul, S.L. Portalupi, M. Jetter, P. Michler, On-chip beamsplitter operation on single photons from quasi-resonantly excited quantum dots embedded in GaAs rib waveguides. Appl. Phys. Lett. **107**, 021101 (2015)
39. M. Schwartz, U. Rengstl, T. Herzog, M. Paul, J. Kettler, S.L. Portalupi, M. Jetter, P. Michler, Generation, guiding and splitting of triggered single photons from a resonantly excited quantum dot in a photonic circuit. Opt. Express **24**, 3089–3094 (2016)
40. S. Unsleber, Y.-M. He, S. Gerhardt, S. Maier, C.-Y. Lu, J.-W. Pan, N. Gregersen, M. Kamp, C. Schneider, S. Höfling, Highly indistinguishable on-demand resonance fluorescence photons from a deterministic quantum dot micropillar device with 74% extraction efficiency. Opt. Express **24**, 8539 (2016)
41. X. Ding, Y. He, Z.-C. Duan, N. Gregersen, M.-C. Chen, S. Unsleber, S. Maier, C. Schneider, M. Kamp, S. Höfling, C.-Y. Lu, J.-W. Pan, On-demand single photons with high extraction efficiency and near-unity indistinguishability from a resonantly driven quantum dot in a micropillar. Phys. Rev. Lett. **116** (2016)
42. S. Bounouar, C. de la Haye, M. Strauss, P. Schnauber, A. Thoma, M. Gschrey, J.-H. Schulze, A. Strittmatter, S. Rodt, S. Reitzenstein, Generation of maximally entangled states and coherent control in quantum dot microlenses. Appl. Phys. Lett. **112**, 153107 (2018)
43. G.P. Agrawal, N.K. Dutta, *Semiconductor Lasers* (Springer US, New York, 1993)
44. T. Numai, *Fundamentals of Semiconductor Lasers* (Springer Japan, Tokyo, 2015)
45. R.J. Glauber, Photon correlations. Phys. Rev. Lett. **10**, 84–86 (1963)
46. R.J. Glauber, The quantum theory of optical coherence. Phys. Rev. **130**, 2529–2539 (1963)
47. R.J. Glauber, Coherent and incoherent states of the radiation field. Phys. Rev. **131**, 2766–2788 (1963)
48. E.C.G. Sudarshan, Equivalence of semiclassical and quantum mechanical descriptions of statistical light beams. Phys. Rev. Lett. **10**, 277–279 (1963)
49. A. Rahimi-Iman, Recent advances in VECSELs. J. Opt. **18**, 093003 (2016)
50. M. Kuznetsov, F. Hakimi, R. Sprague, A. Mooradian, High-power (>0.5-W CW) diode-pumped vertical-external-cavity surface-emitting semiconductor lasers with circular TEM_{00} beams. IEEE Photonics Technol. Lett. **9**(8), 1063–1065
51. U. Keller, A.C. Tropper, Passively modelocked surface-emitting semiconductor lasers. Phys. Rep. **429**, 67–120 (2006)
52. M. Guina, A. Rantamäki, A. Härkönen, Optically pumped vecsels: review of technology and progress. J. Phys. D: Appl. Phys. **50**, 383001 (2017)
53. M. Fujita, A. Sakai, T. Baba, Ultrasmall and ultralow threshold GaInAsP-InP microdisk injection lasers: design, fabrication, lasing characteristics, and spontaneous emission factor. IEEE J. Sel. Top. Quantum Electron. **5**(3), 673–681 (1999)
54. G. Björk, A. Karlsson, Y. Yamamoto, Definition of a laser threshold. Phys. Rev. A **50**, 1675–1680 (1994)
55. F. De Martini, G.R. Jacobovitz, Anomalous spontaneous-stimulated-decay phase transition and zero-threshold laser action in a microscopic cavity. Phys. Rev. Lett. **60**, 1711–1714 (1988)
56. J.-M. Gérard, Solid-state cavity-quantum electrodynamics with self-assembled quantum dots, in *Single Quantum Dots: Fundamentals, Applications, and New Concepts* (Springer, Berlin, 2003), pp. 269–314
57. G. Eisenstein, D. Bimberg, *Green Photonics and Electronics* (Springer, Cham, 2017)
58. S. Noda, Seeking the ultimate nanolaser. Science **314**, 260 (2006)
59. K. Vahala, *Optical Microcavities* (World Scientific, Singapore, 2004)

60. S. Reitzenstein, T. Heindel, C. Kistner, A. Rahimi-Iman, C. Schneider, S. Höfling, A. Forchel, Low threshold electrically pumped quantum dot-micropillar lasers. Appl. Phys. Lett. **93**(6), 061104 (2008)
61. Z.G. Xie, S. Götzinger, W. Fang, H. Cao, G.S. Solomon, Influence of a single quantum dot state on the characteristics of a microdisk laser. Phys. Rev. Lett. **98**, 117401 (2007)
62. D. Al Nakdali, M.K. Shakfa, M. Gaafar, M. Butkus, K.A. Fedorova, M. Zulonas, M. Wichmann, F. Zhang, B. Heinen, A. Rahimi-Iman, W. Stolz, E.U. Rafailov, M. Koch, High-power quantum-dot vertical-external-cavity surface-emitting laser exceeding 8 W. IEEE Photon. Technol. Lett. **26**, 1561–1564 (2014)
63. D. Al Nakdali, M. Gaafar, M. K. Shakfa, F. Zhang, M. Vaupel, K. A. Fedorova, A. Rahimi-Iman, E.U. Rafailov, M. Koch, High-power operation of quantum-dot semiconductor disk laser at 1180 nm. IEEE Photonics Technol. Lett. **27**, 1128–1131 (2015)
64. B. Heinen, T.L. Wang, M. Sparenberg, A. Weber, B. Kunert, J. Hader, S.W. Koch, J.V. Moloney, M. Koch, W. Stolz, 106 W continuous-wave output power from vertical-external-cavity surface-emitting laser. Electron. Lett. **48**, 516–517 (2012)
65. F. Zhang, B. Heinen, M. Wichmann, C. Möller, B. Kunert, A. Rahimi-Iman, W. Stolz, M. Koch, A 23-watt single-frequency vertical-external-cavity surface-emitting laser. Opt. Express **22**, 12817–12822 (2014). https://doi.org/10.1364/OE.22.012817
66. B. Cocquelin, G. Lucas-Leclin, P. Georges, I. Sagnes, A. Garnache, Design of a low-threshold VECSEL emitting at 852 nm for cesium atomic clocks. Opt. Quantum. Electron. **40**, 167–173 (2008)
67. M. Fallahi, L. Fan, Y. Kaneda, C. Hessenius, J. Hader, H. Li, J. Moloney, B. Kunert, W. Stolz, S. Koch, J. Murray, R. Bedford, 5-W yellow laser by intracavity frequency doubling of high-power vertical-external-cavity surface-emitting laser. IEEE Photonics Technol. Lett. **20**, 1700–1702 (2008)
68. E. Kantola, T. Leinonen, S. Ranta, M. Tavast, M. Guina, High-efficiency 20 W yellow VECSEL. Opt. Express **22**, 6372 (2014)
69. M. Scheller, J.M. Yarborough, J.V. Moloney, M. Fallahi, M. Koch, S.W. Koch, Room temperature continuous wave milliwatt terahertz source. Opt. Express **18**, 27112 (2010)
70. J.R. Paul, M. Scheller, A. Laurain, A. Young, S.W. Koch, J. Moloney, Narrow linewidth single-frequency terahertz source based on difference frequency generation of vertical-external-cavity source-emitting lasers in an external resonance cavity. Opt. Lett. **38**, 3654 (2013)
71. M. Lukowski, C. Hessenius, M. Fallahi, Widely tunable high-power two-color VECSELs for new wavelength generation. IEEE J. Sel. Top. Quantum Electron. **21**, 432–439 (2015)
72. M. Lukowski, C. Hessenius, R. Bedford, M. Fallahi, Tunable type II intracavity difference frequency generation at 54μm in a two chip vertical external cavity surface emitting laser. Opt. Lett. **40**, 4174 (2015)
73. R. Bek, S. Baumgärtner, F. Sauter, H. Kahle, T. Schwarzbäck, M. Jetter, P. Michler, Intra-cavity frequency-doubled mode-locked semiconductor disk laser at 325 nm. Opt. Express **23**(15), 19947–19953 (2015)
74. L. Fan, M. Fallahi, J. Hader, A.R. Zakharian, J.V. Moloney, W. Stolz, S.W. Koch, R. Bedford, J.T. Murray, Linearly polarized dual-wavelength vertical-external-cavity surface-emitting laser. Appl. Phys. Lett. **90**, 181124 (2007)
75. M. Wichmann, M. Stein, A. Rahimi-Iman, S.W. Koch, M. Koch, Interferometric Characterization of a Semiconductor Disk Laser driven Terahertz Source. J. Infrared Millim. Terahertz Waves **35**, 503–508 (2014)
76. M. Wichmann, M.K. Shakfa, F. Zhang, B. Heinen, M. Scheller, A. Rahimi-Iman, W. Stolz, J.V. Moloney, S.W. Koch, M. Koch, Evolution of multi-mode operation in vertical-external-cavity surface-emitting lasers. Opt. Express **21**, 31940–31950 (2013)
77. K.A. Fedorova, H. Guoyu, M. Wichmann, C. Kriso, F. Zhang, W. Stolz, M. Scheller, M. Koch, A. Rahimi-Iman, Widely-tunable THz-generating semiconductor disk laser. Phys. Status Solidi RRL **14**, 2000204 (2020). https://doi.org/10.1002/pssr.202000204
78. L. Fan, M. Fallahi, J. Hader, A.R. Zakharian, J.V. Moloney, J.T. Murray, R. Bedford, W. Stolz, S.W. Koch, Multichip vertical-external-cavity surface-emitting lasers: a coherent power scaling scheme. Opt. Lett. **31**, 3612 (2006)

79. F. Zhang, M. Gaafar, C. Möller, W. Stolz, M. Koch, A. Rahimi-Iman, Dual-wavelength emission from a serially connected two-chip VECSEL. IEEE Photonics Technol. Lett. **28**, 927–929 (2016)
80. C. Hessenius, M. Lukowski, M. Fallahi, High-power tunable two-wavelength generation in a two chip co-linear T-cavity vertical external-cavity surface-emitting laser. Appl. Phys. Lett. **101**, 121110 (2012)
81. H. Guoyu, C. Kriso, F. Zhang, M. Wichmann, W. Stolz, K.A. Fedorova, A. Rahimi-Iman, Two-chip power-scalable THz-generating semiconductor disk laser. Opt. Lett. **44**, 4000–4003 (2019)
82. T.A.G. Bondaz, A. Laurain, J.V. Moloney, J.G. McInerney, Generation and stabilization of continuous-wave THz emission from a bi-color VECSEL. IEEE Photonics Technol. Lett. **31**(19), 1569–1572 (2019)
83. M. Wichmann, G. Town, J. Quante, M. Gaafar, A. Rahimi-Iman, W. Stolz, S.W. Koch, M. Koch, Antiphase noise dynamics in a dual-wavelength vertical-external-cavity surface-emitting laser. IEEE Photonics Technol. Lett. **27**, 2039–2042 (2015)
84. M.A. Gaafar, A. Rahimi-Iman, K.A. Fedorova, W. Stolz, E.U. Rafailov, M. Koch, Mode-locked semiconductor disk lasers. Adv. Opt. Photonics **8**, 370–400 (2016)
85. F.F. Voigt, F. Emaury, P. Bethge, D. Waldburger, S.M. Link, S. Carta, A. van der Bourg, F. Helmchen, U. Keller, Multiphoton in vivo imaging with a femtosecond semiconductor disk laser. Biomed. Opt. Express **8**, 3213–3231 (2017)
86. T.W. Hänsch, Nobel lecture: passion for precision. Rev. Mod. Phys. **78**, 1297–1309 (2006)
87. S.M. Link, D.J.H.C. Maas, D. Waldburger, U. Keller, Dual-comb spectroscopy of water vapor with a free-running semiconductor disk laser. Science **356**, 1164 (2017)
88. P. Klopp, U. Griebner, M. Zorn, M. Weyers, Pulse repetition rate up to 92 GHz or pulse duration shorter than 110 fs from a mode-locked semiconductor disk laser. Appl. Phys. Lett. **98**, 071103 (2011)
89. D. Waldburger, S.M. Link, M. Mangold, C.G.E. Alfieri, E. Gini, M. Golling, B.W. Tilma, U. Keller, High-power 100 fs semiconductor disk lasers. Optica **3**, 844–852 (2016)
90. Y.F. Chen, Y.C. Lee, H.C. Liang, K.Y. Lin, K.W. Su, K.F. Huang, Femtosecond high-power spontaneous mode-locked operation in vertical-external cavity surface-emitting laser with gigahertz oscillation. Opt. Lett. **36**, 4581–4583 (2011)
91. L. Kornaszewski, G. Maker, G.P.A. Malcolm, M. Butkus, E.U. Rafailov, C.J. Hamilton, SESAM-free mode-locked semiconductor disk laser. Laser Photon. Rev. **6**, L20–L23 (2012)
92. A.R. Albrecht, Y. Wang, M. Ghasemkhani, D.V. Seletskiy, J.G. Cederberg, M. Sheik-Bahae, Exploring ultrafast negative Kerr effect for mode-locking vertical external-cavity surface-emitting lasers. Opt. Express **21**, 28801–28808 (2013)
93. M. Gaafar, C. Möller, M. Wichmann, B. Heinen, B. Kunert, A. Rahimi-Iman, W. Stolz, M. Koch, Harmonic self-mode-locking of optically pumped semiconductor disc laser. Electron. Lett. **50**, 542–543 (2014)
94. M. Gaafar, D.A. Nakdali, C. Möller, K.A. Fedorova, M. Wichmann, M.K. Shakfa, F. Zhang, A. Rahimi-Iman, E.U. Rafailov, M. Koch, Self-mode-locked quantum-dot vertical-external-cavity surface-emitting laser. Opt. Lett. **39**, 4623–4626 (2014)
95. M. Gaafar, P. Richter, H. Keskin, C. Möller, M. Wichmann, W. Stolz, A. Rahimi-Iman, M. Koch, Self-mode-locking semiconductor disk laser. Opt. Express **22**, 28390–28399 (2014)
96. A. Rahimi-Iman, M. Gaafar, W. Stolz, E.U. Rafailov, M. Koch, Self-mode-locked semiconductor disk lasers. Opt. Photonics News **25**, 28–28 (2014)
97. A. Rahimi-Iman, M. Gaafar, D.A. Nakdali, C. Möller, F. Zhang, M. Wichmann, M.K. Shakfa, K.A. Fedorova, W. Stolz, E.U. Rafailov, M. Koch, Recent advances in the field of vertical-external-cavity surface-emitting lasers, in *Vertical External Cavity Surface Emitting Lasers (VECSELs) V*, vol. 9349. (International Society for Optics and Photonics, 2015), p. 934906
98. A. Rahimi-Iman, M. Gaafar, C. Möller, M. Vaupel, F. Zhang, D. Al-Nakdali, K.A. Fedorova, W. Stolz, E.U. Rafailov, M. Koch, Self-mode-locked vertical-external-cavity surface-emitting laser, in *Vertical External Cavity Surface Emitting Lasers (VECSELs) VI*, vol. 9734. (International Society for Optics and Photonics, 2016), p. 97340M

99. R. Bek, M. Grossmann, H. Kahle, M. Koch, A. Rahimi-Iman, M. Jetter, P. Michler, Self-mode-locked AlGaInP-VECSEL. Appl. Phys. Lett. **111**, 182105 (2017)
100. A. Rahimi-Iman, *Vertical External Cavity Surface Emitting Lasers: VECSEL Technology and Applications*, ed. by M. Jetter, P. Michler, Ch. Self-mode-locked vertical-external-cavity surface-emitting laser, 2nd edn. (Wiley-VCH, Weinheim, 2020) expected release, in production. ISBN 978-3527413621
101. C. Kriso, S. Kress, T. Munshi, M. Grossmann, R. Bek, M. Jetter, P. Michler, W. Stolz, M. Koch, A. Rahimi-Iman, Microcavity-enhanced Kerr nonlinearity in a vertical-external-cavity surface-emitting laser. Opt. Express **27**, 11914–11929 (2019). https://doi.org/10.1364/OE.27.011914
102. A.H. Quarterman, M.A. Tyrk, K.G. Wilcox, Z-scan measurements of the nonlinear refractive index of a pumped semiconductor disk laser gain medium. Appl. Phys. Lett. **106**, 011105 (2015)
103. E.A. Shaw, A.H. Quarterman, A.P. Turnbull, T.C. Sverre, C.R. Head, A.C. Tropper, K.G. Wilcox, Nonlinear lensing in an unpumped antiresonant semiconductor disk laser gain structure. IEEE Photonics Technol. Lett. **28**(13), 1395–1398 (2016)
104. A.H. Quarterman, S. Mirkhanov, C.J.C. Smyth, K.G. Wilcox, Measurements of nonlinear lensing in a semiconductor disk laser gain sample under optical pumping and using a resonant femtosecond probe laser. Appl. Phys. Lett. **109**, 121113 (2016)
105. C. Kriso, S. Kress, T. Munshi, M. Grossmann, R. Bek, M. Jetter, P. Michler, W. Stolz, M. Koch, A. Rahimi-Iman, Wavelength and pump-power dependent nonlinear refraction and absorption in a semiconductor disk laser. IEEE Photonics Technol. Lett. **32**, 85–88 (2020). Published online Dec 2019
106. C. Kriso, M. Stein, T. Haeger, N. Pourdavoud, M. Gerhard, A. Rahimi-Iman, T. Riedl, M. Koch, Nonlinear refraction in $CH_3NH_3PbBr_3$ single crystals. Opt. Lett. **45**, 2431–2434 (2020)
107. C. Möller, C. Fuchs, C. Berger, A. Ruiz Perez, M. Koch, J. Hader, J.V. Moloney, S.W. Koch, W. Stolz, Type-II vertical-external-cavity surface-emitting laser with Watt level output powers at 1.2 μm. Appl. Phys. Lett. **108**, 071102 (2016)
108. C. Möller, C. Fuchs, C. Berger, F. Zhang, A. Rahimi-Iman, M. Koch, A.R. Perez, S.W. Koch, J. Hader, J.V. Moloney, W. Stolz, The development and fundamental analysis of type-II VECSELs at 1.2 μm (Conference Presentation), in *Vertical External Cavity Surface Emitting Lasers (VECSELs) VII*, vol. 10087 (International Society for Optics and Photonics, 2017), p. 100870L
109. C. Möller, F. Zhang, C. Fuchs, C. Berger, A. Rehn, A.R. Perez, A. Rahimi-Iman, J. Hader, M. Koch, J.V. Moloney, S.W. Koch, W. Stolz, Fundamental transverse mode operation of a type-II vertical-external-cavity surface-emitting laser at 1.2 μm. Electron. Lett. **53**, 93–94 (2017)
110. C. Lammers, M. Stein, C. Berger, C. Möller, C. Fuchs, A. Ruiz Perez, A. Rahimi-Iman, J. Hader, J.V. Moloney, W. Stolz, S.W. Koch, M. Koch, Gain spectroscopy of a type-II VECSEL chip. Appl. Phys. Lett. **109**, 232107 (2016)
111. V.K. Singh, A.K. Rai, Prospects for laser-induced breakdown spectroscopy for biomedical applications: a review. Lasers Med. Sci. **26**, 673–687 (2011)
112. J. Kaiser, K. Novotny, M.Z. Martin, A. Hrdlička, R. Malina, M. Hartl, V. Adam, R. Kizek, Trace elemental analysis by laser-induced breakdown spectroscopy–biological applications. Surf. Sci. Rep. **67**, 233–243 (2012)
113. M. Markiewicz-Keszycka, X. Cama-Moncunill, M.P. Casado-Gavalda, Y. Dixit, R. Cama-Moncunill, P.J. Cullen, C. Sullivan, Laser-induced breakdown spectroscopy (libs) for food analysis: a review. Trends Food Sci. Technol. **65**, 80–93 (2017)
114. A.J.R. Bauer, S.G. Buckley, Novel applications of laser-induced breakdown spectroscopy. Appl. Spectrosc. **71**, 553–566 (2017)
115. D. Abdrabou, L.M. Schneider, A. Rahimi-Iman, M.A. Khedr, A.M. Hussein, T.M. El-Sherbini, Study of laser-induced-plasma parameters for molybdenum targets. Plasma Res. Express **1**, 035004 (2019)

116. A. Rahimi-Iman, *On the Condensation of Polaritons*. Springer Series in Optical Sciences, vol. 229 (Springer International Publishing, Cham, 2020), pp. 65–85
117. A. Rahimi-Iman, *Nichtlineare Effekte in III/V Quantenfilm-Mikroresonatoren: Von dynamischer Bose–Einstein-Kondensation hin zum elektrisch betriebenen Polariton-Laser* (Cuvillier Verlag, Göttingen, 2013)
118. C. Schneider, A. Rahimi-Iman, N.Y. Kim, J. Fischer, I.G. Savenko, M. Amthor, M. Lermer, A. Wolf, L. Worschech, V.D. Kulakovskii, I.A. Shelykh, M. Kamp, S. Reitzenstein, A. Forchel, Y. Yamamoto, S. Höfling, An electrically pumped polariton laser. Nature **497**, 348–352 (2013)
119. A. Rahimi-Iman, *Towards Polariton Condensates and Devices*. Springer Series in Optical Sciences, vol. 229 (Springer International Publishing, Cham, 2020), pp. 1–32
120. A. Rahimi-Iman, *The Concept of Polariton Lasing*. Springer Series in Optical Sciences, vol. 229 (Springer International Publishing, Cham, 2020), pp. 87–118
121. A. Rahimi-Iman, *Fundamentals of Polariton Physics*. Springer Series in Optical Sciences, vol. 229. (Springer International Publishing, Cham, 2020), pp. 33–64
122. A. Imamoglu, R.J. Ram, S. Pau, Y. Yamamoto, Nonequilibrium condensates and lasers without inversion: exciton-polariton lasers. Phys. Rev. A **53**, 4250 (1996)
123. A. Kavokin, G. Malpuech, F.P. Laussy, Polariton laser and polariton superfluidity in microcavities. Phys. Lett. A **306**, 187–199 (2003)
124. H. Deng, G. Weihs, C. Santori, J. Bloch, Y. Yamamoto, Condensation of semiconductor microcavity exciton polaritons. Science **298**(5591), 199–202 (2002)
125. H. Deng, G. Weihs, D. Snoke, J. Bloch, Y. Yamamoto, Polariton lasing vs. photon lasing in a semiconductor microcavity. Proc. Natl. Acad. Sci. **100**(26), 15318–15323 (2003)
126. M. Richard, J. Kasprzak, R. Romestain, R. André, L.S. Dang, Spontaneous coherent phase transition of polaritons in CdTe microcavities. Phys. Rev. Lett. **94**, 187401 (2005)
127. J. Kasprzak, M. Richard, S. Kundermann, A. Baas, P. Jeambrun, J.M.J. Keeling, F.M. Marchetti, M.H. Szymańska, R. André, J.L. Staehli, V. Savona, P.B. Littlewood, B. Deveaud, L.S. Dang, Bose-Einstein condensation of exciton polaritons. Nature **443**, 409–414 (2006)
128. S. Christopoulos, G.B.H. von Högersthal, A.J.D. Grundy, P.G. Lagoudakis, A.V. Kavokin, J.J. Baumberg, G. Christmann, R. Butté, E. Feltin, J.-F. Carlin, N. Grandjean, Room-temperature polariton lasing in semiconductor microcavities. Phys. Rev. Lett. **98**, 126405 (2007)
129. G. Christmann, R. Butté, E. Feltin, J.-F. Carlin, N. Grandjean, Room temperature polariton lasing in a GaN/AlGaN multiple quantum well microcavity. Appl. Phys. Lett. **93**, 051102 (2008)
130. D. Bajoni, P. Senellart, E. Wertz, I. Sagnes, A. Miard, A. Lemaître, J. Bloch, Polariton laser using single micropillar GaAs-GaAlAs semiconductor cavities. Phys. Rev. Lett. **100**, 047401 (2008)
131. M. Maragkou, A.J.D. Grundy, T. Ostatnický, P.G. Lagoudakis, Longitudinal optical phonon assisted polariton laser. Appl. Phys. Lett. **97**, 111110 (2010)
132. D. Bajoni, Polariton lasers. Hybrid light-matter lasers without inversion. J. Phys. D: Appl. Phys. **45**(31), 313001 (2012)
133. S. Kim, B. Zhang, Z. Wang, J. Fischer, S. Brodbeck, M. Kamp, C. Schneider, S. Höfling, H. Deng, Coherent polariton laser. Phys. Rev. X **6**, 011026 (2016)
134. M. Ramezani, A. Halpin, A.I. Fernández-DomíDnguez, J. Feist, S.R.-K. Rodriguez, F.J. Garcia-Vidal, J. Gómez Rivas, Plasmon-exciton-polariton lasing. Optica **4**, 31–37 (2017)
135. P. Bhattacharya, B. Xiao, A. Das, S. Bhowmick, J. Heo, Solid state electrically injected exciton-polariton laser. Phys. Rev. Lett. **110**, 206403 (2013)
136. P. Bhattacharya, T. Frost, S. Deshpande, M.Z. Baten, A. Hazari, A. Das, Room temperature electrically injected polariton laser. Phys. Rev. Lett. **112**, 236802 (2014)
137. M. Amthor, S. Weissenseel, J. Fischer, M. Kamp, C. Schneider, S. Höfling, Electro-optical switching between polariton and cavity lasing in an InGaAs quantum well microcavity. Opt. Express **22**, 31146–31153 (2014)
138. M.D. Fraser, S. Höfling, Y. Yamamoto, Physics and applications of exciton-polariton lasers. Nat. Mater. **15**, 1049 (2016)
139. M.D. Fraser, Coherent exciton-polariton devices. Semicond. Sci. Technol. **32**, 093003 (2017)

140. Y.G. Rubo, A. Kavokin, I. Shelykh, Suppression of superfluidity of exciton-polaritons by magnetic field. Phys. Lett. A **358**, 227–230 (2006)
141. S. Utsunomiya, L. Tian, G. Roumpos, C.W. Lai, N. Kumada, T. Fujisawa, M. Kuwata-Gonokami, A. Löffler, S. Höfling, A. Forchel, Y. Yamamoto, Observation of Bogoliubov excitations in exciton-polariton condensates. Nat. Phys. **4**, 700–705 (2008)
142. J. Fischer, S. Brodbeck, A.V. Chernenko, I. Lederer, A. Rahimi-Iman, M. Amthor, V.D. Kulakovskii, L. Worschech, M. Kamp, M. Durnev, C. Schneider, A.V. Kavokin, S. Höfling, Anomalies of a nonequilibrium spinor polariton condensate in a magnetic field. Phys. Rev. Lett. **112**, 093902 (2014)
143. A.V. Chernenko, A. Rahimi-Iman, J. Fischer, M. Amthor, C. Schneider, S. Reitzenstein, A. Forchel, S. Hoefling, Polariton condensate coherence in planar microcavities in a magnetic field. Semiconductors **50**, 1609–1613 (2016). https://doi.org/10.1134/S1063782616120058
144. M. Assmann, J.-S. Tempel, F. Veit, M. Bayer, A. Rahimi-Iman, A. Löffler, S. Höfling, S. Reitzenstein, L. Worschech, A. Forchel, From polariton condensates to highly photonic quantum degenerate states of bosonic matter. Proc. Natl. Acad. Sci. **108**, 1804–1809 (2011)
145. A. Rahimi-Iman, A.V. Chernenko, J. Fischer, S. Brodbeck, M. Amthor, C. Schneider, A. Forchel, S. Höfling, S. Reitzenstein, M. Kamp, Coherence signatures and density-dependent interaction in a dynamical exciton-polariton condensate. Phys. Rev. B **86**, 155308 (2012)
146. J.-S. Tempel, F. Veit, M. Assmann, L.E. Kreilkamp, A. Rahimi-Iman, A. Löffler, S. Höfling, S. Reitzenstein, L. Worschech, A. Forchel, M. Bayer, Characterization of two-threshold behavior of the emission from a GaAs microcavity. Phys. Rev. B **85**, 075318 (2012)
147. A. Rahimi-Iman, C. Schneider, J. Fischer, S. Holzinger, M. Amthor, S. Höfling, S. Reitzenstein, L. Worschech, M. Kamp, A. Forchel, Zeeman splitting and diamagnetic shift of spatially confined quantum-well exciton polaritons in an external magnetic field. Phys. Rev. B **84**, 165325 (2011)
148. A. Rahimi-Iman, *Polaritons in External Fields*. Springer Series in Optical Sciences, vol. 229 (Springer International Publishing, Cham, 2020), pp. 241–262
149. Y. Yamamoto, F. Tassone, H. Cao, *Semiconductor Cavity Quantum Electrodynamics* (Springer, Berlin, 2000)
150. E.M. Purcell, Spontaneous emission probabilities at radio frequencies. Phys. Rev. **69**, 681 (1946)
151. G. Khitrova, H.M. Gibbs, M. Kira, S.W. Koch, A. Scherer, Vacuum Rabi splitting in semiconductors. Nat. Phys. **2**, 81–90 (2006)
152. I.I. Rabi, Space quantization in a gyrating magnetic field. Phys. Rev. **51**, 652–654 (1937)
153. C. Weisbuch, M. Nishioka, A. Ishikawa, Y. Arakawa, Observation of the coupled exciton-photon mode splitting in a semiconductor quantum microcavity. Phys. Rev. Lett. **69**, 3314 (1992)
154. J.P. Reithmaier, G. Sek, A. Löffler, C. Hofmann, S. Kuhn, S. Reitzenstein, L.V. Keldysh, V.D. Kulakovskii, T.L. Reinecke, A. Forchel, Strong coupling in a single quantum dot-semiconductor microcavity system. Nature **432**, 197–200 (2004)
155. T. Yoshie, A. Scherer, J. Hendrickson, G. Khitrova, H.M. Gibbs, G. Rupper, C. Ell, O.B. Shchekin, D.G. Deppe, Vacuum Rabi splitting with a single quantum dot in a photonic crystal nanocavity. Nature **432**, 200–203 (2004)
156. E. Peter, P. Senellart, D. Martrou, A. Lemaître, J. Hours, J.M. Gérard, J. Bloch, Exciton-photon strong-coupling regime for a single quantum dot embedded in a microcavity. Phys. Rev. Lett. 95, 067401 (2005)
157. Y. Yamamoto, A. Imamoglu, *Mesoscopic Quantum Optics* (Wiley, Hoboken, 1999)
158. A. Kavokin, G. Malpuech, *Cavity Polaritons* (Academic, Cambridge, 2003)
159. A.V. Kavokin, J.J. Baumberg, G. Malpuech, F.P. Laussy, *Microcavities* (Oxford University Press, Oxford, 2007)
160. A. Rahimi-Iman, Nonclassical light emission from optically and electrically driven high-Q quantum dot-micropillar cavities. Master's thesis, Würzburg University, 2009
161. D. Press, S. Götzinger, S. Reitzenstein, C. Hofmann, A. Löffler, M. Kamp, A. Forchel, Y. Yamamoto, Photon antibunching from a single quantum-dot-microcavity system in the strong coupling regime. Phys. Rev. Lett. **98**, 117402 (2007)

162. J. McKeever, A. Boca, A.D. Boozer, J.R. Buck, H.J. Kimble, Experimental realization of a one-atom laser in the regime of strong coupling. Nature **425**, 268–271 (2003)
163. S. Reitzenstein, S. Münch, P. Franeck, A. Löffler, S. Höfling, L. Worschech, A. Forchel, I.V. Ponomarev, T.L. Reinecke, Exciton spin state mediated photon-photon coupling in a strongly coupled quantum dot microcavity system. Phys. Rev. B **82**, 121306 (2010)
164. J. Kasprzak, S. Reitzenstein, E.A. Muljarov, C. Kistner, C. Schneider, M. Strauss, S. Höfling, A. Forchel, W. Langbein, Up on the Jaynes-Cummings ladder of a quantum-dot/microcavity system. Nat. Mater. **9**, 304 (2010)
165. S. Reitzenstein, A. Forchel, Quantum dot micropillars. J. Phys. D: Appl. Phys. **43**(3), 033001 (2010)
166. M. Müller, H. Vural, C. Schneider, A. Rastelli, O.G. Schmidt, S. Höfling, P. Michler, Quantum-dot single-photon sources for entanglement enhanced interferometry. Phys. Rev. Lett. **118**, 257402 (2017)
167. X. Lin, X. Dai, C. Pu, Y. Deng, Y. Niu, L. Tong, W. Fang, Y. Jin, X. Peng, Electrically-driven single-photon sources based on colloidal quantum dots with near-optimal antibunching at room temperature. Nat. Commun. **8**, 1132 (2017). https://doi.org/10.1038/s41467-017-01379-6
168. N. Mizuochi, T. Makino, H. Kato, D. Takeuchi, M. Ogura, H. Okushi, M. Nothaft, P. Neumann, A. Gali, F. Jelezko, J. Wrachtrup, S. Yamasaki, Electrically driven single-photon source at room temperature in diamond. Nat. Photonics **6**, 299–303 (2012)
169. M. Nothaft, S. Höhla, F. Jelezko, N. Frühauf, J. Pflaum, J. Wrachtrup, Electrically driven photon antibunching from a single molecule at room temperature. Nat. Commun. **3**, 628 (2012)
170. A. Lohrmann, N. Iwamoto, Z. Bodrog, S. Castelletto, T. Ohshima, T. Karle, A. Gali, S. Prawer, J. McCallum, B. Johnson, Single-photon emitting diode in silicon carbide. Nat. Commun. **6**, 7783 (2015)
171. C. Chakraborty, L. Kinnischtzke, K.M. Goodfellow, R. Beams, A.N. Vamivakas, Voltage-controlled quantum light from an atomically thin semiconductor. Nat. Nanotechnol. **10**, 507–511 (2015)
172. Y.-M. He, G. Clark, J.R. Schaibley, Y. He, M.-C. Chen, Y.-J. Wei, X. Ding, Q. Zhang, W. Yao, X. Xu, C.-Y. Lu, J.-W. Pan, Single quantum emitters in monolayer semiconductors. Nat. Nanotechnol. **10**, 497–502 (2015)
173. M. Koperski, K. Nogajewski, A. Arora, V. Cherkez, P. Mallet, J.-Y. Veuillen, J. Marcus, P. Kossacki, M. Potemski, Single photon emitters in exfoliated WSe_2 structures. Nat. Nanotechnol. **10**, 503–506 (2015)
174. A. Srivastava, M. Sidler, A.V. Allain, D.S. Lembke, A. Kis, A. Imamoglu, Optically active quantum dots in monolayer WSe_2. Nat. Nanotechnol. **10**, 491–496 (2015)
175. P. Tonndorf, R. Schmidt, R. Schneider, J. Kern, M. Buscema, G.A. Steele, A. Castellanos-Gomez, H.S.J. van der Zant, S. Michaelis de Vasconcellos, R. Bratschitsch, Single-photon emission from localized excitons in an atomically thin semiconductor. Optica **2**, 347 (2015)
176. F. Peyskens, C. Chakraborty, M. Muneeb, D. Van Thourhout, D. Englund, Integration of single photon emitters in 2D layered materials with a silicon nitride photonic chip. Nat. Commun. **10**, 4435 (2019). https://doi.org/10.1038/s41467-019-12421-0
177. M. Gschrey, F. Gericke, A. Schüßler, R. Schmidt, J.-H. Schulze, T. Heindel, S. Rodt, A. Strittmatter, S. Reitzenstein, In situ electron-beam lithography of deterministic single-quantum-dot mesa-structures using low-temperature cathodoluminescence spectroscopy. Appl. Phys. Lett. **102**, 251113 (2013)
178. P. Schnauber, J. Schall, S. Bounouar, T. Höhne, S.-I. Park, G.-H. Ryu, T. Heindel, S. Burger, J.-D. Song, S. Rodt, S. Reitzenstein, Deterministic integration of quantum dots into on-chip multimode interference beamsplitters using in situ electron beam lithography. Nano Lett. **18**, 2336–2342 (2018)
179. Q. Shi, B. Sontheimer, N. Nikolay, A.W. Schell, J. Fischer, A. Naber, O. Benson, M. Wegener, Wiring up pre-characterized single-photon emitters by laser lithography. Sci. Rep. **6**, 31135 (2016)

180. M. Colautti, P. Lombardi, S. Pazzagli, S. Nocentini, D. Wiersma, F. Cataliotti, C. Toninelli, 3D laser writing around lifetime-limited quantum emitters, in *2019 Conference on Lasers and Electro-Optics Europe and European Quantum Electronics Conference*, p. ea_p_11 (Optical Society of America, 2019)
181. A. Rahimi-Iman, Advances in functional nano-materials science. Ann. Phys. **532**(9), 2000015 (2020). https://doi.org/10.1002/andp.202000015
182. A. Thoma, P. Schnauber, M. Gschrey, M. Seifried, J. Wolters, J.-H. Schulze, A. Strittmatter, S. Rodt, A. Carmele, A. Knorr, T. Heindel, S. Reitzenstein, Exploring dephasing of a solid-state quantum emitter via time- and temperature-dependent Hong-Ou-Mandel experiments. Phys. Rev. Lett. **116**, 033601 (2016)
183. A. Thoma, P. Schnauber, J. Böhm, M. Gschrey, J.-H. Schulze, A. Strittmatter, S. Rodt, T. Heindel, S. Reitzenstein, Two-photon interference from remote deterministic quantum dot microlenses. Appl. Phys. Lett. **110**, 011104 (2017)
184. S. Bounouar, M. Strauß, A. Carmele, P. Schnauber, A. Thoma, M. Gschrey, J.-H. Schulze, A. Strittmatter, S. Rodt, A. Knorr, S. Reitzenstein, Path-controlled time reordering of paired photons in a dressed three-level cascade. Phys. Rev. Lett. **118**, 233601 (2017)
185. D. Sanvitto, S. Kéna-Cohen, The road towards polaritonic device. Nat. Mater. **15**(10), 1061–1073 (2016)
186. A.V. Kavokin, J.J. Baumberg, G. Malpuech, F.P. Laussy, *Microcavities*, vol. 1 (Oxford University Press, Oxford, 2017)
187. T.C.H. Liew, A.V. Kavokin, I.A. Shelykh, Optical circuits based on polariton neurons in semiconductor microcavities. Phys. Rev. Lett. **101**, 016402 (2008)
188. T. Espinosa-Ortega, T.C.H. Liew, Complete architecture of integrated photonic circuits based on and and not logic gates of exciton polaritons in semiconductor microcavities. Phys. Rev. B **87**, 195305 (2013)
189. A. Amo, T.C.H. Liew, C. Adrados, R. Houdré, E. Giacobino, A.V. Kavokin, A. Bramati, Exciton-polariton spin switches. Nat. Photonics **4**, 361 (2010)
190. M. Assmann, F. Veit, M. Bayer, A. Löffler, S. Höfling, M. Kamp, A. Forchel, All-optical control of quantized momenta on a polariton staircase. Phys. Rev. B **85**, 155320 (2012)
191. I.A. Shelykh, R. Johne, D.D. Solnyshkov, G. Malpuech, Optically and electrically controlled polariton spin transistor. Phys. Rev. B **82**, 153303 (2010)
192. T. Gao, P.S. Eldridge, T.C.H. Liew, S.I. Tsintzos, G. Stavrinidis, G. Deligeorgis, Z. Hatzopoulos, P.G. Savvidis, Polariton condensate transistor switch. Phys. Rev. B **85**, 235102 (2012)
193. M. De Giorgi, D. Ballarini, E. Cancellieri, F.M. Marchetti, M.H. Szymanska, C. Tejedor, R. Cingolani, E. Giacobino, A. Bramati, G. Gigli, D. Sanvitto, Control and ultrafast dynamics of a two-fluid polariton switch. Phys. Rev. Lett. **109**, 266407 (2012)
194. H.S. Nguyen, D. Vishnevsky, C. Sturm, D. Tanese, D. Solnyshkov, E. Galopin, A. Lemaitre, I. Sagnes, A. Amo, G. Malpuech, J. Bloch, Realization of a double-barrier resonant tunneling diode for cavity polaritons. Phys. Rev. Lett. **110**, 236601 (2013)
195. D. Ballarini, M.D. Giorgi, E. Cancellieri, R. Houdré, E. Giacobino, R. Cingolani, A. Bramati, G. Gigli, D. Sanvitto, All-optical polariton transistor. Nat. Commun. **4**, 1778 (2013)
196. F. Marsault, H.S. Nguyen, D. Tanese, A. Lemaître, E. Galopin, I. Sagnes, A. Amo, J. Bloch, Single-wavelength, all-optical switching based on exciton-polaritons. Appl. Phys. Lett **107**, 5263 (2015)
197. C. Sturm, D. Tanese, H. Nguyen, H. Flayac, E. Galopin, A. Lemaitre, A. Amo, G. Malpuech, J. Bloch, All-optical phase modulation in a cavity-polariton Mach-Zehnder interferometer. Nat. Commun. **5**, 3278 (2014)
198. L. Dominici, D. Colas, S. Donati, J.P. Restrepo Cuartas, M. De Giorgi, D. Ballarini, G. Guirales, J.C. López Carreño, A. Bramati, G. Gigli, E. Del Valle, F.P. Laussy, D. Sanvitto, Ultrafast control and Rabi oscillations of polaritons. Phys. Rev. Lett. **113**, 226401 (2014)
199. D. Colas, L. Dominici, S. Donati, A.A. Pervishko, T.C. Liew, I.A. Shelykh, D. Ballarini, M. De Giorgi, A. Bramati, G. Gigli, E. Del Valle, F.P. Laussy, A.V. Kavokin, D. Sanvitto, Polarization shaping of Poincaré beams by polariton oscillations. Light Sci. Appl. **4**, e350 (2015)

200. T.A. Fisher, A.M. Afshar, M.S. Skolnick, D.M. Whittaker, J.S. Roberts, Vacuum Rabi coupling enhancement and Zeeman splitting in semiconductor quantum microcavity structures in a high magnetic field. Phys. Rev. B **53**, R10469 (1996)
201. A. Armitage, T.A. Fisher, M.S. Skolnick, D.M. Whittaker, P. Kinsler, J.S. Roberts, Exciton polaritons in semiconductor quantum microcavities in a high magnetic field. Phys. Rev. B **55**, 16395 (1997)
202. A. Rahimi-Iman, C. Schneider, J. Fischer, S. Holzinger, M. Amthor, L. Worschech, S. Reitzenstein, S. Höfling, A. Forchel, M. Kamp, Magnetic-field interaction of spatially confined quantum-well exciton-polaritons. J. Phys. Conf. Ser. **456**, 012033 (2013)
203. S. Reitzenstein, S. Münch, P. Franeck, A. Rahimi-Iman, A. Löffler, S. Höfling, L. Worschech, A. Forchel, Control of the strong light-matter interaction between an elongated $In_{0.3}Ga_{0.7}As$ quantum dot and a micropillar cavity using external magnetic fields. Phys. Rev. Lett. **103**, 127401 (2009)
204. C. Kistner, T. Heindel, C. Schneider, A. Rahimi-Iman, S. Reitzenstein, S. Höfling, A. Forchel, Demonstration of strong coupling via electro-optical tuning in high-quality QD-micropillar systems. Opt. Express **16**, 15006–15012 (2008)
205. X. Liu, T. Galfsky, Z. Sun, F. Xia, E.C. Lin, Y.H. Lee, S. Kéna-Cohen, V.M. Menon, Strong light-matter coupling in two-dimensional atomic crystals. Nat. Photonics **9**, 30–34 (2014). https://doi.org/10.1038/nphoton.2014.304
206. S. Dufferwiel, S. Schwarz, F. Withers, A.A. Trichet, F. Li, M. Sich, O. Del Pozo-Zamudio, C. Clark, A. Nalitov, D.D. Solnyshkov, G. Malpuech, K.S. Novoselov, J.M. Smith, M.S. Skolnick, D.N. Krizhanovskii, A.I. Tartakovskii, Exciton-polaritons in van der Waals heterostructures embedded in tunable microcavities. Nat. Commun. **6**, 8579 (2015)
207. L.C. Flatten, Z. He, D.M. Coles, A.A. Trichet, A.W. Powell, R.A. Taylor, J.H. Warner, J.M. Smith, Room-Temperature exciton-polaritons with two-dimensional WS_2. Sci. Rep. **6**, 33134 (2016)
208. S. Wang, S. Li, T. Chervy, A. Shalabney, S. Azzini, E. Orgiu, J.A. Hutchison, C. Genet, P. Samorì, T.W. Ebbesen, Coherent coupling of WS_2 monolayers with metallic photonic nanostructures at room temperature. Nano Lett. **16**(7), 4368–74 (2016)
209. K.F. Mak, J. Shan, Photonics and optoelectronics of 2D semiconductor transition metal dichalcogenides. Nat. Photonics **10**, 216–226 (2016)
210. F.H. Koppens, D.E. Chang, F.J. García De Abajo, Graphene plasmonics: a platform for strong light-matter interactions. Nano Lett. **11**, 3370–3377 (2011)
211. L.M. Schneider, J. Kuhnert, S. Schmitt, W. Heimbrodt, U. Huttner, L. Meckbach, T. Stroucken, S.W. Koch, S. Fu, X. Wang, K. Kang, E.-H. Yang, A. Rahimi-Iman, Spin-layer and spin-valley locking in CVD-grown AA'- and AB-stacked tungsten-disulfide bilayers. J. Phys. Chem. C **123**, 21813–21821 (2019). https://doi.org/10.1021/acs.jpcc.9b07213
212. M. Shah, L.M. Schneider, A. Rahimi-Iman, Observation of intralayer and interlayer excitons in monolayered WSe_2/WS_2 heterostructure. Semiconductors **53**(16), 2140–2146 (2019). https://doi.org/10.1134/S1063782619120273
213. T. Karzig, C.E. Bardyn, N.H. Lindner, G. Refael, Topological polaritons. Phys. Rev. X **5**(3), 031001 (2015)
214. J.R. Schaibley, H. Yu, G. Clark, P. Rivera, J.S. Ross, K.L. Seyler, W. Yao, X. Xu, Valleytronics in 2D materials. Nat. Rev. Mater. **1**, 16055 (2016)
215. K.F. Mak, D. Xiao, J. Shan, Light-valley interactions in 2D semiconductors. Nat. Photonics **12**, 451–460 (2018)
216. A.A. Maksimov, I.I. Tartakovskii, E.V. Filatov, S.V. Lobanov, N.A. Gippius, S.G. Tikhodeev, C. Schneider, M. Kamp, S. Maier, S. Höfling, V.D. Kulakovskii, Circularly polarized light emission from chiral spatially-structured planar semiconductor microcavities. Phys. Rev. B **89**, 045316 (2014)
217. S.V. Lobanov, T. Weiss, N.A. Gippius, S.G. Tikhodeev, V.D. Kulakovskii, K. Konishi, M. Kuwata-Gonokami, Polarization control of quantum dot emission by chiral photonic crystal slabs. Opt. Lett. **40**, 1528 (2015)

218. A.A. Demenev, V.D. Kulakovskii, C. Schneider, S. Brodbeck, M. Kamp, S. Höfling, S.V. Lobanov, T. Weiss, N.A. Gippius, S.G. Tikhodeev, Circularly polarized lasing in chiral modulated semiconductor microcavity with GaAs quantum wells. Appl. Phys. Lett. **109**, 171106 (2016)
219. Y.J. Chen, J.D. Cain, T.K. Stanev, V.P. Dravid, N.P. Stern, Valley-polarized exciton-polaritons in a monolayer semiconductor. Nat. Photonics **11**, 431–435 (2017)
220. S. Dufferwiel, T.P. Lyons, D.D. Solnyshkov, A.A. Trichet, F. Withers, S. Schwarz, G. Malpuech, J.M. Smith, K.S. Novoselov, M.S. Skolnick, D.N. Krizhanovskii, A.I. Tartakovskii, Valley-addressable polaritons in atomically thin semiconductors. Nat. Photonics **11**, 497–501 (2017)
221. N. Lundt, C. Schneider, P. Nagler, A. Nalitov, S. Klembt, M. Wurdack, S. Stoll, T.H. Harder, S. Betzold, V. Baumann, A.V. Kavokin, C. Schüller, T. Korn, S. Höfling, Valley polarized relaxation and upconversion luminescence from Tamm-plasmon trion-polaritons with a $MoSe_2$ monolayer. 2D Mater. **4**, 025096 (2017)
222. N. Lundt, S. Stoll, P. Nagler, A. Nalitov, S. Klembt, S. Betzold, J. Goddard, E. Frieling, A.V. Kavokin, C. Schüller, T. Korn, S. Höfling, C. Schneider, Observation of macroscopic valley-polarized monolayer exciton-polaritons at room temperature. Phys. Rev. B **96**, 241403 (2017)
223. P. Qing, J. Gong, X. Lin, N. Yao, W. Shen, A. Rahimi-Iman, W. Fang, L. Tong, A simple approach to fiber-based tunable microcavity with high coupling efficiency. Appl. Phys. Lett. **114**, 021106 (2019). https://doi.org/10.1063/1.5083011
224. F. Wall, O. Mey, L.M. Schneider, A. Rahimi-Iman, Continuously-tunable light-matter coupling in optical microcavities with 2D semiconductors. Sci. Rep. **10**, 8303 (2020). https://doi.org/10.1038/s41598-020-64909-1
225. M. Hertzog, M. Wang, J. Mony, K. Börjesson, Strong light-matter interactions: a new direction within chemistry. Chem. Soc. Rev. **48**, 937–961 (2019)
226. S.R.K. Rodriguez, Classical and quantum distinctions between weak and strong coupling. Eur. J. Phys. **37**, 25802 (2016)
227. R. El-Ganainy, K.G. Makris, M. Khajavikhan, Z.H. Musslimani, S. Rotter, D.N. Christodoulides, Non-Hermitian physics and PT symmetry. Nat. Phys. **14**(1), 11–19 (2018)
228. W.D. Heiss, The physics of exceptional points. J. Phys. A: Math. Theor. **45**, 444016 (2012)
229. B. Peng, Ş. K. özdemir, M. Liertzer, W. Chen, J. Kramer, H. Yılmaz, J. Wiersig, S. Rotter, L. Yang, Chiral modes and directional lasing at exceptional points. Proc. Natl. Acad. Sci. **113**, 6845–6850 (2016)
230. H. Xu, D. Mason, L. Jiang, J.G.E. Harris, Topological energy transfer in an optomechanical system with exceptional points. Nature **537**, 80–83 (2016)
231. W. Chen, Ş.K. özdemir, G. Zhao, J. Wiersig, L. Yang, Exceptional points enhance sensing in an optical microcavity. Nature **548**, 192–196 (2017)
232. L.C. Flatten, D.M. Coles, Z. He, D.G. Lidzey, R.A. Taylor, J.H. Warner, J.M. Smith, Electrically tunable organic-inorganic hybrid polaritons with monolayer WS_2. Nat. Commun. **8**, 14097 (2017). https://doi.org/10.1038/ncomms14097
233. J.D. Thompson, B.M. Zwickl, A.M. Jayich, F. Marquardt, S.M. Girvin, J.G.E. Harris, Strong dispersive coupling of a high-finesse cavity to a micromechanical membrane. Nature **452**, 72–75 (2008)
234. N.E. Flowers-Jacobs, S.W. Hoch, J.C. Sankey, A. Kashkanova, A.M. Jayich, C. Deutsch, J. Reichel, J.G.E. Harris, Fiber-cavity-based optomechanical device. Appl. Phys. Lett. **101**, 221109 (2012)
235. S. Brodbeck, J.-P. Jahn, A. Rahimi-Iman, J. Fischer, M. Amthor, S. Reitzenstein, M. Kamp, C. Schneider, S. Höfling, Room temperature polariton light emitting diode with integrated tunnel junction. Opt. Express **21**, 31098–31104 (2013)
236. J. Fischer, I.G. Savenko, M.D. Fraser, S. Holzinger, S. Brodbeck, M. Kamp, I.A. Shelykh, C. Schneider, S. Höfling, Spatial coherence properties of one dimensional exciton-polariton condensates. Phys. Rev. Lett. **113**, 203902 (2014)

237. K. Winkler, C. Schneider, J. Fischer, A. Rahimi-Iman, M. Amthor, A. Forchel, S. Reitzenstein, S. Höfling, M. Kamp, Electroluminescence from spatially confined exciton polaritons in a textured microcavity. Appl. Phys. Lett. **102**, 041101 (2013)
238. K. Winkler, J. Fischer, A. Schade, M. Amthor, R. Dall, J. Geßler, M. Emmerling, E.A. Ostrovskaya, M. Kamp, C. Schneider, S. Höfling, A polariton condensate in a photonic crystal potential landscape. New J. Phys. **17**, 023001 (2015)
239. S. Klembt, T.H. Harder, O.A. Egorov, K. Winkler, R. Ge, M.A. Bandres, M. Emmerling, L. Worschech, T.C.H. Liew, M. Segev, C. Schneider, S. Höfling, Exciton-polariton topological insulator. Nature **562**, 552–556 (2018)
240. A. Rahimi-Iman, *Technological Realization of Polariton Systems*. Springer Series in Optical Sciences, vol. 229 (Springer International Publishing, Cham, 2020), pp. 139–166

Chapter 5
Optical Measurement Techniques

Abstract The characterisation as well as improvement of various photonic devices, functional nanomaterials and quantum structures heavily relies on advanced optical tools. These tools typically incorporate suitable light sources and detection schemes to probe properties of the target system optically. Through special measurement techniques, one can for instance obtain optical images with high spatial resolution to map features of a sample, or read out spectral properties with sufficient temporal resolution in order to achieve insights into charge-carrier dynamics. While in many situations, the basic material response is of interest when irradiating light onto a sample, some techniques address a material's nonlinearities or composition, the emission's coherence or angle-dependencies, or a structure's phonon modes or magneto-optical properties. In this chapter, numerous optical measurement techniques are summarised which can be useful in the field of semiconductor photonics, particularly with an emphasis on the characterisation of two-dimensional semiconductors and quantum structures. Here, practical examples from the author's works may encourage further reading of expert literature on these subjects.

5.1 Advanced Optical Tools

Optical measurement techniques are widely developed and have been employed in many disciplines, ranging from material, structure and device characterisation to sample imaging and (non-destructive) testing. Optical tools have also heavily impacted on science in the field of astronomy and medicine. Among those techniques of great significance for scientists, microscopy and spectroscopy play a major role in providing insights into the structure of both macroscopic as well as microscopic objects. For such techniques, often lasers are indispensable tools.

Optical microscopy is widely used for imaging purposes, while derivatives of the microscope can fulfil very special roles, for instance fluorescence or Raman microscopy. Among optical microscopes, different operation modes can provide different information about thin films, microscopic objects, height or distribution pro-

A. Rahimi-Iman, *Semiconductor Photonics of Nanomaterials and Quantum Structures*, Springer Series in Solid-State Sciences 196,
https://doi.org/10.1007/978-3-030-69352-7_5

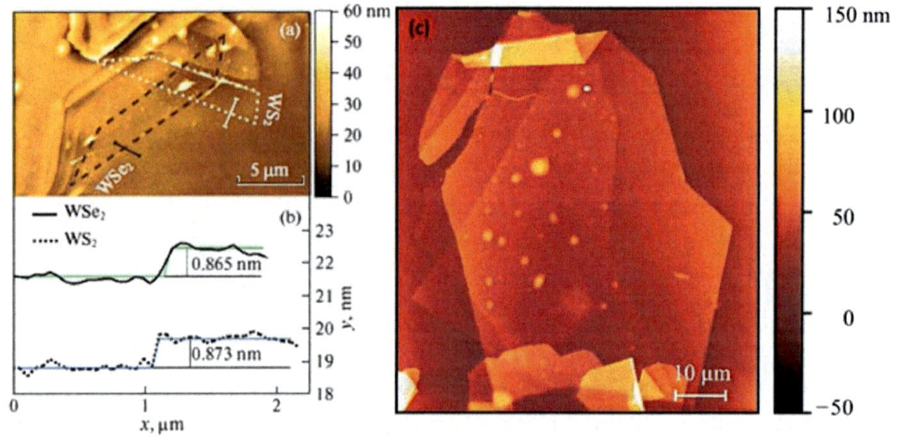

Fig. 5.1 Two AFM examples showing the topography of TMDC heterostructures. **a** False-colour AFM image of a WS_2/WSe_2 type-II monolayer–monolayer heterostructure on hBN buffer (on SiO_2). Heterostructures comprising monolayer TMDCs are commonly revealed by the peculiar height steps in AFM line profiles, as shown in **b**. **a,b** Reproduced with permission. [2] Copyright 2020 Springer Nature. **c** High-quality 2D-material stack of an hBN-encapsulated WSe_2 monolayer. The central area with a few bright spots represents the monolayer with mostly clean interfaces towards the thin hBN layers (bottom 10 nm, top 30 nm). The total thickness amounts to about 40 nm. These AFM images were recorded with the help of the group of G. Witte at the Physics Department in Marburg. **c** Reproduced under the terms of the CC-BY Creative Commons Attribution 4.0 International Licence (http://creativecommons.org/licenses/by/4.0/). [3] Copyright 2020 The Author(s), published by Springer Nature

files, such as differential-interference-contrast and digital holographic microscopy, light-sheet microscopy, or multi-photon microscopy.

Owing to the optical resolution's dependency on the light wavelength, naturally, limitations are imposed with regard to the minimum object sizes resolvable under a microscope. In fact, after centuries of use, new inventions allow further penetration into the sub-micro domain, such as super-resolution microscopy techniques (see for instance Nobel lecture by S. W. Hell in [1]). Further resolution improvement down to the scale of an atom is only provided so far by electron microscopes (e.g., STEM, scanning transmission electron microscopes) and scanning-tunneling microscopes (STM), whereas scanning-near-field optical microscopy (SNOM) and atomic-force microscopy (AFM) can reach tens of nanometers in lateral resolution. In addition, sub-nanometer height profiles become measurable with the use of the AFM's force tip (see for instance Fig. 5.1).

Spectroscopy with its many variations has become a fundamental method for the investigation of all kinds of materials. Particularly, the effects of quantisation can be easily measured for quantum structures when addressing the optical response of the system. This can be directly done by luminescence, reflectance, absorption or transmission studies. While steady-state measurements reveal the energetics of a system, time-resolved variations of the same spectroscopy methods can provide additional information about the dynamics. Thus, time-resolved photoluminescence (commonly μ-PL) and transient absorption have become typical tools for the investigation of nanomaterials and quantum structures.

Conveniently, combining microscopy with spectroscopy allows one to obtain spectral—if time-resolved methods used, then even temporal—information about various microscopic samples and structures with even sub-micrometer lateral resolutions. The major difference of the techniques lies in the underlying measurement principles, and the spatial, momentum, spectral, and temporal resolution. Combined with scanning-probe techniques, spectroscopic methods can even reach resolutions on the nanoscale.

In the following, a selection of useful methods for the characterisation of typical quantum structures and functional nanomaterials is summarised (also see [4] for methods explained in the context of polariton research). Where appropriate, examples related to the research presented in this work are provided. A more complete picture regarding these subjects can be acquired through the vast literature and a thorough discussion of these topics remains out of the scope here.

5.2 Microscopy and Spectroscopy

To perform imaging and spectroscopy with microscopic resolution, an optical microscope extended with a spectroscopy apparatus is required. Optical microscopy itself remains very attractive for sample monitoring and surface imaging. Remarkably, 2D materials as thin as one monolayer usually still provide enough optical contrast to be recognised visually due to their pronounced light–matter interactions. With an excitation source directed into the microscope and a monochromator attached on the diagnostics side, various spectroscopy modes can be used to characterise such materials (see for instance [5]) even during processes such as van-der-Waals-mediated layer stacking (e.g., shown in the Supporting Information section of [6]).

5.2.1 Monitoring and Imaging

Spectroscopy of samples first of all requires knowledge about the sample location, its spatial properties and its environment. In the first place, microscopy is used to investigate microscopic samples with features which can be still resolved optically. In addition, it can be also used to monitor and analyse samples comprising smaller structures not visible under the microscope, but accessible (while invisible) under the microscope for spectroscopic studies through the knowledge of the environment and position markers, for instance when dealing with buried quantum-dot structures in planar or patterned substrates.

Nevertheless, microscopy can also serve in conjunction with spectroscopy as a means of strong spatial selection of a seemingly homogeneous sample area. For

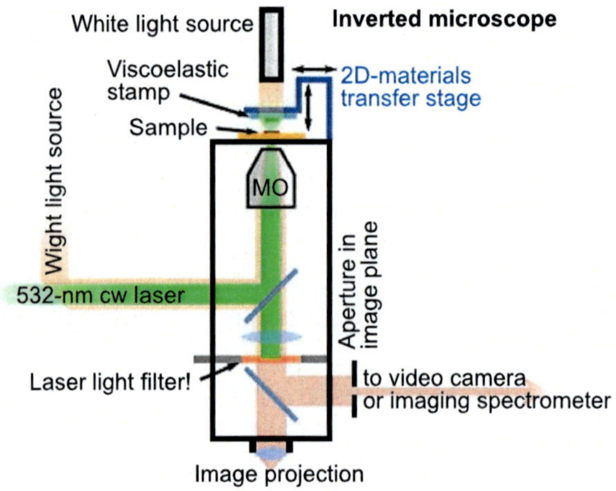

Fig. 5.2 Schematic diagram of a modified (commercial) invert microscope platform with attached stacking unit for controlled 2D heterostructuring, optical control, imaging and spectroscopy capabilities, employed by Mey et al. in their works on 2D–grating structures [6]. Drawn in a similar fashion as in [6] based on an author-team member's visual interpretation. In-coupling of white thermal or 532-nm laser light from a side port allows one to monitor 2D materials during the dry-stamping process and to excite monolayer luminescence optically for the clear identification of direct-gap semiconductor flakes during production, respectively. In addition, a top illumination scheme can be used. Note that laser light can permanently damage the eyes, if they are exposed to it. Thus, proper and accurate eye-safety measures are mandatory and it is crucial to fully block the laser light for eye safety on the imaging side of the device, in compliance with laser safety norms, and to filter out the laser light from digital recordings by an imaging camera or spectrometer

instance, such combination can be used in order to reduce/exploit dependencies on the sample location, or to enable spatially-selective excitation and detection. This in turn can be used for raster scanning of signal correlated to a distinct sample position.

In most cases, microscopy is needed to image structures, surfaces and to monitor the sample position. Particularly, its role in the materials-sciences and spectroscopy communities has been revived in the era of 2D-materials stacking, for which visualisation of 2D materials under the microscope is essential for the manual stacking process of layers.

As a practical microscope setup, one can typically use commercial platforms such as inverted or conventional microscopes (cf. [6]) and self-constructed ones (cf. [7]). For an extension of measurement options, input/output ports and a certain modularity in commercial microscopes are required for the manipulation of optical pathways and couplings, and the insertion of various types of filters.

A schematic overview on two different Optical setup is given in Figs. 5.2 and 5.3. The setup with commercial inverted microscope features a 2D-materials stamping section and a light-source in-coupling as well as signal detection section [6]. The example of a home-built microscope features flexible incoupling pathways and a two-times 4-f imaging configuration behind a microscope objective with multiple exit opportunities towards different diagnostics tools [5, 7].

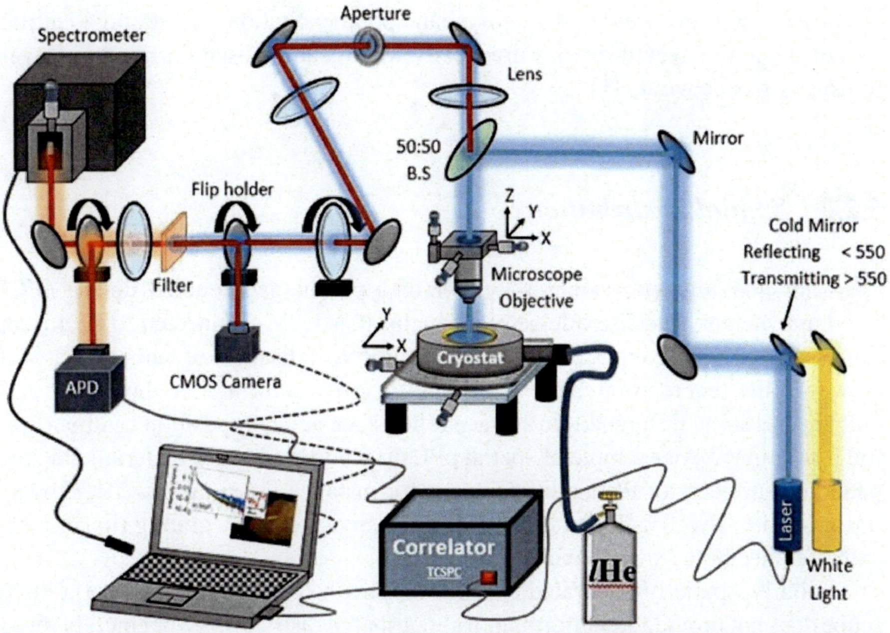

Fig. 5.3 Sketch of a micro-spectroscopy apparatus. This schematic representation was chosen by the author as a typical setup example owing to the fact that it displays his team member's own visual interpretation (with very minor editing remarks by the author himself) of the confocal ($2 \times 4f$ Fourier-space imaging capable) microscopy setup constructed by the author in the initial phase of 2D-materials research for his team. Reproduced under the terms of the CC-BY 3.0 Licence (http://creativecommons.org/licenses/by/3.0/) from Lippert et al., 2D Mater., "Influence of the substrate material on the optical properties of tungsten diselenide monolayers", https://doi.org/10.1088/2053-1583/aa5b21 [7] Copyright 2017 IOP Publishing Ltd

In such setups, white light from a tungsten lamp or laser light can be used for different characterisation aims. The respective light beam is focused onto the sample using a microscope objective (MO) with magnifications typically below 60x and numerical apertures (NA) as high as 0.7. Here, the working distance of the objective can be crucial for the use with optical cryostats. Their optical-window thickness and the sample-to-window distance pose strict limitations regarding the use of highly-magnifying high-NA optics. For optical control, an imaging CCD or CMOS camera is typically used. For micrometer-precise spatial selection of the sample's reflected or emitted signal, an iris aperture or pinhole is placed in the first available focal plane in the path of the collected beam, resembling the configuration of a conventional confocal microscope [7]. With a fully-closeable iris, the confocal spatial selection enables transmission of signal from a sample spot of roughly 1 μm.

One major advantage of imaging 2D flakes under the microscope reveals itself when placing 2D semiconductors under laser illumination (with protection from the laser light for the operator). Based on the direct-to-indirect band gap transition around a layer number of two, the monolayers of common TMDCs reveal themselves as particularly shiny areas under above-band photoexcitation, facilitating identification of the monolayered regions drastically which can be used during monolayers-transferring endeavours [6].

5.2.2 *Spatial Distribution*

One-shot imaging with cameras—either of type charged-coupled device (CCD) or complementary metal-oxide-semiconductor (CMOS)—connected to microscope setups can provide easy access to high-quality visualisation of sample materials. However, the recorded intensity profile only carries information about the spatial domain and some differences in surface reflectance and composition (without spectral resolution).[1] An example of spatial μPL distributions for 2D-materials stacks on patterned surfaces for PL modifications is for instance given in Fig. 5.4, shown in linear (right) as well as logarithmic (left) intensity scale for a grating (upper panel) and reference hole (lower panel).

Similarly, spatially-integrated spectroscopy with or without the use of a microscope does not provide any information about spatial distributions, but merely reveals spectral signatures. Thus, raster scanning the focal spot of the given light source of a spectroscopy method is used to obtain spatially-resolved spectral information. Hereby, the resolution is determined by the magnification of the optics and spatial filtering methods. Raster scans can be performed both in 1D and 2D,[2] resulting in hyperspectral imaging (see Fig. 5.5). In addition, spatially-resolved spectra can be obtained when the real-space image is projected via lenses onto the entrance slit of an imaging spectrometer's monochromator, which delivers spatial information in the direction of the slit wavelength-resolved on a monochromator-attached imaging CCD. By raster scanning in the perpendicular direction, one can also obtain 2D spatial distributions from such spatially-1D single-shot acquisition mode.

Typically, Raman and fluorescence (luminescence) spectroscopy with spatial resolution is widely used to investigate structured surfaces and microscopic materials, such as 2D-material flakes and their artificial stacks [2, 7–10]. For example, comparisons of Raman and PL spectra for monolayer and few-layer systems can already be well performed by hyperspectral line scans (see for instance the Supporting Information section of [8]), for which the sample is scanned in one direction using a high-precision translation stage, e.g., driven by piezo actuators. Spatial information can further be obtained by sheet-/surface-sensitive nonlinear frequency conversion techniques, the signal of which can be mapped [8] (see Fig. 5.6).

Another approach to gain information about spatial profiles and optical properties can be obtained by SNOM and scattering-type SNOM. While the former collects

[1]Amendment: Note that also imaging spectrometer cameras in μPL setups can often well resolve projected sample images, when the spectrometer with fully-open slit is operated in non-dispersive mode.

[2]Amendment: Or even 3D, focal plane by plane scanned in confocal microscopy.

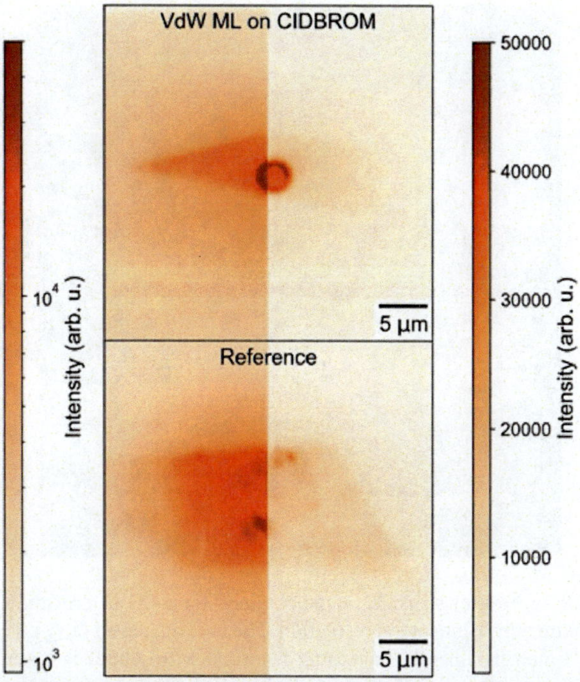

Fig. 5.4 Off-resonantly excited sample emission from a monolayer-WS_2–hBN stack on photonic nanostructures. The wavelength-integrated signal is spatially resolved on the spectrometer's CCD camera for direct luminescence mapping, as the false-colour intensity profiles in logarithmic (left) and linear (right) intensity scale show. Top: enhanced local PL from the stack on a circular in-plane DBR-based optical microcavity (CIDBROM). Bottom: Signal from a similar stack on a reference hole. Data replotted from author's work published in [6]

emission signal locally with the help of a probe tip, s-SNOM detects the scattered signal from the surface of a sample at the laser-light-irradiated tip's position. In fact, s-SNOM measures two effects simultaneously, local electric fields and differences of the material properties caused by corresponding variations of the scattering efficiency [11]. Thus, its measurement result can also have some spectroscopic material-sensing contribution.

A home-built s-SNOM setup can be based on a modified AFM near-field microscope [12]. It uses a metallized AFM tip oscillating at a fundamental frequency. Its apex is illuminated by a laser beam, which is focused down by a paraboloidal mirror from the side. Usually, the tip is kept at a constant time-averaged distance from the surface while the sample is raster scanned with a piezo-based XY-translation stage. Owing to the scanning-probe scheme using an AFM tip, s-SNOM measurements simultaneously yield the topography of the specimen by the setup's AFM functionality and the spectroscopic near-field information via the tip-scattered light. This method can be for instance used to characterise spatial field distributions in

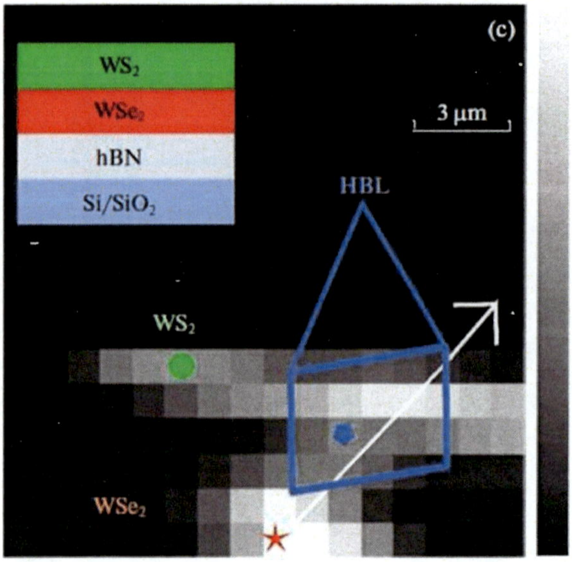

Fig. 5.5 Example of a raster scan PL image obtained for a 2D heterostructure (sketched in the inset) with approximately 1-μm lateral resolution. The here displayed false-colour map with spatial intensity profile (intensity linearly increasing from black to white) is based on the spectrally-integrated emission covering all output from WS_2 and WSe_2 monolayers and heterobilayers (HBL). The HBL region is boxed as a guide to the eyes. The arrow indicated one direction, from which spot-specific spectra were shown in a waterfall diagram in [2] (not shown here). In fact, since spectra are available for every spot, this method is known as hyperspectral imaging. Reproduced with permission. [2] Copyright 2020 Springer Nature

sub-micron optical structures combined with 2D materials [6] (see Fig. 5.7, and for cross-sectional line-outs see Fig. 5 of [6]).

In s-SNOM measurements, the probe tip acts as a nano-antenna. It scatters a part of the near-field wave into the far-field. The scattered light, which is nonlinearly modulated at the frequency of the tip oscillation, is then collected by the paraboloidal mirror and detected using a photodetector in the intensity detection scheme. For extraction of the near-field signature, the detected signal is demodulated at the third harmonic of the tip's oscillation frequency. The spatial resolution of such apparatus is of the order of 20–50 nm [13].

5.2.3 *Time-Integrated Detection*

Commonly, microscopy and spectroscopy are performed in a time-integrating detection scheme, which acquires signal over time following an excitation process. This detection method averages over time and provides time-integrated information about a system, with the spectral resolution and signal sensitivity depending on the dis-

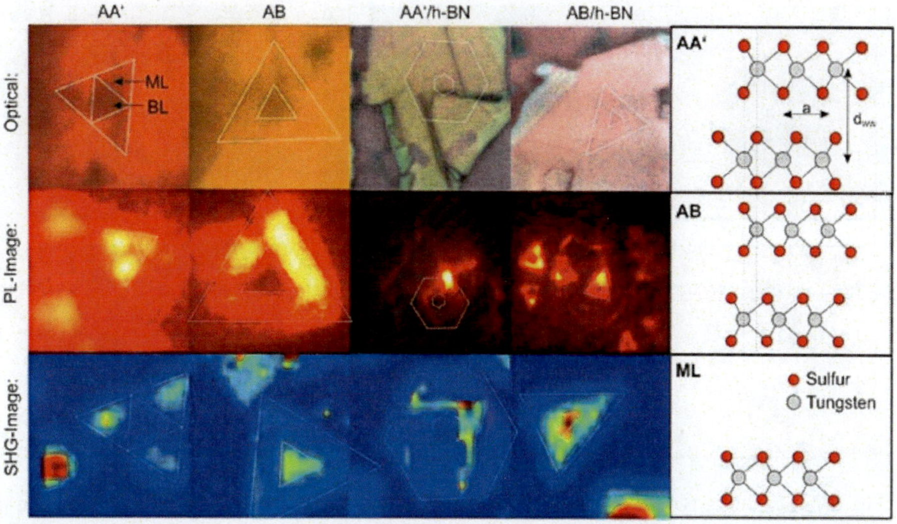

Fig. 5.6 The value of imaging techniques can be evidenced in another example, which combines optical micrographs with PL as well as SHG mapping. These methods can be very helpful in analysing vdW heterostructures, bilayers and monolayers. Here, the first row shows microscope images for AA' and AB stacked homobilayers (stacking configurations sketched in the insets) of WS_2 on bare substrates or hBN. The second row displays corresponding false-colour PL images recorded in a micro-PL (μPL) setup. Since bilayers become indirect semiconductors, their luminescence yield is drastically lower than the monolayer parts, which show spatial inhomogeneity due to quality fluctuations. The third row shows raster-scan SHG micrographs, where the symmetry situation determines the signal strength of the interface-sensitive SHG. The rotational symmetric AA' bilayer areas show no signal, whereas the symmetry-broken AB configuration shows increased signal for the bilayer region. The white dashed lines indicate the edges of the WS_2 layers in the bilayer structures and are guides to the eyes. Reproduced with permission. [8] Copyright 2019 American Chemical Society

persive element's refraction (prism) or diffraction (grating) strengths and the camera chip's responsivity, respectively.

Assuming a non-deformation regime under steady exposure with light and in ambient environment, sample imaging with time integration well displays a specimen's attributes such as size and orientation (cf. [6–8]). In video mode, even changes of sample shape and position with time can be observed, whereas temporal resolution depends on the frame-per-second rate recordable. Combined with pulsed illumination and stroboscopic detection, even fast processes such as membrane vibrations can be made visible by imaging techniques.

Similarly, spectroscopic information can be gathered in the steady-state regime under constant cw excitation, or time-averaged under repeated pulsed excitation. These two cases represent prominent forms of luminescence acquisition from a sample. When exciting the sample optically, one measures PL, when exciting the sample electrically, then electroluminescence (EL). Indeed, other luminescence types exist, such as chemiluminscence, cathodoluminescence and so forth. Combining pulsed

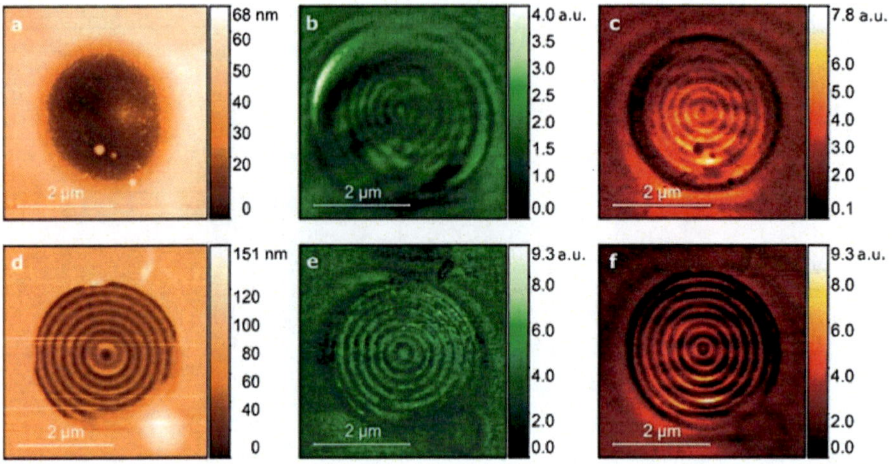

Fig. 5.7 Scattering-type SNOM (s-SNOM) measurements on a TMDC-covered and an open in-plane microcavity structure in GaP substrate. Practically, near-field measurements with s-SNOM setups typically yield topography information due to the AFM functionality of the technique. **a** Topography of the cavity covered by a monolayer-on-hBN stack. Corresponding false-colour near-field intensity maps of the covered cavity at 532 nm **b** and at 850 nm **c** are displayed in greenish and reddish colours, respectively. **d–f** Analogue recordings for an open uncovered cavity structure for a direct comparison of the s-SNOM signals to the covered case in **a–c**. Reproduced with permission. [6] Copyright 2019 American Chemical Society

excitation with a stroboscopic detection scheme can further provide access to the dynamics of a system, as summarised in the next section.

One major advantage of time-integrated acquisition is the possibility of recording high-enough signal through the expansion of the measurement time, enabling detection of weak signals and improvement of signal-to-noise ratios. However, this requires that the measurement conditions remain as stable as possible over the integration time, since thermal and position instabilities can spoil the spectrum for different reasons. Temperature-dependent wavelength shifts for resonances can lead to broadened signatures in recorded spectra, while position fluctuations can easily translate to inhomogeneous broadening and signal-strength fluctuations or complete loss of signal from a particular spot.

Time-integrated measurements also bear the advantage that the system is analysed in its steady-state configuration. Although no information about the dynamics can be obtained, a homogeneously-broadened line can still provide a good estimate about the emitter's lifetime or the linewidth of a laser can indicate the coherence length of a laser.

5.2.4 Time-Resolved Measurements

The dynamics in electronic systems can be studied by time-resolved spectroscopy. For this purpose, different techniques have been established over the last decades, whereas the main difference originates from the time scales accessible and the equipment used to resolve fast to ultrafast processes. With time-resolved (TR) measurements (cf. Fig. 5.8), one can determine carrier lifetimes, as well as characterise the recombination mechanisms in the investigated structure.

One common technique to slice the time in order to have temporal resolution in signal acquisition is given by the employment of gated detectors such as intensified CCDs. A trigger pulse coinciding with the excitation pulse is used to define an electronic delay for the detection time window. Thereby, an electronic shutter determines from which time frame the detector obtains signal. Then, the time trace is established by scanning through the delay time electronically. Typically, this technique is useful for processes which are longer than nanoseconds, such as excitonic decay in colloidal quantum dots [15]. The repetition rate of the laser must be correspondingly low in order to allow decay processes lasting up to the micro- or milliseconds to be investigated.

A much better temporal resolution down to below 1 ps for optimized devices is provided by the streak-camera technique, which works the same way as a stroboscope. Spectrally-resolved incident photons impinge on a photocathode from which they eject electrons. These electrons accelerated in a field experience a controlled deflection periodic in time based on a sweep circuit. A trigger signal fed into the circuit sets the rising flank of a sinusoidal AC bias with its most linear part around the zero delay time with regard to the optical excitation pulse. Based on their time of ejection, the deflected photoelectrons end up on different parts of a phosphorous screen behind a multi-channel plate, resulting in temporal resolution of spectroscopic data. The signal on the phosphorous screen is then imaged by a CCD-type camera.

From the acquired 2D image (an example is given in Fig. 5.9), transients (see right panel) and spectra (see upper panel) can be retrieved by integration over the desired energy or time range, respectively. In case of a mono-exponential decay,[3] a typical fit of the form

$$I(t) = I_0 \exp\left(-\frac{t - t_0}{\tau}\right) \tag{5.1}$$

[3]Note that, while an exponential decay is typically the case for excitons (after being initially prepared by an ultrashort excitation to a starting population N_0), for which a simple rate equation applies, the transient of uncorrelated electron–hole pairs in a plasma follows a different trend. In some cases, where density-dependent Auger-like processes such as exciton–exciton annihilation occur, the mono-exponential fit also cannot hold. Some systems may feature short and long decay time constants, which might require a biexponential fitting. In the case of a reservoir feeding a lower-lying state, such as for cavity–polaritons, more elaborate rate equations which take into account various scattering and decay channels may reproduce the dynamics of a system appropriately (see e.g., [16]). Similarly, the transfer between free and bound excitons could require a specific rate equation (described e.g., for GaSe and GaTe in [17]). For further insights on charge-carrier dynamics in semiconductors, the interested reader is referred to semiconductor theory textbooks such as [18].

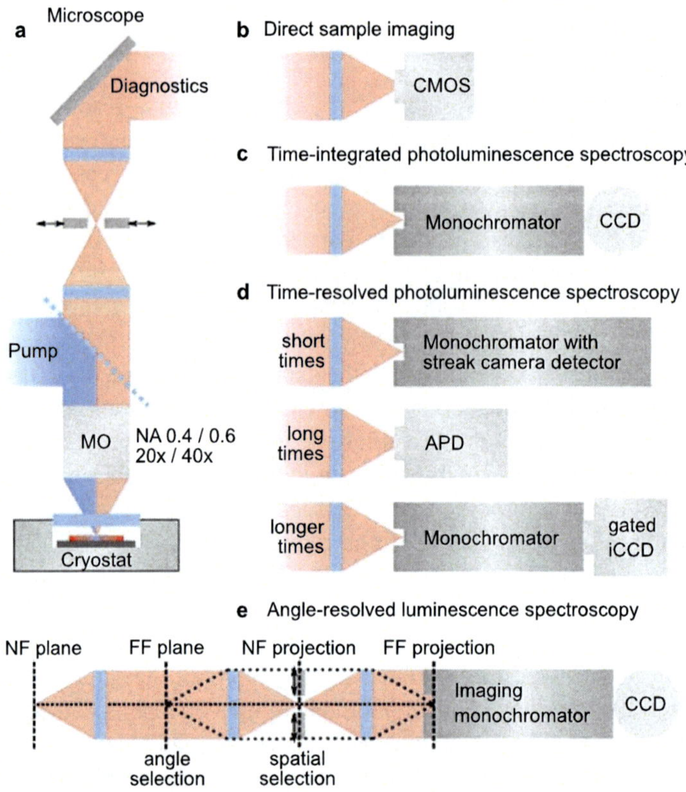

Fig. 5.8 Time-integrated and time-resolved spectroscopy in an optical microscopy setup typically used for micro-PL measurements. **a** The pump light from a laser system is coupled into the setup and focused onto the sample inside an optical cryostat through a microscope objective (e.g., with 20x magnification and NA 0.4, resulting in an excitation spot-size of about 4 μm). The collected signal through the same objective is then directed towards diagnostics. On the optical axis, spatial selection of the detection area can be achieved by an iris aperture in the sample projection plane. **b** By employing a (flexibly removable) lens and flip mirror in the detection path, a CMOS camera can be used to image the sample (also see Fig. 5.3). **c** If the light is focused onto the spectrometer slit, time-integrated PL spectra can be acquired with the monochromator's CCD (e.g., a nitrogen-cooled Si CCD or Peltier-cooled iCCD). If an imaging monochromator is used, 1D-spatially-resolved spectroscopy is enabled. **d** For dynamics studies such as lifetime measurements on different time scales, either a gated detector (e.g., iCCD), an APD or a streak camera can be used to obtain time-resolved PL for rather long (ns–μs), intermediate (ps–ns) or short (ps) times, respectively, based on the devices' temporal resolution (common ranges indicated in parentheses). Note that the APD can be similarly placed behind a monochromator, if not placed behind other appropriate spectral filters. **e** In addition, angle-resolved spectroscopy is shown, which gives direct access to the Fourier-space (phase-space/far-field) plane and enables single-shot acquisition of far-field spectra with an imaging spectrometer. Due to a microcryostat setup, measurements can be performed both at room temperature and at cryogenic temperatures down to a few Kelvins or shortly below 100 K with liquid helium or nitrogen, respectively. Drawn in a similar fashion as the author's diagram used in [5] and amended according to representations of the Fourier-space spectroscopy technique after [4, 14]

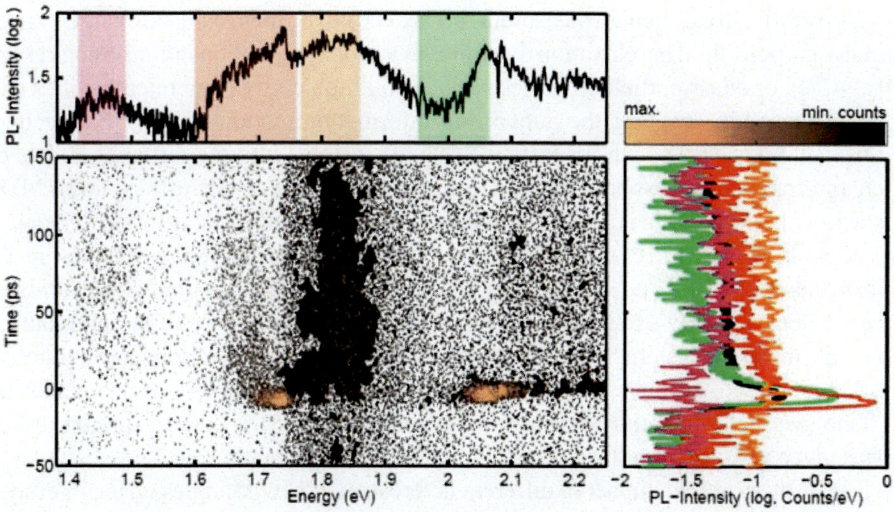

Fig. 5.9 PL of a WS_2/WSe_2 heterobilayer on hBN recorded with a streak camera with picoseconds time resolution. Here, an off-resonant excitation at 2.7 eV at an irradiance of $54\,\mu J\,cm^{-2}$ was used. This false-colour logarithmically-scaled intensity contour diagram shows TRPL data from four energy ranges stitched together (from dark to bright: minimum to maximum counts, white: background). For the peaks representing intralayer excitons of the two TMDC materials, a fast decay is seen on the time scale of a few ps (WSe_2 about twice as slow as WS_2). The slightly shorter instrument response time for the used setting amounts to 2 ps. After the initial picoseconds, the WSe_2 peak exhibits a red-shift and the decay in this spectral region continues with a longer time constant. Top panel: Temporally-integrated spectrum obtained from the time–energy spectrum. Side panel: PL transients were extracted by spectral integration (indicated as colour-coded shaded area in the top panel), that is 1.42–1.49 eV (interlayer excitons, pink line), 1.60–1.76 eV (WSe_2 and lower-energetic moiré states, thick red line), 1.77–1.86 eV (higher-energetic moiré states, orange line), and 1.95–2.06 eV (WS_2, thick green line). Apparently, the low-energy interlayer population builds up after about 50 ps, being electronically 'fed' by higher energy states, and decays slowly on the 100-ps time scale, whereas the spectrally centered so-called bilayer moiré states decay even much slower. The spectrally fully-integrated signal is shown as black transient (thick line) for comparison. Transient counts are in semilogarithmic representation normalised to the energy integration range. Data obtained by M. Shah and L.M. Schneider in the author's team, for the bilayer system studied time-integrated in [2]

can deliver the time constant τ for the emission process starting at t_0, which commonly coincides for ultrafast excitation with the arrival of the pump pulse. Typically, picosecond- and sub-picosecond-pulsed lasers with synchronized repetition rate of typically 80 MHz (corresponding to a pulse separation of 12.5 ns) are employed for such time-resolved detection scheme. For further details, the interested reader is referred to the guide on the webpage of a renowned company, which produces scientific streak camera systems [19]. Note that the evaluation of the data may require some background, noise or artifact correction, as well as the consideration of wavelength-dependent responsivity of the detection apparatus. Moreover, for decay times in the ns range, the backsweep artifact should be taken into account, which typically reveals itself in the recordings as a signal preceeding the excitation pulse.

Typically, streak camera measurements are employed to investigate charge-carrier, quasi-particle, fluid or relaxation dynamics in various types of quantum structures (cf. [20–24]), or quasi-particle lifetimes such as polariton decay from microcavities (see for instance [25, 26], and the Supporting Information section of [27]). Picosecond-resolved transient PL can for instance provide insights into the different effects on charge-transfer processes in organic semiconductor heterosystems [28] and TMDC stacks [5], or the influence of growth parameters on quantum structures [29, 30].

Streak camera traces for monolayer WSe_2 and heterostructures of it were for instance studied in [5, 31] in order to measure the very fast temporal luminescence decays of 2D excitons and density-dependent exciton–exciton annihilation[4] in this material system for different environmental configurations (substrate- or hBN-supported, hBN-capped or hBN-encapsulated monolayers, and monolayer–monolayer heterostructures on hBN). Thereby, an effective time resolution of 3 ps was given which cannot be easily achieved with other means.

Even the possible impact of different heterostructure twist angles on the interlayer coupling situation can be deduced from spectrally-resolved lifetime measurements (cf. [5, 32]). Similar streak camera measurements have been performed for such samples regarding high-charge-carrier-density effects and dynamics on different substrates, and the influence of hBN as well as the bilayer stacking order on the spectral and dynamics properties of 2D WSe_2 emission [33]. Recently, works by the author and co-workers have indicated by streak camera TRPL data that interlayer excitons and possible moiré features in stacked monolayer–monolayer TMDC heterolayers (cf. [2]) with nearly aligned/anti-aligned twist angle (estimation was only possible modulo 60°) feature very long decay times on the ns scale (see Fig. 5.9). This is in contrast to the ultrafast picoseconds decay of the intralayer counterparts in the same structure.

Alternatively, time-correlated single-photon counting with ultrafast APDs[5] can be used to build histograms of detection events with a time-tagging mode based on the measured delay between start and stop signal. Conveniently, a pulse from the excitation laser provides the start signal via a photodiode (or an electronic trigger signal is used) and the emitted sample photon detected by the APD delivers the stop time.

Transient PL recorded with the time-tagging mode is assembled through a plot of photon counts as a function of delay time. The temporal resolution is limited by the instrument response function of the photodetector, typically ranging between tens of picoseconds to hundreds of picoseconds, whereas the histogram resolution depends

[4]The exciton–exciton annihilation rate was modelled in [5] using a rate equation model $\frac{dN}{dt} = G(t) - \frac{N}{\tau} - \xi N^2$, with source term $G(t)$, linear decay term with time constant (low-density PL lifetime) τ and quadratic term with annihilation rate ξ. Two possibilities for modelling were pursued in [5] (also see references therein for details about the models): (1) exciton diffusion, (2) Förster transfer (long-range dipole–dipole interactions). The former one is intuitive, with diffusion constant, mean-free path and mean exciton size (Bohr radius) relevant for possible annihilation processes due to short distance interactions (collision of excitons), the latter one with Förster radius is considered a long-range annihilation mechanism, reminiscent of the situation for exciton–polaritons.

[5]Acronym for ***a**valanche **p**hoto**d**etectors*.

on the time binning of the time-correlator (TCSPC)[6] unit, which can be as good as a few picoseconds.

Such single-photon counting method can be applied for instance to detect very weak signals from single quantum emitters, 2D materials and indirect excitons (see [7] for use with 2D materials). Comparably, a streak camera can be operated in a single-photon counting mode (see for instance [34]). In addition, the streak camera's CCD-type camera can also enable the measurement of (1D) spatially-resolved PL transients if the projection of the sample surface is directly imaged on the entrance slit of the streak camera, i.e. without dispersive element before streak system.

For time-resolved PL, the resolution is limited by the excitation pulse duration τ_p and the temporal resolution of the detection τ_d, which results in the overall time-resolution of $\Delta t \approx \sqrt{\tau_p^2 + \tau_d^2}$. The "faster" the excitation source (in terms of pulse duration), the sharper the events can be recorded around $t = 0$. However, the shorter the excitation pulses in time, the broader their spectral bandwidth, whereby spectral selectivity in resonant excitation schemes gets reduced due to the Fourier relationship $\gamma \geq 1/\tau_p$. Similarly, temporal resolution comes at the cost of spectral resolution through the uncertainty principle. A too strong grating dispersion in the monochromator before the streak camera affects the time trace quality. Remarkably, state-of-the art streak cameras provide sub-200-fs resolution.

A very good temporal resolution for PL measurements can be obtained by a special technique called PL up-conversion. Employing a nonlinear crystal and a reference beam from the excitation laser, the sample signal is overlapped with the reference for sum-frequency generation (SFG). This corresponds to cross correlation between the probe laser pulse and the fluorescence signal. Thus, SFG signal is only generated within the time period when the probe laser pulse is present in the crystal. Thereby, the laser pulse exposure acts as an optical gating process [35]. By adjusting a temporal delay between emitted pulse and reference pulse, the transient PL can be sampled in time. While this technique provides temporal resolution on the order of the pulse duration, recordings of transients from pulsed sample signal become particularly challenging at longer delays when the emission signal becomes very weak owing to the strong intensity dependence of SFG.

In fact, optical gating is the standard technique for the characterisation of ultrashort pulses, using intensity autocorrelation in a nonlinear crystal for second-harmonic generation (SHG). In this scheme, the pulse probes itself, and the temporal resolution is given by the time duration of the pulse. Derivatives of the intensity autocorrelator that provide a more complete characterisation of light pulses use spectral analysis of the SHG signal for the extraction of phase information. FROG and GRENOUILLE are two commonly used tools for this purpose with similar background but different setup realisation. For a broader overview on pulse characterisation and commercial tools, the interested reader is referred to the tutorials webpage of a prominent producer of such pulse-analysis tools [36].

---

[6]Time-correlated single-photon counting.

The linear-absorption counterpart of the PL-up-conversion technique which does not rely on nonlinear conversion on the other side is referred to as optical pump–probe spectroscopy, also known as transient absorption spectroscopy (the probe signal can be analysed in reflection or transmission[7]). The transient reflection technique was for instance used to examine the gain dynamics of type-I and type-II VECSEL chips [37]. Derivatives of it are transient reflection contrast, popular in the domain of 2D materials [9, 38], optical-pump–THz-probe [39, 40] and optical-pump–IR-probe [41], to name but a few examples.

All pump–probe experiments have in common that a pulsed optical excitation with ultrashort duration takes place, while a probe pulse with variable delay time with respect to the pump pulse samples the situation in the studied material for numerous temporal settings in order to acquire a transient signal (the time trace).

If the probe in such experiment is monochromatic and resonant, a simple photodetection scheme can be used to read out the signal modulation as a function of delay time. If the probe is broadband such as a white-light supercontinuum, the whole spectrum can be retrieved for every time step.

If the probe is done via a broadband THz pulse, intraexcitonic transitions can be probed in common III/V semiconductors after optical excitation. This can for instance be used to investigate the exciton formation times [39] or the fraction between dark excitons in reservoir states and condensed ground-state polaritons in optical microcavities [40]. Recently, also indirect excitons in type-II heterostructures were probed by such method [42]. Note that for large-binding energy materials such as TMDCs, the probe pulse typically is in the IR range (see for instance [41]).

Opposite to transient absorption or reflection spectroscopy, which can be used for instance to see ultrafast Rabi oscillations arising from the strong-coupling regime in quantum-well microcavities [43], luminescence digital holography can be employed to resolve Rabi oscillations in the emission behaviour on ultrashort time scales [44]. In this scheme, a laser beam is split into two paths, one excitation path and one reference path, which are superposed to map the phase relation of the sample signal with the help of the reference signal on a monochromator imaging CCD.[8] Involving additional pulses or polarisation schemes, the polariton can be manipulated optically and various states on the Bloch sphere of the cavity–polaritons set [44, 45]. Extending these manipulation schemes to THz radiation could show the direct impact of strong transient electric fields on the coherent states of the system, offering a new level of control for light–matter coupled devices.

The coherent detection scheme provided by the digital holography method and previously used in [44, 45] for cavity–polaritons in combination with polarisation sensitivity could give interesting insights into coherent oscillations of hybridised

[7]For absorbance A, reflectance R and transmittance T are needed, as $A = 1 - R - T$.

[8]Amended information: In off-axis digital holography, the recorded real-space interferogram for every delay step undergoes in the evaluation part a 2D Fourier-transform to the phase space, where the diagonal phase components apart from zero momentum are isolated (truncating the information in the phase map) and re-transformed into real space. Thereby, sample signal is deconvoluted from the overall signal with strong reference laser contribution. For details, see [44] and its Supporting Information.

excitons in the valley landscape of monolayer TMDCs (cf. [46] and references therein), promising a better understanding of valley depolarisation in TMDCs through valley dynamics studies as a function of the quasi-particles' centre-of-mass (in-plane) momentum.

Using another time-resolved investigation method without spectral resolution, which utilized four-wave mixing (FWM), Rabi-flopping signatures of quantum-well excitons under pulsed THz illumination in the time domain were demonstrated to become irreversible for strong "ionising" THz fields. Whereas, weak THz radiation quasi-resonantly drives intraexcitonic transitions between the $1s$ and higher-order states [47]. Here, typically large laser spots on the sample from an amplifier laser are used owing to the THz exposed areas being well above 300 µm in diameter. To employ similar techniques in combination with TMDCs successfully, the ability to focus strongly onto micro-sized mono-/few-layer flakes or the availability of large-area mono-/few-layers is needed, e.g., for two-pulse or three-pulse (laser-induced dynamic gratings) FWM experiments. In such experiments, the laser pulses induce polarisations in the crystal which interfere and form lattices (gratings). The refracted signal for different time delays between the pulses is acquired for a specific diffraction order with a photodiode, from which a FWM transient can be obtained that is analysed with respect to dephasing processes. In addition, to address intraexcitonic transitions, the high binding energies of TMDC excitons shifts the manipulation-pulse's wavelength from the THz to the IR (cf. optical-pump–IR-probe in [41]). First FWM attempts by the author's team on CVD samples showed the challenges in recording monolayer signal. The interested reader is referred to a shortly later published comprehensive article on FWM microscopy with monolayer $MoSe_2$ in the literature [48], nicely showing how FWM can give access both to dynamics (radiative lifetime T_1) and the coherence (dephasing times T_2, homogeneous linewidth $\gamma = 2\hbar/T_2$) measurements, with radiatively-limited dephasing $T_2 = 2T_1$ demonstrated at a temperature of 6–10 K ($\gamma_{<10\,\mathrm{K}} \approx 1\,\mathrm{meV}$).

A different example of time-resolved experiments is given by the measurement of a laser's output characteristics, which can be analysed using fast photodetectors. Intensity as a function of time is recorded and the time trace Fourier-transformed to obtain a noise spectrum. The frequency bandwidth of the detector limits the maximum-measurable frequency components present in relative intensity noise (RIN). Such a RIN characterisation of dual-wavelength laser emission from a VECSEL revealed power-dependend antiphase-noise dynamics between the two wavelengths simultaneously oscillating in the resonator [49].

Another interesting time-resolving experiment reveals the photon statistics of an emitter and is referred to as Hanbury–Brown-and-Twiss (HBT) experiment [51] (see Fig. 5.10). With this method, photon bunching, antibunching and random emission (coherent state) can be distinguished via temporal photon statistics, by obtaining the second-order temporal autocorrelation function $g^{(2)}(\tau = 0) = \langle : \hat{n}^2 : \rangle / \langle \hat{n} \rangle^2$ at zero time delay ($\tau = 0$) where the correlation between photons is maximum. For increased time intervals between photons, their correlation decays. $\hat{n}$ is the photon counting operator. Colons indicate normal ordering of the photon field operators.

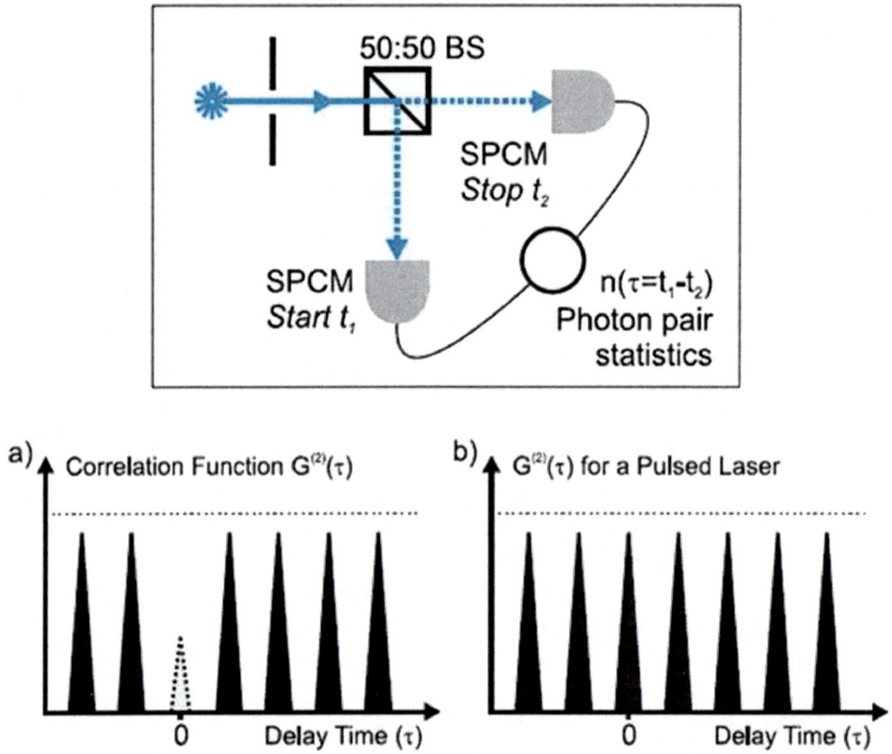

Fig. 5.10 Top: Sketch of the principle of photon statistics measurements in a Hanbury–Brown-and-Twiss experiment. **a** Schematic diagram of the photon second-order temporal autocorrelation function $G^{(2)}(\tau)$ for a pulsed single-photon Fock state (with no peak occuring at $\tau = 0$) as well as a sub-Poissonian state (exhibiting an additional peak at $\tau = 0$, here dash-lined grey-shaded), i.e. with $G^{(2)}(0) = 0$ and $G^{(2)}(0) < 1$, respectively. Dashed horizontal lines are guides to the eyes indicating a constant long-delay count level. **b** Schematic $G^{(2)}(\tau)$ for a pulsed coherent state with Poissonian distribution. Reproduced with permission. [50] Copyright 2009 Arash Rahimi-Iman

This function, which represents the correlation of intensities, is used as a measure of the second-order temporal coherence of emitted light. It is often represented by the following equation:

$$g^{(2)}(\tau) = \frac{\langle I(t)I(t+\tau)\rangle}{\langle I(t)\rangle^2} = \frac{\langle I_R(t)I_T(t+\tau)\rangle}{\langle I_R(t)\rangle\,\langle I_T(t+\tau)\rangle}, \tag{5.2}$$

with $I_R(t)$ and $I_T(t+\tau)$ the measured incidence intensities of the reflected and transmitted detection beam from a beam splitter (in the HBT setup), respectively, at two times separated by a delay time τ. Quantum-mechanically represented, it becomes $g^{(2)}(\tau) = \langle : \hat{n}_T\hat{n}_R : \rangle / (\langle \hat{n}_T\rangle \langle \hat{n}_R\rangle) = \left\langle : \hat{a}_T^\dagger \hat{a}_R^\dagger \hat{a}_T \hat{a}_R : \right\rangle / (\left\langle \hat{a}_T^\dagger \hat{a}_T\right\rangle \left\langle \hat{a}_R^\dagger \hat{a}_R\right\rangle)$, with photon creation and annihilation operators $\hat{a}^\dagger$ and $\hat{a}$, respectively.

In contrast to $g^{(2)}(\tau)$, which addresses the correlation between intensities, the first-order correlation function $g^{(1)}(\tau)$, which can be for instance obtained through Michelson interferometry, gives the correlation of light fields.

For a theoretical view on this subject, the interested reader is referred to textbooks such as [52]. In practice, the experimental value $G^{(2)}(0)$ obtained from pulsed light represents a time-averaged $g^{(2)}(0)$ for every pulse [53], as used for instance in polariton condensate research to investigate the degree of temporal coherence and the intensity noise for different regimes of microcavity emission.

The HBT experiment has become an important tool for the characterisation of single-photon sources such as quantum dots in optical micropillars [54], or for laser thresholds characterisation, where a transition from bunching to a coherent state is expected, as demonstrated on semiconductor microcavity lasers [55] (see also electrically-driven low-threshold nanolasers [56]). Later, it was introduced in efforts towards the characterisation of a double-threshold behaviour from polariton systems [26], whereas it had been generally used for density-dependent photon statistics measurements for polariton condensation effects [53, 57–62]. In combination with disturbing transient electric fields, pulsed HBT measurements may give additional insights into the external manipulation of polariton systems below and above a condensation threshold and the impact on intensity noise.

5.3 Basic Material Response

Typically, novel materials and quantum structures are studied using their absorption, reflection, transmission, emission, or vibrational behaviour. In the following, a brief introduction to basic spectroscopy modes is given together with examples, and some of their strengths and drawbacks are summarised.

5.3.1 Absorbance

While laser excitation enables luminescence studies, white-light illumination in reflection geometry allows one to probe the absorption-behaviour of samples by recording reflected signal and comparing it to a reference of the light source. Absorption in the low-intensity regime belongs to the field of linear optics, where the material polarisation is a linear response to the incident light field. In contrast to nonlinear optics, the) (macroscopic polarisation $\mathbf{P} = \epsilon_0 \chi(\omega, \mathbf{k})\mathbf{E}$ (with susceptibility $\chi = \epsilon_r - 1$, a complex function with real and imaginary part) can be approximated by the term linear in electric field strength, neglecting contributions of higher-order terms.[9] Features with a high oscillator strength typically exhibit a stronger optical absorption due to a more pronounced light–matter interaction.

By measuring the reflectance R and the transmission T of a sample, the absorbance $A = 1 - T - R$ can be obtained. Spectrum acquisition can be either done with the use of tunable monochromatic light at suitable controlled irradiances for serial frequency sampling or with the use of broadband white light. However, not all investigated samples have the desired transparency in the relevant spectral region, typically due to their substrate material. This makes for instance transmission experiments in the visible range on 2D material flakes isolated on SiO_2/Si substrates impossible as long as opaque silicon is attached on the reverse side of the sample. Similarly, gain media deposited onto highly-reflective dielectric/Bragg mirrors cannot be probed in transmission geometry.

The equipment used for measurements of this kind at low temperature and detection of signal is not different from that of steady-state PL spectroscopy (see for instance [7]). Similarly, polarisation or time-dependent experiments can be performed. Yet, transmission experiments in a cryostat need optical access from opposite sides. Furthermore, the employed broadband spectrum should not be blocked in the desired spectral range by the used optical windows. Naturally, the detector's responsivity over the whole spectral range and its dynamic range should be sufficiently high.

Typically, 2D materials deposited onto substrates are characterised in the reflection-contrast spectroscopy mode to determine excitonic resonances and to probe their Rydberg-like energy series (see [64, 65], and for instance the monolayer and bilayer study of [8] or the PL/absorption enhancement study of [6]). This requires a comparison of the spectrum acquired on the flake to a position off the flake on the same substrate in order to remove (i.e. cancel out) contributions from the supporting substrate.

White-light spectroscopy at reasonably low incident powers generates such a low amount of excited charge carriers that the spectra resemble a zero-density-regime probe of a system's resonance. This can be for instance useful to compare optical dispersion measurements obtained in PL and reflection contrast with respect to density-dependent effects in the linear excitonic regime [66], or to characterise cavity–polariton-mode anticrossing behaviour in basically-unexcited optical microcavities (e.g., shown in [27]). Note that high-enough spectral density at shorter wavelength can also lead to charge-carrier excitation during white-light exposure, particularly in combination with a strong focus achieved by microscope objectives.

At higher intensities (i.e. irradiances), saturable absorption leads to nonlinear spectroscopy, where the active medium becomes transparent. This is needed to characterise for instance saturable-absorber mirrors used for mode-locked lasers by the respective time-resolved method. On the other hand, an exciting light–matter cou-

[9]The resonance(s) described by such (complex) dielectric function $\epsilon_r(\omega, \mathbf{k})$ in the ideal case feature a Lorentzian lineshape, which is representative for homogeneously-broadened resonances (linewidth linked to lifetime by $\gamma \approx 1/\tau_0$). In contrast, inhomogeneously-broadened spectral lines (e.g., statistically-broadened by size fluctuations) are well described by a Gaussian lineshape. A convolution of both is given by the Voigt profile.

pling phenomenon can be observed when a strong mono-mode resonant CW light field is incident on an ideal two-level system: periodic absorption (excitation) and stimulated emission (deexcitation) occur within the frame of so-called Rabi oscillations,[10] known for instance from atom–cavity experiments (cf. [52, 69, 70]).

In some structures, an additional pump can be used to probe low-signal gain with white-light spectroscopy such as in the time-resolved optical pump–probe experiment in reflection geometry on VECSEL chips [37]. A steady-state spectroscopy of gain mirrors is similarly possible with pumped chips probed by white light.

Novel semiconductor materials such as 2D materials have been also investigated in the literature with nonlinear spectroscopy using two-photon absorption, with which optically dark states for one-photon transition processes ("dark" in terms of inactive, due to optical selection rules) can be addressed [71].

5.3.2 Photoluminescence

Luminescence spectroscopy can be clearly separated into two different excitation modes that are (i) resonant or quasi-resonant excitation, (ii) non-resonant or above-band excitation. A mix of both is obtained in a third variation that is (iii) energy-tunable excitation (PL excitation spectroscopy, delivering relative excitation efficiency as a function of excitation photon energy). The former two cases are distinctions of static-excitation-energy spectroscopy (widely used to obtain emission spectra), whereas the latter gives a probing mechanism using the excitation energy and the response of the system. Exciton Rydberg states have been for instance observed for monolayer TMDCs with PL excitation spectroscopy in [72]. Thereby, it allows one to scan over transitions usually hidden in time-averaged above-band excitation spectroscopy, the spectra of which are dominated by the most shiny (most probable) transitions (due to much higher oscillator strengths). This method and reflection contrast in combination with the two-photon absorption principle can provide complementary information about higher-order excitonic states (cf. [64, 65, 71, 72]).

Still, fixed-wavelength excitation bears an important role in PL studies. An example spectrum for a TMDC heterobilayer is shown in Fig. 5.11 together with a reflection contrast measurement of that heterostructure. Naturally, non-resonant excitation, be it through single-photon or two-photon excitation, introduces excess energy into the electronic system which is dissipated through multiple particle–particle or particle–quasi-particle scattering processes, such as electron- or phonon-mediated relaxation. This produces heat and *hot* non-thermalized charge carriers, which can reach the lower states through different channels and irradiate light by recombination at different stages on the way to the system's excitation ground-state. More importantly, emission from the exciton mode is likely not anymore representing recombination from Coulomb-bound electron–hole pairs, as the fraction of electron–hole plasma

[10]The related Rabi flopping behaviour in the time domain and Rabi splitting phenomenon in the frequency domain are named in reference to historic atom–field experiments [67, 68].

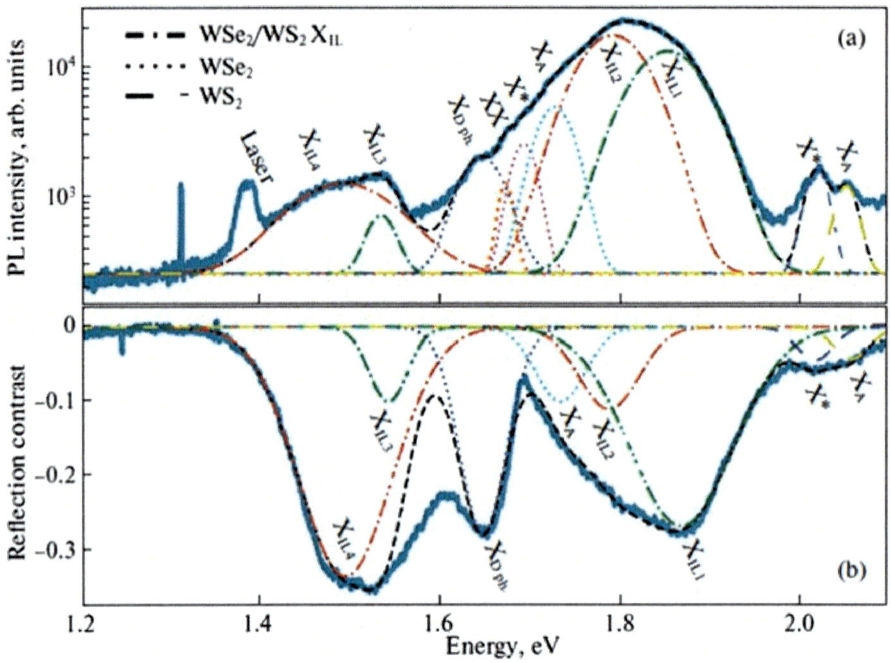

Fig. 5.11 Example measurement showing the linear optical response for a WS_2/WSe_2 hetero-bilayer structure (on hBN) as studied in [2]. **a** PL spectrum recorded spatially-resolved for the heterobilayer (here spatially integrated over a certain heterobilayer region) exhibiting numerous features at 10 K, in semilogarithmic representation. Next to monolayer A-excitonic signatures for the individual TMDCs, pronounced interlayer features labelled IL1–4 are obtained. **b** Corresponding reflection contrast spectrum for the same flake (spatially-integrated over a certain heterobilayer region from spatially-resolved data), showing absorption signatures and strong energetic congruence of features from **a** and **b**. Remarkably, the energy-split indirect excitons at long wavelengths indicate a noticeable influence of moiré patterns according to the literature (cf. [63]). Note the linear scale here, indicating a pronounced oscillator strength of these optical IL transitions. Additional interlayer features (IL1,2) between the monolayer resonances have been attributed to moiré states in good spectral agreement with the aforementioned literature. Also see corresponding lifetimes obtained from preliminary investigations shown in Fig. 5.9. **a, b** Reproduced with permission. [2] Copyright 2019 Springer Nature

(ionised excitons) becomes dominant[11] which can radiate out of the exciton mode as well (cf. [66], which discusses off-resonantly and resonantly excited Fourier-space-resolved emission spectra, and references therein).

Typical non-resonantly CW-excited PL spectra show multiple emission modes and background features (see for instance [46, 66], and inset in Fig. 5.17(d) in the section below), whereas pulsed measurements can reveal the decay times of bright species (cf. [5, 7, 31, 33], and see Fig. 5.9). Additionally, time-resolved spectra can even indicate transfer rates between populations in excitation modes that feed each others.

[11] The interested reader is referred to the semiconductor luminescence equations, see e.g., [18].

Quasi-resonant excitation in contrast to off-resonant excitation can directly excite a species, and it creates a coherent (macroscopic) polarisation of matter (coherent excitons, also understood as exciton–polaritons [69]). The macroscopic polarisation then dephases over time and forms an incoherent population of Coulomb-bound electron–hole pairs (microscopic polarisation). Such incoherent population can be for instance probed through intraexcitonic transitions (cf. [39, 41]). Furthermore, for high intensities of the driving light field, resonant excitation result in coupled-oscillator effects such as the aforementioned Rabi oscillations.

However, in time-averaged spectroscopy, spectral resonance of the excitation source and the emitter drastically complicates detection of luminescence signal under resonant pumping. Typically, (back-) scattered laser light collected by the detection optics is the dominant signal contribution and even filtering using cross-polarisation techniques may not always be sufficient to cancel out signal, although applicable in some cases such as for cavity–polariton studies under optical side pumping [59]. In contrast to CW acquisition, resonant ultrashort-pulsed excitation in time-resolved spectroscopy can be filtered out temporally.

Recent investigations on the optical dispersion of 2D excitons in WSe_2 by means of angle-resolved (Fourier-space) spectroscopy applied quasi-resonant excitation of the $1s$ species with a photon energy well-enough below the $2s$ resonance and far-enough away from the $1s$ resonance to be able to spectrally filter it out with the help of a premium edge filter and by a monochromator [66]. Furthermore, it could be recently experimentally demonstrated that the dispersion feature, which also occurs in white-light reflection contrast spectra, is detuning dependent and its occurrence weakens with increased pump energy (above the $2s$ resonance) until disappearance for continuum pumping (well above the gap for the TMDC's A exciton, and above the B-exciton level) [46]. Simultaneously, these measurements revealed a reduction in the valley-polarisation degree and change in the $1s$-exciton's pseudo-spin texture within the light cone as a function of the excitation detuning.

One major advantage of ultrashort-pulsed excitation in time-averaged acquisition is the reduced heat transfer to the lattice at simultaneously very-high peak intensities, for which under CW pumping the sample would not survive exposure due to the thermal load incorporated into the investigated structure, allowing to address a wider range of charge-carrier densities in optical experiments. Such strong irradiances with short temporal footprint allow one to reach higher-density regimes in (2D) semiconductors on ultrashort time scales that give rise to the formation and observation of various exciton complexes including biexcitons,[12] charged biexcitons and other multi-particle states even formed across (TMDC) valleys (cf. [3, 7, 73–75]), and the observation of out-of-plane "grey" excitons (see [3, 76]). Indeed, some of these exciton complexes are also obtained under CW pumping, but their occurrence shows strong excitation energy dependencies [46] and the obtained densities are much lower than peak densities in pulsed experiments.

---

[12] Biexcitons are exciton molecules similar to the hydrogen or positronium molecule, which require sufficiently high exciton populations to be established, whereas trions that are charged excitons, or even charged biexcitons, need an (additional) excess in free carriers.

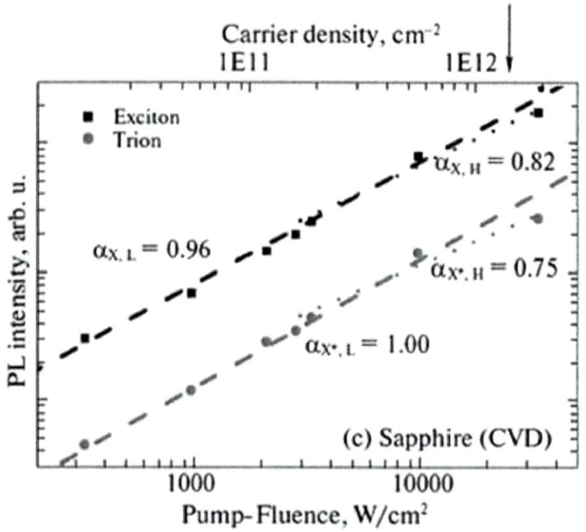

Fig. 5.12 Example input-output diagram showing PL intensity as a function of the irradiance for the emission from a CVD monolayer grown on a sapphire substrate. The linearity factors α for the exciton (X) and trion (X*) are given in the low- and high-density regimes (L/H) according to the pseudo-linear fit of the double-logarithmically plotted data (dashed/dotted lines). The estimated carrier excitation density is given on the top axis, with an arrow indicating the Mott density. Reproduced with permission. [31] Copyright 2018 Springer Nature

Analysing the linearity factor α of an emitter according to

$$I = P^{\alpha} \tag{5.3}$$

(see Fig. 5.12), i.e. the proportionality of measured intensity I to excitation power P, can be helpful to shed light on the nature of the optical transition. Charge-carrier-density-dependent PL signal for different species can provide a hint regarding qualitatively different dynamics and types of resonances. Typically, input–output measurements are plotted with double-logarithmic scale to indicate the exponent via the slope. A linear regime indicates free excitons [7, 77, 78], a sublinear behaviour indicates defect or bound states [77]. However, this type of experiment is not unambiguous. For instance, biexcitons and plasma share the same predicted linearity factor [73, 79].

For monolayer WSe_2, also hBN-encapsulated monolayers, PL measurements of the linearity factor as a function of temperature indicate a transition from a linear to a sublinear regime by a temperature-dependent decrease of the linearity factor above 100 K [5]. A similar temperature-dependent PL experiment focused on linewidth evaluation showed that at temperatures below 100 K, the linewidth changes linearly, but above 100 K, an exponential increase of the linewidth with temperature was obtained [80, 81]. Such behaviour has been attributed to the formation of phonon sidebands (PSBs) [80, 82, 83], obtained for both bright and dark excitonic species (cf. Fig. 2.7a, b), and the exciton–polaron-formation picture has been supported by temperature-dependent dispersion measurements [66]. Furthermore, the linearity factor can also show a change within density-dependent measurements at higher excitation densities near to the Mott transition for TMDCs (Fig. 5.12). This gets manifest as a kink in the (double-logarithmical) input–output curve, which can be caused by a different band renormalisation at K and Σ-point [31, 84]. Thus, complementary information for a material system can be obtained by different considerations. This becomes further clear in another PL-measurement example related to interactions with the phonon bath.

In addition, for different stacks of WSe_2 and hBN (i.e. substrate-supported, hBN-supported, hBN-capped and hBN-encapsulated), temperature-dependent energy shifts were in parallel to the linearity factor evaluated in a series between 10 K and 300 K [5], indicating reduced exciton–phonon coupling strengths for capped and encapsulated monolayers. To link the shifts with the role of optical phonon modes at elevated temperatures, the phenomenological O'Donnell model can be for instance useful [85]:

$$E(T) = E_0(T = 0) - S \langle \hbar\omega \rangle \left(\frac{\coth \hbar\omega}{2k_B T} - 1 \right), \tag{5.4}$$

with the two parameters $\langle \hbar\omega \rangle$ and S denoting the mean energy of optical phonons and the dimensionless coupling constant, respectively. The latter is also referred to as Huang-Rhys factor. Interestingly, the extracted trion–phonon coupling parameter was systematically weaker than the exciton–phonon counterpart ($S_{X^\pm} < S_{X_0}$). Note that with regard to such temperature-series evaluation, if the emphasis is not on the discussion of exciton–phonon coupling, also the phenomenological Varshni equation [86] typically describes the data well (also the case for [5]). Therefor, the second term $-S\ldots$ in (5.4) is replaced with $-\alpha T^2/(T + \beta)$, with parameters α and β.

In 2D semiconductors with excitons arising from different bands such as the A and B (and C) excitons of TMDCs [87], an additional possibility of injecting excitons is given by resonantly addressing the A or B type, for instance. However, decay channels to lower-lying excitons cause depopulation and introduce additional depolarisation pathways in valley-sensitive materials. Unless the free-carrier energy gap is above the addressed species, resonance excitation of higher energy excitons automatically leads to direct-gap hot carrier excitation. This might become even more relevant when combining TMDC monolayers to form heterostructures (cf. [2, 5, 88–90]).

Last, but not least, excitation and detection can be purposefully adjusted to be in a certain polarisation of light. This can be particularly useful when dealing with valleytronic materials such as the valley-polarisable TMDCs (see for instance [91–93]). TMDCs as monolayers exhibit considerable polarisation anisotropy, which can be in the case of bilayers further dependent on the bilayer configuration (for instance discussed in [8]). In [8], superachromatic wave-plates and polarisers were used for polarisation-resolved measurements. Typically, co- and cross-/counter- polarisation are of interest to determine the degree of linear/circular polarisation. The degree of linear or circular polarisation is given by $\rho = (I_{Co} - I_{Contra})/(I_{Co} + I_{Contra})$. Generally, the polarisation-filtering behaviour of the optics and the monochromator should be taken into account for such measurements.

5.3.3 Photocurrent Measurements

Photocurrent spectroscopy can provide additional information about semiconducting behaviour for materials through measurements of the electrical response as a function of optical irradiation. For instance, this method was applied in the last century to characterise bulk TMDC crystals [94]. It is also widely used in the field of phototransistor studies, including recent works with graphene or ultrathin TMDCs (see for instance [95, 96] and references therein) or photodetecting devices involving both 2D material types (e.g., in [97]).

The use of a monochromator behind a thermal white-light source (a wavelength sweep) can provide a simple means of wavelength tunable photoexcitation of the target material placed for instance in a specially-prepared microscope setup with contacting needles. The photoinduced current (measured with external bias applied) as a function of the photon energy allowed for instance the characterisation of the photoexcitation wavelength (peak in the photocurrent spectrum) for a new material with semiconductor properties in [98] (further details on the method are summarised in the Supporting Information sections). The setup employed for this experiment was placed in a nitrogen-filled box to minimise exposure to moisture and oxygen.

In this context, also current–voltage (I/V) characteristics can shed light on electrical properties when comparing for instance white-light exposed samples with their unexcited case in a dark environment. As a reference system, a prominent optoelectronic material with a direct band gap such as GaAs can be employed. Such studies for instance added to the characterisation of a new chemical compound $K_2Hg_2Se_3$ developed by the Dehnen group in Marburg with interesting optoelectronic, photophysical, and thermoelectric properties, which was thereby identified as photoconducting material with a direct band gap around 1.4 eV [98]. Similar experiments also led to the characterisation of a new compound from the family of organic–inorganic hybrid materials based on a porphyrin diacid with interesting colour-change effect [99], synthesised and explored in the Heine group in Marburg.

In addition to I/V curves, typical optoelectronic characterisation furthermore involves the on/off behaviour of a photoconducting material. This can demonstrate the reversibility of the light-induced effects, as well as provide information on the recovery timescales and possible switching times.

5.3.4 Raman Signatures

Raman spectroscopy employs typical PL equipment and a very narrow-line excitation laser, which typically impinges the sample surface after passing a well-focusing microscope objective. In Raman measurements, the back-scattered light from the sample is collected and analysed in order to identify the Raman shifts due to scattering processes between incident photons (wavelength λ_0) and lattice vibrations. Depending on the absorption or emission of a phonon's energy by the photons, the back-scattered photon either obtains more or less energy, respectively. This gives the lines referred to as anti-Stokes (AS) and Stokes (S) lines, respectively. Their lines appear at characteristic wave-numbers $\Delta k = 2\pi(\lambda_0^{-1} - \lambda_{(A)S}^{-1})$ which are related to the phonon energies involved. The intensity spectrum as a function of the wave-number acts as a fingerprint for a sample's material.

Commercial Raman microscopes are widely used in the field of materials characterisation. In lab applications, self-built confocal microscopes for μ-PL spectroscopy can be equipped with a Raman functionality. Since the modes are located very closely around the excitation laser's position, a very narrow linewidth of the laser light and a very strong filtering of it with a sharp-edged filter are necessities in this spectroscopy mode. Holographic and interference filters are commonly employed to suppress the back-scattered Rayleigh line. Compared to the Rayleigh signal, the targeted Raman modes are orders of magnitude weaker and long integration times are typically needed to obtain sufficient signal-to-noise ratios in Raman spectra. Moreover, the optical power has to be kept below the sample damage threshold and cannot be freely used for intensity scaling. Further details about Raman spectroscopy can be found for instance in laser spectroscopy textbooks such as [100].

In combination with 2D materials, Raman spectroscopy can provide useful information about the layer thickness and strain (e.g., used in [5, 7, 31]). When for instance verification is needed that the data was obtained for sample locations which exhibit monolayer or few-layer material, the characteristic modes can be analysed. This all-optical combination of PL and Raman spectroscopy can even compensate for the lack of atomic-force microscopy (see Fig. 5.13 and cf. [7, 8]), which is alternatively used to determine sub-nm height steps occurring at the transition from one layer terrace to the next (e.g., used in [2, 3, 6]). For WSe_2 samples, a shift of the main (monolayer material specific) Raman mode towards higher wave-numbers can be clearly observed on various substrates with increasing layer number, i.e. thickness [7, 31].

For instance, exfoliated WSe_2 monolayer samples feature one important Raman peak around 252 cm^{-1} with a weak shoulder slightly higher than 260 cm^{-1}. However, a bilayer or multilayer of this TMDC exhibits the weak shoulder below 260 cm^{-1} [101]. The monolayer Raman mode for CVD-grown monolayers can be slightly shifted to higher wave-numbers when compared to the exfoliated counterparts, which can be an indicator for strain [102]. Such shift's magnitude can be comparable or larger than the shift induced by a transition from monolayer to bilayer TMDC.

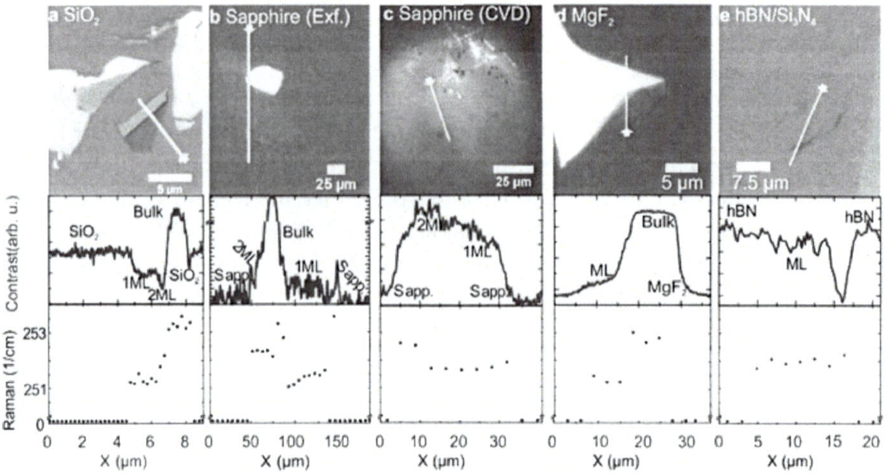

Fig. 5.13 All-optical identification of monolayer WSe_2 with the help of optical contrast analysis of micrographs and micro-Raman line scans. The measurements are further detailed in [7]. The microscope images for the measured WSe_2 monolayers deposited on different substrates **a**–**e** are shown together with brightness cross sections (along yellow line, shown in the central row). **a**, **b**, and **d** Monolayer flake (mechanically exfoliated and) transferred onto the bare substrate material (labelled), **c** as-grown monolayers on sapphire (by CVD), **e** hBN-buffered monolayer on substrate using the dry-stamping technique for 2D stacking. The last row shows extracted Raman peak positions obtained along the yellow line by a line scan. Data points at zero in the graphs correspond to locations, for which no WSe_2 peak was obtained. All monolayers can be identified by the wave-number below $252\,\mu m^{-1}$. Reproduced under the terms of the CC-BY 3.0 Licence (http://creativecommons.org/licenses/by/3.0/) from Lippert et al., 2D Mater., "Influence of the substrate material on the optical properties of tungsten diselenide monolayers", https://doi.org/10.1088/2053-1583/aa5b21 [7] Copyright 2017 IOP Publishing Ltd

Raman spectra can also be useful for the characterisation of bilayers. No change in the energetic position is expected for the A_g^1 and E_{1g}^1 when comparing both stacking configurations according to previous density-functional theory (DFT) calculations and experimental studies on other TMDC-bilayer stacks [103, 104]. However, shear and breathing modes and out-of-plane modes at the M point [105, 106] are expected to change. The shifts in these modes that appear at low wave-numbers and the two-phonon processes that give rise to the shoulders of the in-plane Raman modes can be attributed to the predicted change in the interlayer distance and interaction [103].

It was indicated by the Raman measurements in [8] that the stacking type influences the Raman intensities and energies enough to assume distinguishability between AA' and AB bilayers, particularly when taking a resonant Raman effect into account. As the Raman laser can be nearly in resonance with the B exciton, an enhanced electron–phonon coupling with strong Raman signal is observed as previously reported [107]. Relative red-shifts of the B exciton based on different stacking configurations can then cause strong intensity differences of a particular Raman mode due to a detuning of the resonance Raman effect.

5.4 Nonlinearities

While many standard characterisation efforts rely on linear spectroscopy or imaging modes, nonlinear optical techniques can provide a unique access to nonlinear material properties else inaccessible to the basic measurement techniques. How some of these nonlinear methods have been employed in this work is briefly outlined in the following with corresponding examples. For introductory purposes, the basic principle behind nonlinear optics based on the nonlinear polarisation is highlighted, the details of which will be beyond the scope of this work.

It shall be recalled that the complex dielectric function $\epsilon(\omega)$ and the susceptibility $\chi(\omega)$ are linked by

$$\epsilon(\omega) = 1 + \chi(\omega), \tag{5.5}$$

which sets in relation a polarisable medium's dielectric displacement $\mathbf{D} = \epsilon_0 \epsilon(\omega) \mathbf{E}$ and an applied electric field $\mathbf{E}$. The light-field induced polarisation $\mathbf{P}_i$ for a field with electric field amplitude $\mathbf{E}_i$ and oscillation frequency ω_i can be expressed, with $\chi(\omega_i, \mathbf{E}_i) = \epsilon(\omega_i, \mathbf{E}_i) - 1$, as a power series of the incident field amplitudes (see [69] and references therein):

$$\mathbf{P}_i = \epsilon_0 \left[\sum_j \chi_{ij}^{(1)} \mathbf{E}_j + \sum_{j,k} \chi_{ijk}^{(2)} \mathbf{E}_j \mathbf{E}_k + \sum_{j,k,l} \chi_{ijkl}^{(3)} \mathbf{E}_j \mathbf{E}_k \mathbf{E}_l + \ldots \right], \tag{5.6}$$

whereas oscillating electric fields can be represented as $\mathbf{E}_i = \mathbf{e} E_i \exp(-\mathrm{i}\omega_i t) + \text{c.c.}$ (c.c. denotes the complex conjugated part,[13] and $\mathbf{e}$ the unit vector of the field's direction). While linear optics neglects the nonlinear terms due to their often negligible influence for small fields (thus, linear optics, i.e. in the regime of the harmonic oscillator), for strong fields (or high intensities), the higher-order terms can play an important role. These higher-order terms matter for instance for frequency conversion (e.g., $\chi^{(2)}$ processes), Kerr lensing or two-photon absorption (e.g., $\chi^{(3)}$ processes). For equal frequencies $\omega_i = \omega$, a similar expansion is given for the complex refractive index.

$$\tilde{n}(\omega, I) = \tilde{n}_0(\omega) + \tilde{n}_2(\omega) I + \ldots, \tag{5.7}$$

which is the basis for nonlinear lensing and absorption effects ($\chi^{(3)}$ nonlinearities) in the regime of high intensities ($I = |E|^2$). Note that the first term is the linear optics term, which is not intensity dependent. The effect of the second term is for instance evidenced in Z-scan experiments described below [108–110] (see e.g., Fig. 5.14).

Concerning frequencies, without going into many details, one can discriminate two basic situations. First, the case with equal frequencies, which leads to a $\mathbf{P}^{(2)}(t) = \epsilon_0 \chi^{(2)} E^2(t)$ that results in an expression such as

---

[13] That is with complex conjugated amplitudes E_i^*.

$$\mathbf{P}^{(2)}(t) = \frac{1}{2}\epsilon_0 \chi^{(2)} E_0^2 \left(1 - \cos\left(2\omega t\right)\right), \tag{5.8}$$

which describes second-harmonic generation, i.e. frequency conversion from ω to 2ω due to the fast oscillating term of the polarisation. Second, one can describe a more general case with different frequencies, e.g., ω_1 and ω_2 (and their fields $\mathbf{E}_{1,2}(t) \propto \exp(\mathrm{i}\omega_{1,2}t)$), for which the $\mathbf{P}^{(2)}$ contains more terms summed up briefly as the SHG terms for each frequency, an optical rectification term, a sum-frequency term

$$\mathbf{P}^{(2)}(t) \propto \chi^{(2)} E_1 E_2 \exp\left((\omega_1 + \omega_2)t\right), \tag{5.9}$$

and a difference-frequency term

$$\mathbf{P}^{(2)}(t) \propto \chi^{(2)} E_1 E_2^* \exp\left((\omega_1 - \omega_2)t\right). \tag{5.10}$$

The latter term is for instance used for THz generation in periodically- or aperiodically-poled lithium niobate (nonlinear) crystals placed inside a dual-wavelength laser's resonator, such as in a VECSEL [111–114] (see e.g., Fig. 5.15).

Note that for efficient frequency conversion, nonlinear crystals with a high nonlinear susceptibility are employed and that phase-matching conditions should be fulfilled for effective outcoupling, otherwise the energy of the propagating signal is transferred back to the fundamental modes. Typically, for SHG such as in intensity autocorrelators, or phase-resolving autocorrelation techniques such as FROGs or GRENOUILLEs [36], the crystal thickness is optimised with regard to signal strengths, phase-matching and dispersion.

5.4.1 Z-Scans

One effective tool to probe intensity-dependent nonlinear lensing, as well as nonlinear absorption, in a medium is given by the Z-scan technique. Kerr-lensing, which is a nonlinear lensing phenomenon, is for instance the pulse forming mechanism in mode-locked Ti:sapphire lasers (cf. [115, 116]). Z-scan measurements allow one to directly measure the nonlinear refractive index changes associated with the Kerr effect.

While the Z-scan technique [117] is the most prominent one due to its high sensitivity and simplicity, several measurement schemes have been developed in the past for the characterisation of the nonlinear refractive index [118–120]. With the rise of novel material classes like graphene and other 2D semiconductors, which often exhibit a very strong refractive nonlinearity [121–123], the Z-scan technique continues to be of experimental value.

The effective nonlinear refractive index n_2 of a medium is defined by the relationship

$$n(I_0) = n_0 + n_2 I_0 \tag{5.11}$$

(an approximation neglecting higher-order terms). $n(I_0)$ is the total refractive index, n_0 the intensity independent refractive index and I_0 the optical intensity, e.g., at the peak power of an ultrashort pulse. Such definition of n_2 comprises both non-instantaneous contributions, i.e. those induced by free-carrier nonlinearities (FCN), and instantaneous contributions, i.e. from the bound-electron Kerr effect (BEKE). In principle, the relative contributions from FCN and BEKE to the nonlinear refractive index could be unraveled with the help of time-resolved investigations.

Similarly, an intensity-dependent absorption term can be defined (in similar approximation neglecting higher-order terms) with a nonlinear absorption coefficient β by the relationship

$$\alpha(I_0) = \alpha_0 + \beta I_0, \tag{5.12}$$

with α_0 the linear absorption coefficient. Thus, $\alpha(I_0)$ is the total absorption coefficient. In contrast to the effective nonlinear refractive index n_2, β only comprises ultrafast $\chi^{(3)}$ effects. This is understandable by the fact that carrier-related absorption would lead to saturation and thus to a more complex relation on incident intensity.

To perform the Z-scan measurement, the sample is step-wise moved finely in the beam direction through the focal spot of the focused 100-fs-pulsed probe beam (see Fig. 5.14a). The simultaneously (behind a beam splitter) recorded sample-transmitted as well as sample-reflected signal on an open-aperture and closed-aperture power meter detector are then used to obtain a nonlinear absorption trace (c) as well as an absorption-corrected nonlinear lensing trace (b), respectively, as shown for example in Fig. 5.14. A typical setup is employed and described for instance in [108].

One practical example of the determination of the effective n_2 for a sample is given by a study of a VECSEL chip motivated by the observation of self-mode-locking behaviour reminiscent of that in a Ti:sapphire laser [108, 109]. This clearly indicated nonlinear lensing as a possible driving mechanism for mode-locking. In a follow-up work to the one without charge-carrier excitation [108], even close-to-operation conditions with an optically-pumped chip were investigated [109]. The spectroscopic part enters the investigation, when the dependence on the probe wavelength is analysed (in both works). A microcavity effect can for instance drastically shape the wavelength-dependent nonlinear response of such a semiconductor disk laser, as discussed in [108]. Moreover, Z-scans readily deliver information about the nonlinear absorption coefficient β and reveal to which extent intensity-dependent absorption can act as a possible loss mechanism.

In fact, such technique can be also used to study 2D materials. Endeavours to achieve clear results by the author's team have not been completed yet and the limiting factor remain the size of the specimen and insufficient signal-to-noise ratio when probing sub-nm-thin microscale flakes with a strong optical focus (particularly for displacements away from the focal spot). Thus, further improvements on the side of the sample production and experiment are generally required for such studies. On the other hand, bulk crystals such as a single-crystalline perovskite material were

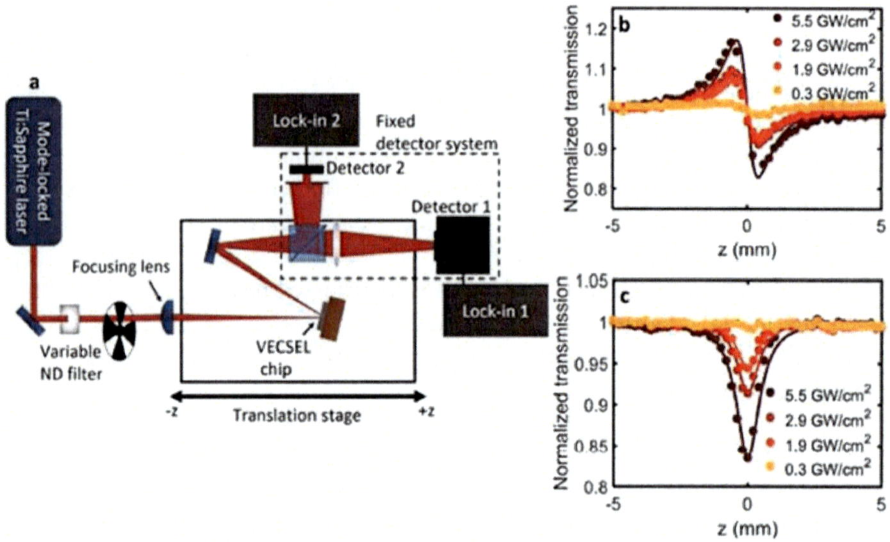

Fig. 5.14 **a** Typical Z-scan setup based on a mode-locked Ti:sapphire laser system and two intensity detectors (power meter heads) analysing the reflected beam from a translated sample. Here, the experiment is set up in reflection geometry to meet the requirements of VECSEL chip investigations [108]. Two lock-ins are used for signal acquisition in combination with a chopper wheel. The irradiation is adjusted by neutral-density (ND) filters. **b** Z-scan measurement at a fixed probe wavelength for different irradiances, measured for an incidence angle of 20°. The corresponding open scan **c** recorded by detector 1 shows nonlinear absorption. A partially closed aperture in front of detector 2 is used to obtain nonlinear lensing signatures in the trace. To obtain the Z-scan trace, the closed scan is corrected for nonlinear absorption by dividing it by the open scan, a background subtraction with a low-intensity scan is performed and the transmission at z positions far away from the focus region is normalised to a transmittance of 1. Reproduced with permission. [108] Copyright 2019 Optical Society of America

recently investigated with regard to their nonlinear refraction and absorption using the Z-scan method [110].

5.4.2 *Nonlinear Frequency Conversion*

Nonlinear processes can enrich the study of materials and structures. Among these processes are particularly famous SHG [124] and difference-frequency generation (DFG). In addition, the Kerr effect and other nonlinear processes such as optical rectification can be used to generate strong fields of electro-magnetic radiation (cf. [125, 126]) or white-light supercontinuum spectra (e.g., in fibers, see for instance [127]) which are both desired for different spectroscopic applications.

On the one hand, higher-harmonic generation such as SHG and third-harmonic generation (THG) can be used to probe crystal symmetries of a given sample due to the dependence of nonlinear frequency conversion efficiencies on the tensor-elements of the second-order susceptibility of a medium $\chi^{(2)}$. In bulk, the microscopic dipole fields contributing to the nonlinear polarisation signal cancel each other out, but the symmetry-break at the surface of a highly symmetric crystal leads to considerable nonlinear frequency conversion. Thus, as a surface- and interface-sensitive tool, it can become useful for the study of 2D materials. Particularly, when the incident polarisation is tuned, different crystalline domains of different orientation in poly-crystalline monolayers and their grain boundaries can become visible, for example demonstrated by imaging microscopy in [128].

On the other hand, frequency conversion can be used to generate new wavelengths from incident light on a nonlinear crystal, such as SHG (frequency doubling), DFG and sum-frequency generation (SFG). SHG is particularly desired as a means to expand the available frequencies of a pulsed laser towards higher energies for off-resonant excitation schemes or excitation of high-gap materials. For example, in [129], a mode-locked ultrafast VECSEL's output was frequency doubled externally to achieve a high-repetition-rate pump for the excitation of a quantum-dot single-photon source (see Fig. 4.2), thereby obtaining a world-record single-photon flux rate (see Fig. 4.7).

Similarly, DFG can be attractive to access spectral regions, for which light sources are rare, for instance in the previously labeled THz gap, and to produce emitters useful for applications in spectroscopy or industry. One example employs intracavity DFG in an open cavity of a VECSEL to provide a strong coherent THz output in the sub-mW to mW range [111]—a well-studied concept of multi-mode TEM_{00} THz generation [112, 130], which has been recently made power-scalable using a two-chip design [113]. This approach was previously explored and demonstrated with the help of intracavity SHG and SFG for tunable frequency conversion [131] using angle-dependencies of the chip resonances [132]. Typically, periodically-poled lithium niobate crystals are employed for efficient DFG designed for 1 THz output. In fact, the use of such crystals with aperiodical poling enables even tunable THz generation as demonstrated recently in the range of 0.79-1.11 THz [114] (Fig. 5.15).

SHG imaging, which has been described above as a means to obtain information about symmetries in crystalline systems, has not only become useful for the characterisation of monolayers on substrates, but also of the bilayer configurations and their twist angles [8, 133]. A comparison of PL and SHG images for instance enables one to identify monolayer, bilayer or few-layer TMDCs and their stacking symmetry, as the example of a homobilayer-system study in [8] shows (see Fig. 5.6). For PL images, the laser radiation is filtered for instance by a holographic notch filter, whereas for SHG, the fundamental light is blocked by a short-pass wavelength filter. Both spatially-resolved signals can be well recorded using a scientific intensified CCD. For the CW excitation of PL, a frequency-stabilised 532-nm Nd:YAG-based laser of a μPL-Raman setup with emission above material band gap can be employed. For spectrally-integrated SHG measurements, a pulsed Ti:sapphire laser (with 100 fs pulse duration) of a time-resolved PL setup at wavelengths below the band gap serves well. Both lasers can be conveniently used for common 2D semiconductors due to the energetic position of their excitons.

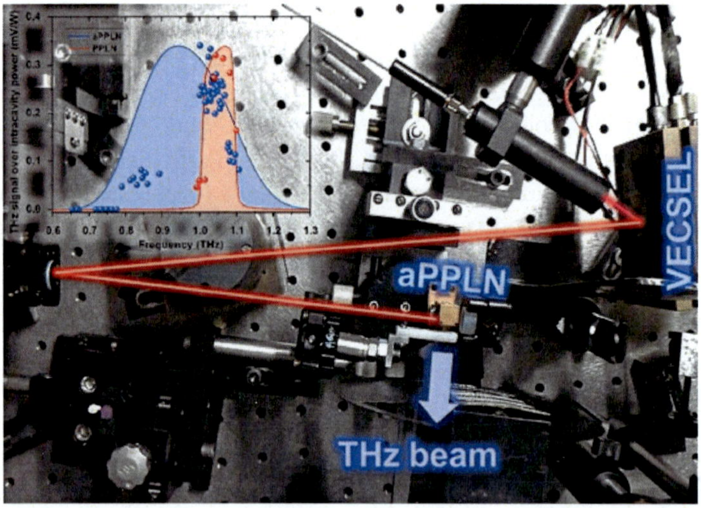

Fig. 5.15 Photo of a THz-generating dual-wavelength semiconductor disk laser device with nonlinear crystal, also referred to as THz-emitting external-cavity surface-emitting laser (TECSEL). Inset: Comparison of the THz frequency tunability for an aperiodically-poled and a periodically-poled lithium niobate (aPPLN/PPLN) crystal designed for 1-THz output. The laser operates on two laser modes simultaneously at around 1 μm wavelength with the help of an intracavity etalon and with a wavelength spacing around 3.4 nm, corresponding to about 1 THz. Intracavity difference-frequency generation exploits high intensities inside the laser resonator for effective THz emission, which is designed to take place perpendicular to the optical pump light (i.e. resonator mode beam). Due to a chirp of the periods in the ferroelectric crystal, the aperiodic poling enables phase-matching for a wider range of wavelengths than periodical poling, which allows phase-matching only for a specific wavelength. Adapted under the terms of the CC-BY 4.0 Licence (http://creativecommons.org/licenses/by/4.0/). [114] Copyright 2020 The Author(s), published by Wiley VCH

Bilayer areas reveal themselves in PL maps by their significantly reduced emission, whereas the twist angle of bilayers determines the strength of nonlinear frequency conversion and, thus, the recorded signal in SHG maps. In general, the SHG response of both layers will cancel each other in an inversion symmetric case. In contrast, the response will increase nearly 4-fold [133] in the scenario with broken inversion symmetry.

Correspondingly, for AA' and AB bilayers, destructive and constructive SHG behaviours are observed [8]. The signal of the symmetry-dependent SHG process cancels out for the symmetric AA' bilayer (type abundant in nature as the energetically favourable stacking configuration) and results in a dark bilayer area surrounded by SHG from monolayer regions (for the studied CVD-grown stacks). In contrast, for the AB-type bilayer (energetically less favoured configuration), the symmetry-break in the stacking configuration provides increased signal in comparison to a single layer, as the individual layers' contributions constructively add up. Thus, for samples which were grown to result in bilayers of different configuration, dark PL with bright SHG indicates the AB-type bilayer, whereas bright SHG and bright PL is associated with the monolayer parts. Indeed, a trilayer of naturally-stacked TMDC material should be dark in PL owing to its indirect gap, but not dark in SHG due to the symmetry configuration, whereas a naturally-stacked (AA') bilayer remains dark in both signal maps (i.e. PL and SHG).

5.5 Fourier-Space Spectroscopy

Fourier-space imaging has become helpful with regard to single-shot acquisition of angle-resolved emission. In fact, momentum-space-resolved spectroscopy can be useful for many studies in which an energy–momentum dispersion can be probed, such as for exciton–polaritons in optical microcavities. This usefulness can be also evidenced from recent experiments on 2D semiconductor monolayers highlighted below. The underlying Fourier-space projection method is explained in more detail for instance in [4] in the context of polariton research.

5.5.1 Angle-Resolved Detection

Angle-resolved detection can be achieved by various means (for a more detailed overview see [4]). The simplest approach employs a goniometer arm probing different angles. Alternatively, one can spatially scan an aperture through the far field of an objective-collected signal or directly map the Fourier-space plane on an imaging camera.

Angle-variation experiments are in different ways also applicable to laser performance characterisation (as done for instance in [108, 132, 134]), but the main focus remains here on the characterisation of angle-resolved PL and reflection contrast (such as in [3, 46, 66]).

For angle-resolved PL (Fig. 5.16), measurements were done in [66] and follow-up work using a conventional 4-f μPL setup based on a self-built confocal optical microscope (cf. Fig. 5.3) with Fourier-space imaging capabilities (similar to those established by the author for polariton research in [14, 27, 59, 135, 136]). To reach very low temperatures, the sample is typically mounted in a continuous-flow cryostat at high vacuum.[14] Commonly, a microscope objective with long-enough working distance such as a 40x (NA 0.6) objective can be used to focus light onto the cooled sample. If discussed at the example of WSe_2 investigations (with its monolayer neutral exciton located slightly below 720 nm at 10 K), a (pulsed) Ti:sapphire laser tuned to its short-wavelength edge enables quasi-resonant excitation of monolayer excitons. For 100-fs-pulsed light, a short-pass filter for 700 nm was used to shape the pulse in front of the sample [3]. To spectrally filter the excitation beam out of the collection path, a

[14]To reduce the dependence on cryogens, closed-cycle cryostats are nowadays a popular option for the replacement of flow cryostats. However, the noise specifications due to mechanical vibrations have to be taken into account for high-precision measurements.

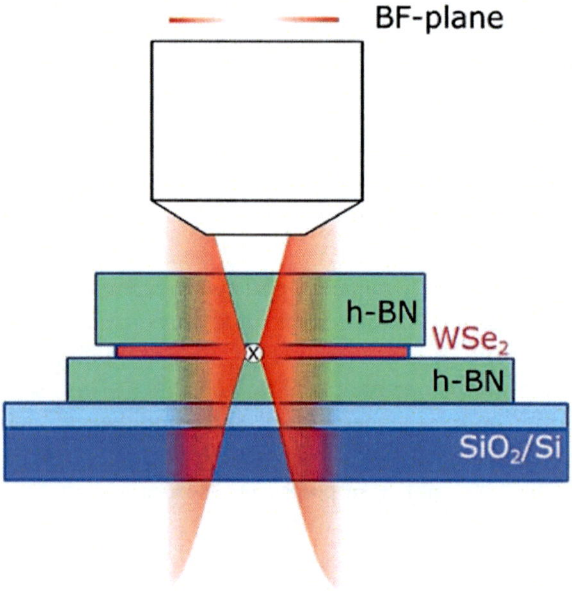

Fig. 5.16 Sketch of a typical angle-resolved PL measurement of hBN-encapsulated monolayer TMDCs with a microscope objective. In a simplified drawing, the sideways emitting light from an out-of-plane dipole is projected angle-resolved in the backfocal (BF) plane of the objective. From there, projection onto an imaging spectrometer's CCD allows one to map the radiation profile with spectral resolution, as discussed for bright and "grey" excitons in [3]. Reproduced under the terms of the CC-BY 4.0 Licence (http://creativecommons.org/licenses/by/4.0/). [3] Copyright 2020 The Author(s), published by Springer Nature

high aspect-ratio long-pass filter with specified edge at 700 nm (and in some scenarios a polariser) can be inserted between the sample and detector. Filtered μPL sample signal is then detected regularly by a monochromator with research-grade imaging camera (typically a nitrogen-cooled CCD chip). For polarisation-resolved measurements, additional quarter-wavelength or half-wavelength plates can be incorporated into the setup, as done in [46]. At room temperature, the experimental constraints for angle-resolved measurements are much more relaxed, as short working distances can be used to maximise the numerical aperture for wide-angle signal collection, or even immersion liquids can be employed between sample and objective for high-NA measurements (see for instance [137]).

5.5.2 *Dispersion Measurements*

For optical dispersion measurements, time-integrated spectra were recorded in [46, 66] in Fourier-space imaging configuration, which projects the angle information one-to-one from the microscope objective's backfocal plane via a set of lenses onto the entrance slit of the imaging spectrometer's monochromator (cf. [4, 14, 66, 138] for illustrations of this scheme, and Fig. 5.8e). In order to measure the emission

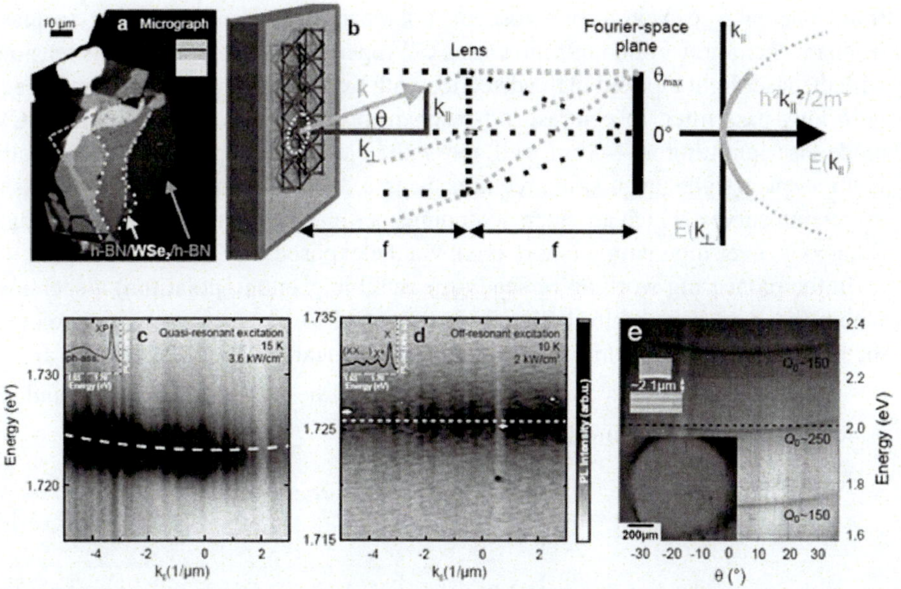

Fig. 5.17 **a** Optical micrograph of an hBN-encapsulated WSe_2 monolayer, which provides excitonic signatures with spectrally very narrow optical modes close to the homogeneous linewidth limit. Arrows indicate the flakes according to the stacking sequence (see inset, and label). The bottom layer is marked by dashed lines, the sandwiched monolayer semiconductor by dotted lines. **b** Schematic drawing of a Fourier-space projection which enables in-plane-momentum ($k_{\|}$) resolved signal acquisition. In a simplified sketch, a parabolic free-particle dispersion for in-plane confined (centre-of-mass) propagation is set in relation to the detectable angle range (limited by the numerical aperture, i.e the maximum collectable angle in air). Parallel optical rays are indicated in grey and dark grey for a finite and zero angle θ, respectively. Such setup has given unique insight into excitonic (exciton–polariton) dispersions within the measurable light cone. **c**, **d** Comparison of the dispersions recorded for the TMDC in **a** at quasi-resonant (**c**) and off-resonant (**d**) excitation, discussed in [66]. Insets: Angle-integrated spectra (XP: exciton–polariton upper branch; ph.ass: phonon-assisted emission from dark states; X, X* and XX... various excitonic resonances discussed and labelled in [3, 46]). **a**–**d** Adapted with permission. [66] Copyright 2019 Optical Society of America. **e** Angle-resolved white-light reflection spectrum (false-colour diagram, from black to white increasing intensity in arb.u.) from a homemade planar tunable Fabry–Pérot-type open microcavity with photonic dispersion of three cavity modes (exhibiting quality factors of up to 250). The dashed line indicates the WS_2 exciton resonance for targeted coupling experiments. Insets: sketch of the cavity (top inset) comprising a dielectric mirror and a silver-coated SU-8-polymer micropost (top facet shown in the bottom inset) with slight concaveness and indicated 2-µm mirror separation. In fact, bare dielectric mirrors are very useful test samples in Fourier-space spectroscopy setups. Data obtained by F. Wall in the author's team, for experiments towards the achievement of tunable coupling strength studied in [139]

from coherent free excitons (macroscopic polarisation) in monolayer WSe_2, a quasi-resonant excitation around 695 nm with a CW-operated Ti:sapphire laser is feasible, with the laser light blocked in the detection path by the aforementioned high-aspect-ratio long-pass filter. In contrast, off-resonant pumping was achieved with a CW diode laser emitting at 445 nm well above the electronic gap of WSe_2. Naturally, Fourier-space signal acquisition requires using a 2D chip read-out, giving access to dispersion curves, which are intensity (counts per integration time) over wavelength (long axis) over momentum (short axis), via the exposed CCD area.

In the pioneering work on optically-measurable exciton(–polariton) dispersions within the light cone of [66] (Fig. 5.17), care had to be taken with the choice of the measurement parameters. For a maximised optical collection of emitted angles $\theta = \arcsin\left(\frac{\sin(\theta_0)}{n}\right)$ (θ_0 is the angle in vacuum, n the material's refractive index), which correspond to the in-plane momentum by

$$k_{\parallel} = \frac{2\pi n}{\lambda_0} \tan\left(\arcsin\left(\frac{\sin(\theta_0)}{n}\right)\right), \tag{5.13}$$

a cryostat-compatible numerical aperture (NA $= 0.6$) had to be chosen for the employed microscope objective, as the samples were placed behind a cryostat window. μPL signal detection from the sample took place behind a spatially-filtering aperture in the real-space projection plane of the confocal microscope, which selected a spot of $\approx 1\,\mu$m diameter. The laser-spot diameter amounted to approximately $2\,\mu$m. By Fourier-space projection (Fig. 5.17), the imaging optics of the setup project angles up to $\theta_0 = \pm 37°$ onto the imaging spectrometer, whereas the monochromator disperses the signal behind the entrance slit horizontally and preserves the angle information in the vertical direction. The light cone in the medium (vacuum) amounts to $|k_{\parallel}| \approx 34$ (8.7) μm^{-1}, the maximum detectable angle of $\pm 37°$ corresponds to $|k_{\parallel}| \approx 5.2\,\mu m^{-1}$ in WSe_2.

It is explained in [66] that in contrast to angle-resolved photo-electron spectroscopy (ARPES), which resolves mainly the electronic band structure but can provide indirect evidence of the exciton dispersion [140] (not easily understood), Fourier-space-resolved μPL spectroscopy provides an effective access to the optical dispersion within the measurable light cone (Fig. 5.17), i.e. the polaritonic exciton's dispersion.[15] In addition, electron energy-loss spectroscopy could also measure exci-

[15]Note that as far as exciton emission is concerned which arises from the coupling of the induced macroscopic polarisation in matter with the light field, the oscillator strength results in hybridisation and radiation out of the upper polariton branch within the light cone, or from the lower branch outside the light cone through phonon scattering processes [69]. In a crystal, polarisation waves propagate, whereas outside the crystal beyond the matter–vacuum interface photons represent the propagating electromagnetic wave carrying the information of the hybrid states according to energy and momentum conservation. Incoherent excitons (microscopic polarisation, with effective masses on the order of electron masses) and electron–hole plasma (due to statistics in the emission processes, no centre-of-mass information is contained in the emission's momentum distribution) do not feature a measurable angle-dependent energy as discussed in [66].

ton dispersions using a fundamentally different approach, but no measurement on a monolayer has been obtained yet, according to [66].

An exciton–polariton branch or the literature-predicted big interaction-mediated exciton splitting at finite k within the light cone (cf. [141–145]) can only be detected when the spectral lines are very narrow (i.e. with a small dephasing rate) and the optical resolution in E and k is sufficiently high. In fact, optical measurements allowed to evidence a pronounced dispersion feature in Fourier-space-resolved spectra on the order of meV with sub-meV precision in the determination of angle-dependent energy changes for WSe_2 and $MoSe_2$ (the latter with less curvature). Furthermore, an underlying exchange-energy-affected dispersion feature of TMDC excitons with predicted meV changes within the light cone could also facilitate observation of the polariton-branch curvature (see considerations in [66]). By merely evaluating the effective curvature of the measurable average dispersion feature, which due to contributions of incoherent excitons and e–h plasma can be (partially) washed out, through a parabola approximation, mean effective masses of as little as 10^{-4} can be extracted in both PL and reflection contrast [66]. Both, pump-power dependencies [66] (the investigated excitation densities were set by a neutral density filter wheel in discrete steps in the linear regime) and detuning dependencies [46] showed changes in the measurable dispersion. Moreover, the latter (i.e. excitation detuning) modified the overall (as well as the momentum-dependent) helicity degree of the excitonic emissions obtained in experiments.

5.6 Additional Methods

At the end of this chapter, a few more optical methods are briefly described which can be used for the characterisation of photonic devices, semiconductor structures or even bare materials.

5.6.1 Beam Characterisation

For the characterisation of laser output beam profiles, typically the beam quality factor M^2 is obtained. This is the parameter, which reflects the degree of variation of a beam compared to an ideal Gaussian beam of the same wavelength. A definition of this beam quality factors and how to derive it on the basis of measuring second moment widths $D4\sigma$, which are the beam diameters along beam-propagation direction, is for instance given in [146, 147]. The closer a beam (profile) is to an ideal Gaussian beam, the closer the M^2 values in the two cross-sectional directions x and y become to one. For most common wavelength ranges, there are commercial solutions available to measure the beam quality. Nevertheless, the quality of measurements is not reduced when employing homemade tools evaluated without commercial software solutions, such as used to measure the beam quality factor of a type-II VECSEL [148].

In contrast to the optical wavelength region, there are only few of such measurements (e.g., [149]) in the THz frequency region. Recently, beam profile measurements in the THz domain were for instance used to provide additional information about the beam quality of a THz-generating VECSEL (TECSEL) [130], which featured M^2 values of about 1.4 and 1.7 in x and y direction, respectively. Indeed, a simple knife-edge method can already reveal the intensity profile of a THz source's output beam, as shown in [111] for the first TECSEL—something more trivial in the optical domain when imaging cameras are available (intensity profiles acquired from a single shot). With the rise of THz imaging cameras, such M^2 measurements are facilitated. Nevertheless, care has to be taken when extracting values from raw data, because offset values and noise strongly influence the calculation of the second-moment width D4σ derived from beam profile measurements. Thus, an additional processing of the measured raw data was applied in [130], with the aim to correct for the background offset signal, the camera non-uniformity, the background noise, where there was no THz signal present, and smoothing of the signal. If no imaging device is available, the knife-edge method in combination with a beam-waist scanning can also enable the determination of beam quality factors. Therefor, a knife edge is translated in front of a photodetector and sequential knife-edge scans at each position along the beam are acquired.

Important information about a light source can also be obtained interferometrically. Characterisation of a semiconductor-disk-laser-driven terahertz source by Michelson interferometry in the THz frequency domain, similar to Fourier-transform infrared-spectroscopy (FTIR), can reveal the emission frequency and bandwidth [112]. Such experiment by Wichmann et al. facilitated the estimation of the number of difference-frequency-generated THz modes obtained from a TECSEL.

White-light interferometry using a comparison of balanced and unbalanced Michelson interferometer arms can provide useful insights into the group-delay dispersion of a transmissive sample. Therefor, the frequency-dependent phase[16] is deduced via a fast Fourier transform of the interferogram from the time domain.[17] By applying a second derivative (with respect to frequency) to this frequency-dependent phase, the group-delay dispersion within a VECSEL introduced by the gain chip or saturable-absorber mirror can be estimated for chirp compensation approaches. This is currently pursued to improve self-mode-locked VECSELs with regard to pulse stability and duration for ultrashort-pulse operation.

Recently, angle-resolved PL spectroscopy, which projects the far-field of an emission signal onto the spectrometer, was even used to identify excitonic species in an encapsulated WSe_2 monolayer based on their spectrally-resolved angle-dependent emission characteristics. Thereby, one type of dark exciton, which can only emit due to symmetry-dependent selection rules in the plane but not out of the plane and which is referred to as grey exciton (Γ_4 state with out-of-plane dipole orientation in WSe_2), was clearly singled out from the pool of excitonic lines due to its radiation profile [3].

---

[16] Part of the complex spectrum with amplitude and phase.

[17] The interferogram results from the scanning of a length-change induced temporal delay in one arm.

Other dark excitons (Γ_3 state or dipole-forbidden transitions) are still observable in emission spectra when phonon-scattering processes are involved. For WSe_2, the correspondingly long-living Γ_3 dark state lies energetically below the bright states and acts as a nonradiative decay channel for Γ_6 excitons. This peculiarity distinguishes WSe_2 from its $MoSe_2$ counterpart. For group-theory analysis and early experiments on this matter see [150–152].

In fact, far-field beam characterisation is else used for laser beam divergence studies, e.g., from VCSELs and edge emitters, and it can also reveal a phase transition in bosonic emitters from a Maxwell–Boltzmann distributed regime to a Bose–Einstein-like condensate with narrow momentum distribution (commonly achieved by Fourier-space imaging, cf. [4, 153, 154]).

5.6.2 *Time-Domain Spectroscopy*

Time-domain spectroscopy (TDS) is widely known from THz spectroscopy, which often derives material information from amplitude and phase changes of transmitted or backreflected THz pulses. In the literature, THz spectroscopy has been widely used to characterise materials and photoconductivity with sub-ps resolution without physical contact and non-destructively. Using pump–probe schemes, even excitonic and free-charge-carrier behaviour can be distinguished through the analysis of the frequency-dependent complex conductivity. More importantly, it has become a commercially available method after initial developments in this field took shape around two decades ago [155]. It covers the far-infrared spectral range between approximately 3 to 300 wave-numbers, corresponding to frequencies between 0.1 to 10 THz. In the following, a few examples are briefly provided.

For instance, THz spectroscopy allows one to study fundamental physical mechanisms such as the formation and decay of excitons in various semiconductor systems [39, 42, 156]. In addition, it was recently demonstrated that the formation of crystals out of a solution could be monitored with THz (attenuated-total-reflection) spectroscopy [157]. Note that the optics of THz waves are material-wise different than those for visible light, which required all kinds of developments to obtain the simplest optics, such as lenses, wave-plates, and also beam splitters [158]. In the field of non-destructive testing, THz experimentalists from the Semiconductor Photonics Group in Marburg recently combined a THz system with a robotic arm to achieve a novel THz tomography system for analysis of arbitrarily shaped samples [159].

Furthermore, THz spectroscopy can enable the exploration of carbon nanomaterials with regard to applications at THz frequencies [160, 161]. In addition, recent optical ultrafast conductivity studies on graphene nanoribbons were carried out using THz spectroscopy which showed how the conductivity varies with the precise structure of graphene nanoribbons [162–164].

Often, THz-TDS systems are employed for material studies which are typically composed of THz emitting antennae photoexcited by fs-pulsed laser pulses and THz detecting antennae optically gated by delayed laser pulses from the same fs-laser source, thereby generating and detecting ps-short THz pulses in the 1 THz spectral region. To achieve the desired performance of THz antennae, the type and density of defects is usually optimized in low-temperature-grown GaAs (LT-GaAs) THz antennae, but to further enhance the performance after growth, one can decorate their surface in the metal-antenna gap region with nanoparticles, as demonstrated independently in [165, 166].

5.6.3 *Laser-Induced Plasma/Breakdown Spectroscopy*

Laser-Induced Plasma/Breakdown Spectroscopy (LIBS/LIPS) is an element spectroscopy technique, which is based on the analysis of the spectral emission from laser-induced plasmas produced by high-power ultrashort laser pulses applied to the surface of a target material. An optically induced plasma is formed on the surface of or in the sample when the laser power irradiance exceeds the breakdown threshold of the medium. When a plasma is ignited, a plume of ionised matter emits radiation on different time scales on the way to its normal configuration, with the plasma cooling and electrons consecutively relaxing into bound element energy states. Typically, characteristic element lines can be spectrally resolved for low-repetition-rate pulsed laser-induced breakdown after μs delay times using a gated intensified-CCD (iCCD) camera. The laser fluence is commonly in the range of J cm^{-2}.

Fortunately, LIPS is minimally destructive because the amount of sample consumed is very small (nanograms). Thus, this method is suitable where only small amounts of material are available. Although the technique is usually qualitative or at best semi-quantitative, it opens up a unique pathway to the characterisation or identification of materials, or even the contamination in environmental or biomedical samples by, for instance, metals and plastics. However, the abundance of carbon in organic material can impose strong constraints when it comes to the analysis of carbon-dominated materials in a carbon-rich environment. With a spatial resolving power and spot sizes of the order of 100 µm, micro-regions can be analysed. This limit can be pushed considerably down by implementing a microscope-based LIPS technique that offers a rapid and simple analysis. Multiple elements in a sample can be determined simultaneously, giving it an advantage over many other analysis techniques, which often even require extensive sample preparation. Particularly, spatial profiling is the main advantage of such microscope-based technique. This enables imaging of samples with microscopic resolution of the distribution of contaminants. Versatile sampling of all sorts of media including biomaterials/organic samples promises successful application of LIPS for the analysis of contamination levels, but also the discovery of micro-plastics in environmentally-relevant samples.

In an exploratory effort driven by the desire to identify pollutants in certain tissue types and polymer-based debris in soil, the μLIPS technique (Fig. 5.18) was explored by the author and co-workers in conjunction with machine-learning approaches to identify different types of carbon-rich materials (such as by a comparison of different plastics among each other). In addition, the microscale localisation of debris

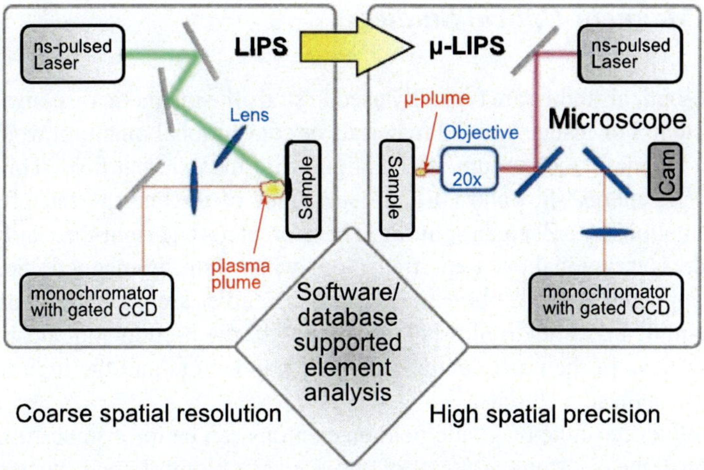

Fig. 5.18 Schematic drawing of a typical LIBS setup in comparison to a spatially-selective microscope-coupled experiment acquiring element signal from micro-sized plasma plumes

in suitable static lifescience samples was pursued. In this context, also a commercial (external) platform in its development phase was tested for this purpose, with the μLIPS test system utilising UV pulses in a conventional optical microscope with microprecise control over the sample position. In fact, this non-semiconductor study motivated further investigations and methods development which require more concentrated efforts and long-term investments to become fruitful. Particularly, the differentiation between organic materials, such as polymers, soil, cells, food etc. remains a fundamental challenge for future endeavours in this field, which could be best tackled with artificial-intelligence schemes.

The interested reader is referred to topical reviews in the LIPS literature, e.g., on food analysis [167], biomedical or biological applications [168, 169], and other novel applications [170].

In a different study, plasma parameters were studied for molybdenum samples to improve the measurement capabilities by Abdrabou and co-workers [171] based on the expertise in the El-Sherbini group in Cairo: Firstly, optimised measurement parameters were identified through variations of the laser fluencies, delay time and distance from the target surface in a 10-Hz 2-ns-pulsed laser spectroscopy setup (with frequency-doubled Q-switched Nd:YAG, UV-compatible quartz lenses, and iCCD). Secondly, the molybdenum-I lines were corrected against self-absorption and the corrected values of the plasma's temperature and the electron density using the presence of the hydrogen H_α-line were obtained.

5.6.4 Magneto-Optical Studies

Magneto-optical studies are typically used to study the magnetic response of an excitonic system. For instance in 2D materials or conventional quantum wells, excitons exhibit a magnetic momentum which aligns with the external field. This leads to a diamagnetic energy shift and a lifted degeneracy of the energy states for different spin orientations by a Zeeman splitting. Therefore, left- and right-circularly polarised light address individual excitonic transitions when a strong magnetic field of a few Tesla is applied. Currently, these magnetic-field studies are particularly of interest in the domain of 2D valleytronics [91, 93] owing to the pseudo-spin attributed to the distinct valleys in the most common 2D semiconductors and the high polarisation anisotropy in these materials.

The influence of the magnetic field on excitons can be for instance used not only to determine the electronic g-factor of individual Coulomb-bound species [172], but also the exciton Bohr radius change as a function of the dielectric environment [173]. Recent studies with very-strong pulsed magnetic fields have also provided estimates of the 2D exciton's effective reduced mass in a monolayer material [174].

In the domain of polariton research using quantum-well microcavity systems, strong magnetic fields were used to study magneto-optical effects in polariton gases. One such experiment investigated the non-equilibrium spin-Meissner effect and polarisation of condensed polaritons, referred to as spinor condensates [175], while another study analysed coherence properties of (spin-resolved) polariton condensates by second-order temporal-autocorrelation-function measurements in fields up to 5 T [61].

However, one major motivation to apply magnetic fields was given by the need to clearly evidence the matter component of the usually hybrid emitters in a polariton system when driven above the condensation threshold. The challenging task to prove the existence of light–matter-coupled quasi-particles in the nonlinear emission regime, particularly when the peculiar dispersion characteristics of cavity–polaritons is absent, could only be addressed by Zeeman splitting investigations [14, 27]. This magneto-optical method was previously applied and verified using polariton traps formed by buried modulations of the cavity length in the semiconductor microcavity structure, which gave rise to quantised polariton states due to the laterally-confining potentials provided [136]. The coexistence of polaritonic planar-resonator modes and photonic-quantum-box polariton modes allowed a systematic estimation of the matter fractions for the detectable modes, both, through the Zeeman splitting and the diamagnetic shift [136, 176].

References

1. S.W. Hell, Nobel lecture: nanoscopy with freely propagating light. Rev. Mod. Phys. **87**, 1169–1181 (2015)
2. M. Shah, L.M. Schneider, A. Rahimi-Iman, Observation of intralayer and interlayer excitons in monolayered WSe_2/WS_2 Heterostructure. Semiconductors **53**(16), 2140–2146 (2019).

https://doi.org/10.1134/S1063782619120273
3. L.M. Schneider, S.S. Esdaille, D.A. Rhodes, K. Barmak, J.C. Hone, A. Rahimi-Iman, Direct measurement of the radiative pattern of bright and dark excitons and exciton complexes in encapsulated tungsten diselenide. Sci. Rep. **10**, 8091 (2020). https://doi.org/10.1038/s41598-020-64838-z
4. A. Rahimi-Iman, *Spectroscopy Techniques for Polariton Research*. Springer Series in Optical Sciences, vol. 229 (Springer International Publishing, Cham, 2020), pp. 167–193
5. L.M. Schneider, S. Lippert, J. Kuhnert, O. Ajayi, D. Renaud, S. Firoozabadi, Q. Ngo, R. Guo, Y.D. Kim, W. Heimbrodt, J.C. Hone, and A. Rahimi-Iman, The influence of the environment on monolayer tungsten diselenide photoluminescence. Nano-Struct. Nano-Objects, **15**, 84–97 (2018). https://doi.org/10.1016/j.nanoso.2017.08.009
6. O. Mey, F. Wall, L.M. Schneider, D. Günder, F. Walla, A. Soltani, H. Roskos, N. Yao, P. Qing, W. Fang, A. Rahimi-Iman, Enhancement of the monolayer tungsten disulfide exciton photoluminescence with a two-dimensional material/air/gallium phosphide in-plane microcavity. ACS Nano **13**, 5259–5267 (2019). https://doi.org/10.1021/acsnano.8b09659
7. S. Lippert, L.M. Schneider, D. Renaud, K.N. Kang, O. Ajayi, J. Kuhnert, M.-U. Halbich, O.M. Abdulmunem, X. Lin, K. Hassoon, S. Edalati-Boostan, Y.D. Kim, W. Heimbrodt, E.-H. Yang, J.C. Hone, A. Rahimi-Iman, Influence of the substrate material on the optical properties of tungsten diselenide monolayers, 2D Mater. **4**, 025045 (2017). https://doi.org/10.1088/2053-1583/aa5b21
8. L.M. Schneider, J. Kuhnert, S. Schmitt, W. Heimbrodt, U. Huttner, L. Meckbach, T. Stroucken, S.W. Koch, S. Fu, X. Wang, K. Kang, E.-H. Yang, A. Rahimi-Iman, Spin-layer and spin-valley locking in CVD-grown AA'- and AB-stacked tungsten-disulfide bilayers. J. Phys. Chem. C **123**, 21813–21821 (2019). https://doi.org/10.1021/acs.jpcc.9b07213
9. X. Hong, J. Kim, S.-F. Shi, Y. Zhang, C. Jin, Y. Sun, S. Tongay, J. Wu, Y. Zhang, F. Wang, Ultrafast charge transfer in atomically thin MoS_2/WS_2 heterostructures. Nat. Nanotechnol. **9**, 682–686 (2014)
10. H. Fang, C. Battaglia, C. Carraro, S. Nemsak, B. Ozdol, J.S. Kang, H.A. Bechtel, S.B. Desai, F. Kronast, A.A. Unal, G. Conti, C. Conlon, G.K. Palsson, M.C. Martin, A.M. Minor, C.S. Fadley, E. Yablonovitch, R. Maboudian, A. Javey, Strong interlayer coupling in van der Waals heterostructures built from single-layer chalcogenides. Proc. Natl. Acad. Sci. **111**, 6198–6202 (2014)
11. A.J. Huber, D. Kazantsev, F. Keilmann, J. Wittborn, R. Hillenbrand, Simultaneous IR material recognition and conductivity mapping by nanoscale near-field microscopy. Adv. Mater. **19**(17), 2209–2213 (2007)
12. F. Walla, M.M. Wiecha, N. Mecklenbeck, S. Beldi, F. Keilmann, M.D. Thomson, H.G. Roskos, Anisotropic excitation of surface plasmon polaritons on a metal film by a scattering-type scanning near-field microscope with a non-rotationally-symmetric probe tip. Nanophotonics **7**(1), 269–276 (2018)
13. T. Taubner, F. Keilmann, R. Hillenbrand, Nanoscale-resolved subsurface imaging by scattering-type near-field optical microscopy. Opt. Express **13**(22), 8893 (2005)
14. A. Rahimi-Iman, *Nichtlineare Effekte in III/V Quantenfilm-Mikroresonatoren: Von dynamischer Bose–Einstein-Kondensation hin zum elektrisch betriebenen Polariton-Laser* (Cuvillier Verlag, Göttingen, 2013)
15. R. Li, L.M. Schneider, W. Heimbrodt, H. Wu, M. Koch, A. Rahimi-Iman, Gate tuning of Förster resonance energy transfer in a graphene - quantum dot FET photo-detector. Sci. Rep. **6**, 28224 (2016). https://doi.org/10.1038/srep28224
16. H. Deng, H. Haug, Y. Yamamoto, Exciton-polariton Bose-Einstein condensation. Rev. Mod. Phys. **82**, 1489 (2010)
17. R.A. Taylor, J.F. Rayn, Time-resolved exciton photoluminescence in GaSe and GaTe. J. Phys. C: Solid State Phys. **20**, 6175–6187 (1987)
18. H. Haug, S.W. Koch, *Quantum Theory of the Optical and Electronic Properties of Semiconductors*, 5th edn. (World Scientific Publishing Co., 2009)

19. Hamamatsu Photonics, Universal streak camera C10910 series (webpage 2019: hamamatsu.com/resources/pdf/sys/SHSS0016E_C10910s.pdf) (2019)
20. M. Müller, R. André, J. Bleuse, L.S. Dang, A. Huynh, J. Tignon, P. Roussignol, C. Delalande, Non-linear polariton dynamics in II-VI microcavities. Semicond. Sci. Technol. **18**(10), S319 (2003)
21. J. Bloch, B. Sermage, M. Perrin, P. Senellart, R. André, L.S. Dang, Monitoring the dynamics of a coherent cavity polariton population. Phys. Rev. B **71**, 155311 (2005)
22. J.J. Baumberg, P.G. Lagoudakis, Parametric amplification and polariton liquids in semiconductor microcavities. Phys. Status Solidi B **242**(11), 2210–2223 (2005)
23. A. Amo, D. Sanvitto, F. P. Laussy, D. Ballarini, E. d. Valle, M.D. Martín, A. Lemaître, J. Bloch, D.N. Krizhanovskii, M.S. Skolnick, C. Tejedor, L. Viña, Collective fluid dynamics of a polariton condensate in a semiconductor microcavity. Nature **457**, 291–295 (2009)
24. M. De Giorgi, D. Ballarini, P. Cazzato, G. Deligeorgis, S.I. Tsintzos, Z. Hatzopoulos, P.G. Savvidis, G. Gigli, F.P. Laussy, D. Sanvitto, Relaxation oscillations in the formation of a polariton condensate. Phys. Rev. Lett. **112**, 113602 (2014)
25. D. Ballarini, D. Sanvitto, A. Amo, L. Viña, M. Wouters, I. Carusotto, A. Lemaître, J. Bloch, Observation of long-lived polariton states in semiconductor microcavities across the parametric threshold. Phys. Rev. Lett. **102**, 056402 (2009)
26. J.-S. Tempel, F. Veit, M. Assmann, L.E. Kreilkamp, A. Rahimi-Iman, A. Löffler, S. Höfling, S. Reitzenstein, L. Worschech, A. Forchel, M. Bayer, Characterization of two-threshold behavior of the emission from a GaAs microcavity. Phys. Rev. B **85**, 075318 (2012)
27. C. Schneider, A. Rahimi-Iman, N.Y. Kim, J. Fischer, I.G. Savenko, M. Amthor, M. Lermer, A. Wolf, L. Worschech, V.D. Kulakovskii, I.A. Shelykh, M. Kamp, S. Reitzenstein, A. Forchel, Y. Yamamoto, S. Höfling, An electrically pumped polariton laser. Nature **497**, 348–352 (2013)
28. M. Gerhard, A.P. Arndt, I.A. Howard, A. Rahimi-Iman, U. Lemmer, M. Koch, Temperature- and energy-dependent separation of charge-transfer states in PTB7-based organic solar cells. J. Phys. Chem. C **119**, 28309–28318 (2015)
29. R. Woscholski, M.K. Shakfa, S. Gies, M. Wiemer, A. Rahimi-Iman, M. Zimprich, S. Reinhard, K. Jandieri, S.D. Baranovskii, W. Heimbrodt, K. Volz, W. Stolz, M. Koch, Time-resolved photoluminescence of Ga(NAsP) multiple quantum wells grown on Si substrate: Effects of rapid thermal annealing. Thin Solid Films **613**, 55–58 (2016)
30. R. Woscholski, S. Gies, M. Wiemer, M.K. Shakfa, A. Rahimi-Iman, P. Ludewig, S. Reinhard, K. Jandieri, S.D. Baranovskii, W. Heimbrodt, K. Volz, W. Stolz, M. Koch, Influence of growth temperature and disorder on spectral and temporal properties of Ga(NAsP) heterostructures. J. Appl. Phys. **119**, 145707 (2016)
31. L.M. Schneider, S. Lippert, J. Kuhnert, D. Renaud, K.N. Kang, O. Ajayi, M.-U. Halbich, O.M. Abdulmunem, X. Lin, K. Hassoon, S. Edalati-Boostan, Y.D. Kim, W. Heimbrodt, E.H. Yang, J.C. Hone, A. Rahimi-Iman, The impact of the substrate material on the optical properties of 2D WSe_2 monolayers. Semiconductors **52**, 565–571 (2018). https://doi.org/10.1134/S1063782618050275
32. H. Zhu, J. Wang, Z. Gong, Y.D. Kim, J. Hone, X.-Y. Zhu, Interfacial charge transfer circumventing momentum mismatch at two-dimensional van der waals heterojunctions. Nano Lett. **17**, 3591–3598 (2017)
33. L.M. Schneider, J. Kuhnert, S. Lippert, D. Renaud, W. Heimbrodt, O. Ajayi, Y.D. Kim, J.C. Hone, K.N. Kang, E.-H. Yang, A. Rahimi-Iman, Density-dependent excitonic properties and dynamics in 2d heterostructures consisting of boron nitride and monolayer or few-layer tungsten diselenide. in *Micro- and Nanotechnology Sensors, Systems, and Applications X*, vol. 10639 (International Society for Optics and Photonics, 2018), p. 106390V
34. M. Assmann, F. Veit, M. Bayer, M. van der Poel, J.M. Hvam, Higher-order photon bunching in a semiconductor microcavity. Science **325**, 297–300 (2009)
35. M.A. Kahlow, W. Jarzeba, T.P. DuBruil, P.F. Barbara, Ultrafast emission spectroscopy in the ultraviolet by time-gated upconversion. Rev. Sci. Instrum. **59**, 1098 (1988)
36. Swamp Optics, Tutorials (webpage 2019: http://www.swampoptics.com/tutorials.html). Online (2019)

37. C. Lammers, M. Stein, C. Berger, C. Möller, C. Fuchs, A. Ruiz Perez, A. Rahimi-Iman, J. Hader, J.V. Moloney, W. Stolz, S.W. Koch, M. Koch, Gain spectroscopy of a type-II VECSEL chip. Appl. Phys. Lett. **109**, 232107 (2016)
38. A. Chernikov, C. Ruppert, H.M. Hill, A.F. Rigosi, T.F. Heinz, Population inversion and giant bandgap renormalization in atomically thin WS2 layers. Nat. Photon. **9**, 466–470 (2015)
39. R.A. Kaindl, M.A. Carnahan, D. Hägele, R. Lövenich, D.S. Chemla, Ultrafast terahertz probes of transient conducting and insulating phases in an electron-hole gas. Nature **423**, 734–738 (2003)
40. J.M. Ménard, C. Poellmann, M. Porer, U. Leierseder, E. Galopin, A. Lemaître, A. Amo, J. Bloch, R. Huber, Revealing the dark side of a bright exciton-polariton condensate. Nat. Commun. **5**, 4648 (2014)
41. C. Poellmann, P. Steinleitner, U. Leierseder, P. Nagler, G. Plechinger, M. Porer, R. Bratschitsch, C. Schüller, T. Korn, R. Huber, Resonant internal quantum transitions and femtosecond radiative decay of excitons in monolayer WSe_2. Nat. Mater. **14**, 889–893 (2015)
42. M. Stein, C. Lammers, P. Springer, P.-H. Richter, S.W. Koch, M. Koch, M. Kira, Density-dependent exciton dynamics and L-valley anisotropy in germanium. Phys. Rev. B **95**, 155207 (2017)
43. A.D. Jameson, J.L. Tomaino, Y.-S. Lee, G. Khitrova, H.M. Gibbs, C.N. Böttge, A.C. Klettke, M. Kira, S.W. Koch, Direct measurement of light-matter energy exchange inside a microcavity. Optica **1**, 276 (2014)
44. L. Dominici, D. Colas, S. Donati, J.P. Restrepo Cuartas, M. De Giorgi, D. Ballarini, G. Guirales, J.C. López Carreño, A. Bramati, G. Gigli, E. Del Valle, F. P. Laussy, D. Sanvitto, Ultrafast control and Rabi oscillations of polaritons. Phys. Rev. Lett. **113**, 226401 (2014)
45. D. Colas, L. Dominici, S. Donati, A.A. Pervishko, T.C. Liew, I.A. Shelykh, D. Ballarini, M. De Giorgi, A. Bramati, G. Gigli, E. Del Valle, F.P. Laussy, A.V. Kavokin, D. Sanvitto, Polarization shaping of Poincaré beams by polariton oscillations. Light Sci. Appl. **4**, e350 (2015)
46. L.M. Schneider, S.S. Esdaille, D.A. Rhodes, K. Barmak, J.C. Hone, A. Rahimi-Iman, Optical dispersion of valley-hybridised coherent excitons with momentum-dependent valley polarisation in monolayer semiconductor. 2D Mater. **8**, 015009 (2020)
47. M.J. Drexler, R. Woscholski, S. Lippert, W. Stolz, A. Rahimi-Iman, M. Koch, Disturbing the coherent dynamics of an excitonic polarization with strong terahertz fields. Phys. Rev. B **90**, 195304 (2014)
48. T. Jakubczyk, V. Delmonte, M. Koperski, K. Nogajewski, C. Faugeras, W. Langbein, M. Potemski, J. Kasprzak, Radiatively limited dephasing and exciton dynamics in MoS_2 monolayers revealed with four-wave mixing microscopy. Nano Lett. **16**, 5333–5339 (2016)
49. M. Wichmann, G. Town, J. Quante, M. Gaafar, A. Rahimi-Iman, W. Stolz, S.W. Koch, M. Koch, Antiphase noise dynamics in a dual-wavelength vertical-external-cavity surface-emitting laser. IEEE Photon. Technol. Lett. **27**, 2039–2042 (2015)
50. A. Rahimi-Iman, Nonclassical Light Emission from Optically and Electrically Driven High-Q Quantum Dot-Micropillar Cavities, Master's thesis, Würzburg University (2009)
51. R. Hanbury Brown, R.Q. Twiss, Correlation between photons in two coherent beams of light. Nature **177**, 27–29 (1956)
52. M. Orszag, *Quantum Optics: Including Noise Reduction, Trapped Ions, Quantum Trajectories, and Decoherence*, 3rd edn. (Springer International Publishing, 2016)
53. H. Deng, G. Weihs, C. Santori, J. Bloch, Y. Yamamoto, Condensation of semiconductor microcavity exciton polaritons. Science **298**(5591), 199–202 (2002)
54. S. Reitzenstein, A. Forchel, Quantum dot micropillars. J. Phys. D: Appl. Phys. **43**(3), 033001 (2010)
55. S.M. Ulrich, C. Gies, S. Ates, J. Wiersig, S. Reitzenstein, C. Hofmann, A. Löffler, A. Forchel, F. Jahnke, P. Michler, Photon statistics of semiconductor microcavity lasers. Phys. Rev. Lett. **98**, 043906 (2007)
56. S. Reitzenstein, T. Heindel, C. Kistner, A. Rahimi-Iman, C. Schneider, S. Höfling, A. Forchel, Low threshold electrically pumped quantum dot-micropillar lasers. Appl. Phys. Lett. **93**(6), 061104 (2008)

57. J. Kasprzak, M. Richard, A. Baas, B. Deveaud, R. Andre, J.-P. Poizat, L.S. Dang, Second-order time correlations within a polariton bose-einstein condensate in a CdTe microcavity. Phys. Rev. Lett. **100**, 067402 (2008)
58. M. Assmann, J.-S. Tempel, F. Veit, M. Bayer, A. Rahimi-Iman, A. Löffler, S. Höfling, S. Reitzenstein, L. Worschech, A. Forchel, From polariton condensates to highly photonic quantum degenerate states of bosonic matter. Proc. Natl. Acad. Sci. **108**, 1804–1809 (2011)
59. A. Rahimi-Iman, A.V. Chernenko, J. Fischer, S. Brodbeck, M. Amthor, C. Schneider, A. Forchel, S. Höfling, S. Reitzenstein, M. Kamp, Coherence signatures and density-dependent interaction in a dynamical exciton-polariton condensate. Phys. Rev. B **86**, 155308 (2012)
60. M. Amthor, S. Weissenseel, J. Fischer, M. Kamp, C. Schneider, S. Höfling, Electro-optical switching between polariton and cavity lasing in an ingaas quantum well microcavity. Opt. Express **22**, 31146–31153 (2014)
61. A.V. Chernenko, A. Rahimi-Iman, J. Fischer, M. Amthor, C. Schneider, S. Reitzenstein, A. Forchel, S. Hoefling, Polariton condensate coherence in planar microcavities in a magnetic field. Semiconductors **50**, 1609–1613 (2016). https://doi.org/10.1134/S1063782616120058
62. S. Kim, B. Zhang, Z. Wang, J. Fischer, S. Brodbeck, M. Kamp, C. Schneider, S. Höfling, H. Deng, Coherent polariton laser. Phys. Rev. X **6**, 011026 (2016)
63. C. Jin, E.C. Regan, A. Yan, M. Iqbal Bakti Utama, D. Wang, S. Zhao, Y. Qin, S. Yang, Z. Zheng, S. Shi, K. Watanabe, T. Taniguchi, S. Tongay, A. Zettl, and F. Wang, Observation of moiré excitons in WSe_2/WS_2 heterostructure superlattices. Nature **567**, 76–80 (2019)
64. K. He, N. Kumar, L. Zhao, Z. Wang, K.F. Mak, H. Zhao, J. Shan, Tightly bound excitons in monolayer WSe_2. Phys. Rev. Lett. **113**, 026803 (2014)
65. A. Chernikov, T.C. Berkelbach, H.M. Hill, A. Rigosi, Y. Li, O.B. Aslan, D.R. Reichman, M.S. Hybertsen, T.F. Heinz, Exciton binding energy and nonhydrogenic rydberg series in monolayer WSe_2. Phys. Rev. Lett. **113**, 076802 (2014)
66. L.M. Schneider, S.S. Esdaille, D.A. Rhodes, K. Barmak, J.C. Hone, A. Rahimi-Iman, Shedding light on exciton's nature in monolayer quantum material by optical dispersion measurements. Opt. Express **27**, 37131–37149 (2019). https://doi.org/10.1364/OE.27.037131
67. I.I. Rabi, Space quantization in a gyrating magnetic field. Phys. Rev. **51**, 652–654 (1937)
68. M. Brune, F. Schmidt-Kaler, A. Maali, J. Dreyer, E. Hagley, J.M. Raimond, S. Haroche, Quantum rabi oscillation: a direct test of field quantization in a cavity. Phys. Rev. Lett. **76**, 1800–1803 (1996)
69. C.F. Klingshirn, *Semiconductor optics* (Springer, 2012)
70. W. Demtröder, *Laser Spectroscopy 1: Basic Principles*, 5th edn. (Springer, Berlin Heidelberg, 2014)
71. Z. Ye, T. Cao, K. O'Brien, H. Zhu, X. Yin, Y. Wang, S.G. Louie, X. Zhang, Probing excitonic dark states in single-layer tungsten disulphide. Nature **513**, 214–218 (2014)
72. H.M. Hill, A.F. Rigosi, C. Roquelet, A. Chernikov, T.C. Berkelbach, D.R. Reichman, M.S. Hybertsen, L.E. Brus, T.F. Heinz, Observation of excitonic rydberg states in monolayer MoS_2 and WS_2 by photoluminescence excitation spectroscopy. Nano Lett. **15**, 2992–2997 (2015)
73. Y. You, X.-X. Zhang, T.C. Berkelbach, M.S. Hybertsen, D.R. Reichman, T.F. Heinz, Observation of biexcitons in monolayer WSe_2. Nat. Phys. **11**, 477–481 (2015)
74. S.-Y. Chen, T. Goldstein, T. Taniguchi, K. Watanabe, J. Yan, Coulomb-bound four- and five-particle intervalley states in an atomically-thin semiconductor. Nat. Commun. **9**, 1–8 (2018)
75. M. Barbone, A.R.-P. Montblanch, D.M. Kara, C. Palacios-Berraquero, A.R. Cadore, D.D. Fazio, B. Pingault, E. Mostaani, H. Li, B. Chen, K. Watanabe, T. Taniguchi, S. Tongay, G. Wang, A.C. Ferrari, M. Atatüre, Charge-tuneable biexciton complexes in monolayer WSe_2. Nat. Commun. **9**, 1–6 (2018)
76. G. Wang, C. Robert, M.M. Glazov, F. Cadiz, E. Courtade, T. Amand, D. Lagarde, T. Taniguchi, K. Watanabe, B. Urbaszek, X. Marie, In-plane propagation of light in transition metal dichalcogenide monolayers: optical selection rules. Phys. Rev. Lett. **119**, 047401 (2017)
77. I.I. Brener, M. Olszakier, E. Cohen, E. Ehrenfreund, A. Ron, L. Pfeiffer, Particle localization and phonon sidebands in GaAs/AlxGa_{1-x}As multiple quantum wells. Phys. Rev. B **46**, 7927–7930 (1992)

78. A. Chiari, M. Colocci, F. Fermi, Y. Li, R. Querzoli, A. Vinattieri, W. Zhuang, Temperature dependence of the photoluminescence in GaAs/GaAlAs multiple quantum well structure. Phys. Status Solidi B **147**, 421 (1988)
79. J. Hader, J.V. Moloney, S.W. Koch, Supression of carrier recombination in semiconductor lasers by phase-space filling. Appl. Phys. Lett. **87**, 201112 (2005)
80. D. Christiansen, M. Selig, G. Berghäuser, R. Schmidt, I. Niehues, R. Schneider, A. Arora, S.M. de Vasconcellos, R. Bratschitsch, E. Malic, A. Knorr, Phonon sidebands in monolayer transition metal dichalcogenides. Phys. Rev. Lett. **119**, 187402 (2017)
81. M. Selig, G. Berghäuser, A. Raja, P. Nagler, C. Schüller, T.F. Heinz, T. Korn, A. Chernikov, E. Malic, A. Knorr, Excitonic linewidth and coherence lifetime in monolayer transition metal dichalcogenides. Nat. Commun. **7**, 13279 (2016)
82. S. Brem, A. Ekman, D. Christiansen, F. Katsch, M. Selig, C. Robert, X. Marie, B. Urbaszek, A. Knorr, E. Malic, Phonon-assisted photoluminescence from indirect excitons in monolayers of transition-metal dichalcogenides. Nano Lett. **20**(4), 2849–2856 (2020)
83. C.M. Chow, H. Yu, A.M. Jones, J.R. Schaibley, M. Koehler, D.G. Mandrus, R. Merlin, W. Yao, X. Xu, Phonon-assisted oscillatory exciton dynamics in monolayer MoSe2. npj 2D Mater Appl. **1**, 33 (2017)
84. A. Steinhoff, M. Rösner, F. Jahnke, T.O. Wehling, C. Gies, Influence of excited carriers on the optical and electronic properties of MoS_2. Nano Lett. **14**(7), 3743–3748 (2014)
85. K.P. O'Donnell, X. Chen, Temperature dependence of semiconductor band gaps. Appl. Phys. Lett. **58**, 2924–2926 (1991)
86. Y.P. Varshni, Temperature dependence of the energy gap in semiconductors. Physica **34**, 149–154 (1967)
87. K.F. Mak, C. Lee, J. Hone, J. Shan, T.F. Heinz, Atomically thin MoS_2: a new direct-gap semiconductor. Phys. Rev. Lett. **105**, 136805 (2010)
88. A.K. Geim, I.V. Grigorieva, Van der Waals heterostructures. Nature **499**, 419 (2013)
89. K.S. Novoselov, A. Mishchenko, A. Carvalho, A.H. Castro Neto, 2D materials and van der Waals heterostructures. *Science* **353**, aac9439 (2016)
90. U. Wurstbauer, B. Miller, E. Parzinger, A.W. Holleitner, Light matter interaction in transition metal dichalcogenides and their heterostructures. J. Phys. D: Appl. Phys. **50**, 173001 (2017)
91. J.R. Schaibley, H. Yu, G. Clark, P. Rivera, J.S. Ross, K.L. Seyler, W. Yao, X. Xu, Valleytronics in 2D materials. Nat. Rev. Mater. **1**, 16055 (2016)
92. G. Wang, A. Chernikov, M.M. Glazov, T.F. Heinz, X. Marie, T. Amand, B. Urbaszek, Colloquium: excitons in atomically thin transition metal dichalcogenides. Rev. Mod. Phys. **90**, 21001 (2018)
93. K.F. Mak, D. Xiao, J. Shan, Light-valley interactions in 2D semiconductors. Nat. Photon. **12**, 451–460 (2018)
94. K.K. Kam, B.A. Parkinson, Detailed photocurrent spectroscopy of the semiconducting group VIB transition metal dichalcogenides. J. Phys. Chem. **86**, 463–467 (1982)
95. F.H.L. Koppens, T. Mueller, P. Avouris, A.C. Ferrari, M.S. Vitiello, M. Polini, Photodetectors based on graphene, other two-dimensional materials and hybrid systems. Nat. Nanotechnol. **9**, 780–793 (2014)
96. K.F. Mak, J. Shan, Photonics and optoelectronics of 2D semiconductor transition metal dichalcogenides. Nat. Photon. **10**, 216–226 (2016)
97. T. Chen, Y. Sheng, Y. Zhou, R.-J. Chang, X. Wang, H. Huang, Q. Zhang, L. Hou, J.H. Warner, High Photoresponsivity in Ultrathin 2D Lateral Graphene:WS_2: Graphene Photodetectors Using direct CVD growth. ACS Appl. Mater. Interfaces **11**, 6421–6430 (2019)
98. G. Thiele, S. Lippert, F. Fahrnbauer, P. Bron, O. Oeckler, A. Rahimi-Iman, M. Koch, B. Roling, S. Dehnen, $K_2Hg_2Se_3$: large-scalesynthesis of a photoconductor material prototype with a columnar polyanionic substructure. Chem. Mater. **27**, 4114–4118 (2015)
99. B. Wagner, N. Dehnhardt, M. Schmid, B.P. Klein, L. Ruppenthal, P. Müller, M. Zugermeier, J.M. Gottfried, S. Lippert, M.-U. Halbich, A. Rahimi-Iman, J. Heine, Color change effect in an organic-inorganic hybrid material based on a porphyrin diacid. J. Phys. Chem. C **120**, 28363–28373 (2016)

100. W. Demtröder, *Laser Spectroscopy 2: Experimental Techniques*, 5th edn. (Springer, Berlin Heidelberg, 2015)
101. H. Sahin, S. Tongay, S. Horzum, W. Fan, J. Zhou, J. Li, J. Wu, F.M. Peeters, Anomalous Raman spectra and thickness-dependent electronic properties of WSe_2. Phys. Rev. B **87**, 165409 (2013)
102. R. Zhang, V. Koutsos, R. Cheung, Elastic properties of suspended multilayer WSe_2. Appl. Phys. Lett. **108**(4), 042104 (2016)
103. S. Huang, L. Liang, X. Ling, A.A. Puretzky, D.B. Geohegan, B.G. Sumpter, J. Kong, V. Meunier, M.S. Dresselhaus, Low-frequency interlayer raman modes to probe interface of twisted bilayer MoS_2. Nano Lett. **16**(2), 1435–1444 (2016)
104. A.A. Puretzky, L. Liang, X. Li, K. Xiao, B.G. Sumpter, V. Meunier, D.B. Geohegan, Twisted $MoSe_2$ bilayers with variable local stacking and interlayer coupling revealed by low-frequency raman spectroscopy. ACS Nano **10**(2), 2736–2744 (2016)
105. Q.-H. Tan, Y.-J. Sun, X.-L. Liu, Y. Zhao, Q. Xiong, P.-H. Tan, J. Zhang, Observation of forbidden phonons, Fano resonance and dark excitons by resonance Raman scattering in few-layer WS_2. *2D Mater.* **4**, 031007 (2017)
106. D. Nam, J.U. Lee, H. Cheong, Excitation energy dependent Raman spectrum of MoSe2. Sci. Rep. **5**, 1–6 (2015)
107. A. Berkdemir, H.R. Gutiérrez, A.R. Botello-Méndez, N. Perea-López, A.L. Elías, C.-I. Chia, B. Wang, V.H. Crespi, F. López-Urías, J.-C. Charlier, H. Terrones, M. Terrones, Identification of individual and few layers of WS_2 using Raman Spectroscopy. Sci. Rep. **3**, 1755 (2013)
108. C. Kriso, S. Kress, T. Munshi, M. Grossmann, R. Bek, M. Jetter, P. Michler, W. Stolz, M. Koch, A. Rahimi-Iman, Microcavity-enhanced Kerr nonlinearity in a vertical-external-cavity surface-emitting laser. Opt. Express **27**, 11914–11929 (2019). https://doi.org/10.1364/OE.27.011914
109. C. Kriso, S. Kress, T. Munshi, M. Grossmann, R. Bek, M. Jetter, P. Michler, W. Stolz, M. Koch, A. Rahimi-Iman, Wavelength and pump-power dependent nonlinear refraction and absorption in a semiconductor disk laser. IEEE Photon. Technol. Lett. **32**, 85–88 (2020). Published online (2019)
110. C. Kriso, M. Stein, T. Haeger, N. Pourdavoud, M. Gerhard, A. Rahimi-Iman, T. Riedl, M. Koch, Nonlinear refraction in $CH_3NH_3PbBr_3$ single crystals. Opt. Lett. **45**, 2431–2434 (2020)
111. M. Scheller, J.M. Yarborough, J.V. Moloney, M. Fallahi, M. Koch, S.W. Koch, Room temperature continuous wave milliwatt terahertz source. Opt. Express **18**, 27112 (2010)
112. M. Wichmann, M. Stein, A. Rahimi-Iman, S.W. Koch, M. Koch, Interferometric characterization of a semiconductor disk laser driven terahertz source. J. Infrared Milli. Terahz Waves **35**, 503–508 (2014)
113. H. Guoyu, C. Kriso, F. Zhang, M. Wichmann, W. Stolz, K.A. Fedorova, A. Rahimi-Iman, Two-chip power-scalable THz-generating semiconductor disk laser. Opt. Lett. **44**, 4000–4003 (2019)
114. K.A. Fedorova, H. Guoyu, M. Wichmann, C. Kriso, F. Zhang, W. Stolz, M. Scheller, M. Koch, A. Rahimi-Iman, Widely-tunable THz-generating semiconductor disk laser. Phys. Status Solidi RRL **14**, 2000204 (2020). https://doi.org/10.1002/pssr.202000204
115. D.E. Spence, P.N. Kean, W. Sibbett, 60-fsec pulse generation from a self-mode-locked Ti:sapphire laser. Opt. Lett. **16**, 42–44 (1991)
116. S.T. Cundiff, W.H. Knox, E.P. Ippen, H.A. Haus, Frequency-dependent mode size in broadband kerr-lens mode locking. Opt. Lett. **21**, 662–664 (1996)
117. M. Sheik-Bahae, A.A. Said, T. Wei, D.J. Hagan, E.W.V. Stryland, Sensitive measurement of optical nonlinearities using a single beam. IEEE J. Quantum Electron. **26**, 760–769 (1990)
118. M. Moran, C.-Y. She, R. Carman, Interferometric measurements of the nonlinear refractive-index coefficient relative to cs2in laser-system-related materials. IEEE J. Quantum Electron. **11**, 259–263 (1975)
119. D. Weaire, B.S. Wherrett, D.A.B. Miller, S.D. Smith, Effect of low-power nonlinear refraction on laser-beam propagation in InSb. Opt. Lett. **4**, 331–333 (1979)

120. R. Adair, L.L. Chase, S.A. Payne, Nonlinear refractive-index measurements of glasses using three-wave frequency mixing. J. Opt. Soc. Am. B **4**, 875–881 (1987)
121. H. Zhang, S. Virally, Q. Bao, L. Kian Ping, S. Massar, N. Godbout, P. Kockaert, Z-scan measurement of the nonlinear refractive index of graphene. Opt. Lett. **37**, 1856–1858 (2012)
122. X. Zheng, Y. Zhang, R. Chen, X. Cheng, Z. Xu, T. Jiang, Z-scan measurement of the nonlinear refractive index of monolayer WS_2. Opt. Express **23**, 15616–15623 (2015)
123. N. Vermeulen, D. Castelló-Lurbe, M. Khoder, I. Pasternak, A. Krajewska, T. Ciuk, W. Strupinski, J. Cheng, H. Thienpont, J. Van Erps, Graphene's nonlinear-optical physics revealed through exponentially growing self-phase modulation. Nat. Commun. **9**, 2675 (2018)
124. P.A. Franken, A.E. Hill, C.W. Peters, G. Weinreich, Generation of optical harmonics. Phys. Rev. Lett. **7**, 118–119 (1961)
125. K. Suizu, K. Koketsu, T. Shibuya, T. Tsutsui, T. Akiba, K. Kawase, Extremely frequency-widened terahertz wave generation using cherenkov-type radiation. Opt. Express **17**, 6676–6681 (2009)
126. V. Bleko, P. Karataev, A. Konkov, K. Kruchinin, G. Naumenko, A. Potylitsyn, T. Vaughan, Coherent cherenkov radiation as an intense thz source. J. Phys. Conf. Ser. **732**, 012006 (2016)
127. J.M. Dudley, G. Genty, S. Coen, Supercontinuum generation in photonic crystal fiber. Rev. Mod. Phys. **78**, 1135–1184 (2006)
128. J.E. Zimmermann, B. Li, J. Hone, U. Höfer, G. Mette, Second-harmonic imaging microscopy for time-resolved investigations of transition metal dichalcogenides (2016) arxiv:1608.03434
129. A. Schlehahn, M. Gaafar, M. Vaupel, M. Gschrey, P. Schnauber, J.-H. Schulze, S. Rodt, A. Strittmatter, W. Stolz, A. Rahimi-Iman, T. Heindel, M. Koch, S. Reitzenstein, Single-photon emission at a rate of 143 MHz from a deterministic quantum-dot microlens triggered by a mode-locked vertical-external-cavity surface-emitting laser. Appl. Phys. Lett. **107**, 041105 (2015). https://doi.org/10.1063/1.4927429
130. F. Zhang, S. Wang, A. Rehn, M. Wichmann, G. Urbasch, A. Rahimi-Iman, S.W. Koch, M. Koch, Investigation of the beam quality of a terahertz emitting vertical-external-cavity surface-emitting laser. J. Infrared Milli. Terahz Waves **37**, 536–539 (2016)
131. F. Zhang, M. Gaafar, C. Möller, W. Stolz, M. Koch, A. Rahimi-Iman, Dual-wavelength emission from a serially connected two-chip VECSEL. IEEE Photon. Technol. Lett. **28**, 927–929 (2016)
132. F. Zhang, C. Möller, M. Koch, S.W. Koch, A. Rahimi-Iman, W. Stolz, Impact of detuning on the performance of semiconductor disk lasers. Appl. Phys. B **123**, 291 (2017)
133. W.-T. Hsu, Z.-A. Zhao, L.-J. Li, C.-H. Chen, M.-H. Chiu, P.-S. Chang, Y.-C. Chou, W.-H. Chang, Second harmonic generation from artificially stacked transition metal dichalcogenide twisted bilayers. ACS Nano **8**, 2951–2958 (2014)
134. D. Al Nakdali, M. K. Shakfa, B. Heinen, B. Kunert, W. Stolz, S.W. Koch, J. Hader, J.V. Moloney, A. Rahimi-Iman, M. Koch, Analysis of optical scattering losses in vertical-external-cavity surface-emitting lasers. Appl. Phys. B **120**, 41–46 (2015)
135. A. Rahimi-Iman, Winkelaufgelöste Spektroskopie an Exziton-Polaritonen in optisch und elektrisch angeregten III/V Quantenfilm-Mikroresonatoren, Engineering diploma thesis, Julius-Maximilians-Universität Würzburg, Lehrstuhl für Technische Physik (2010)
136. A. Rahimi-Iman, C. Schneider, J. Fischer, S. Holzinger, M. Amthor, S. Höfling, S. Reitzenstein, L. Worschech, M. Kamp, A. Forchel, Zeeman splitting and diamagnetic shift of spatially confined quantum-well exciton polaritons in an external magnetic field. Phys. Rev. B **84**, 165325 (2011)
137. J.A. Schuller, S. Karaveli, T. Schiros, K. He, S. Yang, I. Kymissis, J. Shan, R. Zia, Orientation of luminescent excitons in layered nanomaterials. Nat. Nanotechnol. **8**, 271–276 (2013)
138. C.W. Lai, N.Y. Kim, S. Utsunomiya, G. Roumpos, H. Deng, M.D. Fraser, T. Byrnes, P. Recher, N. Kumada, T. Fujisawa, Y. Yamamoto, Coherent zero-state and π-state in an exciton-polariton condensate array. Nature **450**, 529–532 (2007)
139. F. Wall, O. Mey, L.M. Schneider, A. Rahimi-Iman, Continuously-tunable light-matter coupling in optical microcavities with 2D semiconductors. Sci. Rep. **10**, 8303 (2020). https://doi.org/10.1038/s41598-020-64909-1

140. A. Rustagi, A.F. Kemper, Photoemission signature of excitons. Phys. Rev. B **97**, 235310 (2018)
141. H. Yu, G.-B. Liu, P. Gong, X. Xu, W. Yao, Dirac cones and dirac saddle points of bright excitons in monolayer transition metal dichalcogenides. Nat. Commun. **5**, 3876 (2014)
142. F. Wu, F. Qu, A.H. MacDonald, Exciton band structure of monolayer MoS_2. Phys. Rev. B **91**, 075310 (2015)
143. D.Y. Qiu, T. Cao, S.G. Louie, Nonanalyticity, valley quantum phases, and lightlike exciton dispersion in monolayer transition metal dichalcogenides: Theory and first-principles calculations. Phys. Rev. Lett. **115**, 176801 (2015)
144. M.M. Glazov, T. Amand, X. Marie, D. Lagarde, L. Bouet, B. Urbaszek, Exciton fine structure and spin decoherence in monolayers of transition metal dichalcogenides. Phys. Rev. B **89**, 201302 (2014)
145. T. Deilmann, K.S. Thygesen, Finite-momentum exciton landscape in mono- and bilayer transition metal dichalcogenides. 2D Mater. **6**, p. 035003 (2019)
146. A.E. Siegman, How to (maybe) measure laser beam quality, in *DPSS (Diode Pumped Solid State) Lasers: Applications and Issues* (Optical Society of America, 1998), p. MQ1
147. ISO 11146, Lasers and laser-related equipment – test methods for laser beam parameters – beam width, divergence, angle and beam propagation factor (1999)
148. C. Möller, F. Zhang, C. Fuchs, C. Berger, A. Rehn, A.R. Perez, A. Rahimi-Iman, J. Hader, M. Koch, J.V. Moloney, S.W. Koch, W. Stolz, Fundamental transverse mode operation of a type-II vertical-external-cavity surface-emitting laser at 1.2 μm. Electron. Lett. **53**, 93–94 (2017)
149. H. Richter, N. Rothbart, H.-W. Hübers, Characterizing the beam properties of terahertz quantum-cascade lasers. J. Infrared Milli. Terahz Waves **35**, 686–698 (2014)
150. Y. Song, H. Dery, Transport theory of monolayer transition-metal dichalcogenides through symmetry. Phys. Rev. Lett. **111**, 026601 (2013)
151. H. Dery, Y. Song, Polarization analysis of excitons in monolayer and bilayer transition-metal dichalcogenides. Phys. Rev. B **92**, 125431 (2015)
152. C. Robert, T. Amand, F. Cadiz, D. Lagarde, E. Courtade, M. Manca, T. Taniguchi, K. Watanabe, B. Urbaszek, X. Marie, Fine structure and lifetime of dark excitons in transition metal dichalcogenide monolayers. Phys. Rev. B **96**, 755423 (2017)
153. J. Kasprzak, M. Richard, S. Kundermann, A. Baas, P. Jeambrun, J.M.J. Keeling, F.M. Marchetti, M.H. Szymańska, R. André, J.L. Staehli, V. Savona, P.B. Littlewood, B. Deveaud, L.S. Dang, Bose-Einstein condensation of exciton polaritons. Nature **443**, 409–414 (2006)
154. A. Rahimi-Iman, *On the Condensation of Polaritons*. Springer Series in Optical Sciences, vol. 229 (Springer International Publishing, Cham, 2020), pp. 65–85
155. P. Jepsen, D. Cooke, M. Koch, Terahertz spectroscopy and imaging - Modern techniques and applications. Laser Photon. Rev. **5**, 124–166 (2011)
156. M. Stein, C. Lammers, P.-H. Richter, C. Fuchs, W. Stolz, M. Koch, O. Vänskä, M.J. Weseloh, M. Kira, S.W. Koch, Dynamics of charge-transfer excitons in type-II semiconductor heterostructures. Phys. Rev. B **97**, 125306 (2018)
157. A. Soltani, D. Gebauer, L. Duschek, B.M. Fischer, H. Cölfen, M. Koch, Crystallization caught in the act with terahertz spectroscopy: non-classical pathway for l-(+)-tartaric acid. Chem. - A Eur. J. **23**, 14128–14132 (2017)
158. W. Lai, N. Born, L.M. Schneider, A. Rahimi-Iman, J.C. Balzer, M. Koch, Broadband antireflection coating for optimized terahertz beam splitters. Opt. Mater. Express **5**, 2812–2819 (2015)
159. E. Stübling, Y. Bauckhage, E. Jelli, B. Fischer, B. Globisch, M. Schell, A. Heinrich, J. Balzer, M. Koch, A THz Tomography System for Arbitrarily Shaped Samples. J. Infrared Milli. Terahz Waves **38**, 1179–1182 (2017)
160. M.D. Thomson, W. Zouaghi, F. Meng, M.M. Wiecha, K. Rabia, T. Heinlein, L. Hussein, D. Babu, S. Yadav, J. Engstler, J.J. Schneider, N. Nicoloso, I. Rychetský, P. Kužel, H.G. Roskos, Dielectric properties of vertically aligned multi-walled carbon nanotubes in the terahertz and mid-infrared range. J. Phys. D: Appl. Phys. **51**, 034004 (2017)

161. W. Zouaghi, L. Hussein, M.D. Thomson, Q.-U. Islam, N. Nicoloso, T. Heinlein, J. Engstler, J.J. Schneider, H.G. Roskos, Towards gas sensing with vertically aligned carbon nanotubes interrogated by THz radiation pulses. Lith. J. Phys. **58**, 38–48 (2018)
162. Z. Chen, H.I. Wang, N. Bilbao, J. Teyssandier, T. Prechtl, N. Cavani, A. Tries, R. Biagi, V. De Renzi, X. Feng, M. Kläui, S. De Feyter, M. Bonn, A. Narita, K. Müllen, Lateral fusion of chemical vapor deposited N = 5 armchair graphene nanoribbons. J. Am. Chem. Soc. **139**, 9483–9486 (2017)
163. Z. Chen, H.I. Wang, J. Teyssandier, K.S. Mali, T. Dumslaff, I. Ivanov, W. Zhang, P. Ruffieux, R. Fasel, H.J. Räder, D. Turchinovich, S. De Feyter, X. Feng, M. Kläui, A. Narita, M. Bonn, K. Müllen, Chemical vapor deposition synthesis and terahertz photoconductivity of low-bandgap N = 9 Armchair Graphene Nanoribbons. J. Am. Chem. Soc. **139**, 3635–3638 (2017)
164. I. Ivanov, Y. Hu, S. Osella, U. Beser, H.I. Wang, D. Beljonne, A. Narita, K. Müllen, D. Turchinovich, M. Bonn, Role of edge engineering in photoconductivity of graphene nanoribbons. J. Am. Chem. Soc. **139**, 7982–7988 (2017)
165. O. Abdulmunem, K. Hassoon, M. Gaafar, A. Rahimi-Iman, J.C. Balzer, TiN nanoparticles for enhanced THz generation in TDS systems. J. Infrared Milli. Terahz Waves **38**, 1206–1214 (2017). https://doi.org/10.1007/s10762-017-0412-z
166. S. Lepeshov, A. Gorodetsky, A. Krasnok, N. Toropov, T.A. Vartanyan, P. Belov, A. Alú, E.U. Rafailov, Boosting terahertz photoconductive antenna performance with optimised plasmonic nanostructures. Sci. Rep. **8**, 6624 (2018)
167. M. Markiewicz-Keszycka, X. Cama-Moncunill, M.P. Casado-Gavalda, Y. Dixit, R. Cama-Moncunill, P.J. Cullen, C. Sullivan, Laser-induced breakdown spectroscopy (libs) for food analysis: a review. Trends Food Sci. Technol. **65**, 80–93 (2017)
168. V.K. Singh, A.K. Rai, Prospects for laser-induced breakdown spectroscopy for biomedical applications: a review. Lasers Med. Sci. **26**, 673–687 (2011)
169. J. Kaiser, K. Novotny, M.Z. Martin, A. Hrdlička, R. Malina, M. Hartl, V. Adam, R. Kizek, Trace elemental analysis by laser-induced breakdown spectroscopy–biological applications. Surf. Sci. Rep. **67**, 233–243 (2012)
170. A.J.R. Bauer, S.G. Buckley, Novel applications of laser-induced breakdown spectroscopy. Appl. Spectrosc. **71**, 553–566 (2017)
171. D. Abdrabou, L.M. Schneider, A. Rahimi-Iman, M.A. Khedr, A.M. Hussein, T.M. El-Sherbini, Study of laser-induced-plasma parameters for molybdenum targets. Plasma Res. Express **1**, 035004 (2019)
172. J. Kuhnert, A. Rahimi-Iman, W. Heimbrodt, Magneto photoluminescence measurements of tungsten disulphide monolayers. J. Phys.: Condens. Matter **29**, 08LT02 (2017)
173. A.V. Stier, N.P. Wilson, G. Clark, X. Xu, S.A. Crooker, Probing the Influence of Dielectric Environment on Excitons in Monolayer WSe_2: Insight from High Magnetic Fields. Nano Lett. **16**, 7054–7060 (2016)
174. A.V. Stier, N.P. Wilson, K.A. Velizhanin, J. Kono, X. Xu, S.A. Crooker, Magnetooptics of Exciton Rydberg States in a Monolayer Semiconductor. Phys. Rev. Lett. **120**, 057405 (2018)
175. J. Fischer, S. Brodbeck, A.V. Chernenko, I. Lederer, A. Rahimi-Iman, M. Amthor, V.D. Kulakovskii, L. Worschech, M. Kamp, M. Durnev, C. Schneider, A.V. Kavokin, S. Höfling, Anomalies of a Nonequilibrium Spinor Polariton Condensate in a Magnetic Field. Phys. Rev. Lett. **112**, 093902 (2014)
176. A. Rahimi-Iman, C. Schneider, J. Fischer, S. Holzinger, M. Amthor, L. Worschech, S. Reitzenstein, S. Höfling, A. Forchel, M. Kamp, Magnetic-field interaction of spatially confined quantum-well exciton-polaritons. J. Phys. Conf. Ser. **456**, 012033 (2013)

Chapter 6
Effects of Quantisation

Abstract Fundamental quantum-physical effects are nowadays harnessed in practical devices through different means. On the one hand, quantum potential boxes can be employed to alter electronic or photonic properties depending on the application. From bulk to zero-dimensional structures, the quantisation of energy states for physical entities, such as electrons or photons, gets pronounced and the density of states is strongly modified based on the dimensionality of the host system. Therefor, the size of a confinement potential provided by a host structure is typically adjusted to the length scale corresponding to the respective particle's spatial extent, resulting in a quantum structure. Generally, quantum structures can be beneficial for optoelectronic or quantum devices due to several aspects, such as strong spatial confinement, tailored energy transport or storage, efficient lasing or nonclassical light generation, and altered interactions of their particles with their environment or other particles. On the other hand, the control of quantum coupling phenomena or the use of many-body effects can enable novel devices or open up new pathways to a fundamental exploration of quantum effects, such as hybridisation, superposition, collective phenomena and cavity quantum electrodynamics. Thus, this chapter aims at an introduction to prominent effects of size reduction to the nanoscale for electronics or microscale for photonics, as well as a summary of benefits and applications, giving examples from optoelectronics and nanophotonics.

6.1 Miniaturisation Towards Quantum Structures

The reduction of a structure's size brings obvious advantages with it, as for example miniaturised devices strongly benefit from microscopic building blocks. In addition, it is well understood that clusters of matter, such as nanoparticles or microparticles, can act as noticeable scattering centers for propagating (electromagnetic) waves when the structure size becomes more or less—in terms of the order of magnitude—comparable to the wavelength. Similarly, apertures or device facets on that length scale introduce considerable diffraction effects. However, more intriguing in terms of miniaturisation remains the fact that the device physics gets influenced by the size of a structure which can have a remarkable impact on various properties when

A. Rahimi-Iman, *Semiconductor Photonics of Nanomaterials and Quantum Structures*,
Springer Series in Solid-State Sciences 196,
https://doi.org/10.1007/978-3-030-69352-7_6

entering the regime of quantum structures. In other words, it means that when one of the structure's dimensions becomes comparable to a (quasi-)particle's wavelength[1] or *size* confined in it, classical descriptions of the system's behaviour fail. Such low-dimensional structures are well described in many text books and the interested reader is referred to the literature for a more complete and detailed description of quantum-physical concepts [1–5].

Nowadays, all kinds of properties, such as electronic, optical and mechanical, are tailored by size reductions towards the respective quantisation length scale. Typical examples are quantum wells, wires and dots for electrons and their analogues for light in the form of microresonators (e.g. [6]) and waveguides (e.g. [7]), widely used in optoelectronic and quantum-technological devices (cf. [8–10]). Simultaneous electronic and optical confinement is for instance important for strong light–matter interactions in optical microcavities, giving rise to polariton physics, and the study of quantum coupling effects, as well as many-body phenomena in solids (cf. [11–15]).

6.2 From Bulk to Zero-Dimensional Structures

The reduction of size is a powerful tool to achieve spatial confinement (of a physical entity, i.e. a fundamental or composite particle) in any of the desired spatial dimensions. Such a reduction in different dimensions is sketched in Fig. 6.1. Its consequences are directly noticeable in the reciprocal space, the phase space, in which the possible particle states are defined in terms of energy and momentum. While in some cases, the sole purpose of size reduction is given by an enhanced spatial confinement of particles, trapping and waveguiding, other scenarios strongly harness the effects of quantisation introduced typically on the nanoscale for electronic and on the microscale for photonic features. In the following, the effects of spatial confinement shall be highlighted regarding the density of particle states and the energy spectrum, and some of their uses as examples. However, also the trivial effect of localisation and its use is summarised.

6.2.1 Spatial Confinement

Spatial confinement can serve different purposes and opens up many possibilities to design and alter device physics. The most typical purpose remains the confinement of charge carriers in a region of interest in terms of localisation, while different dimensionalities can still give rise to a certain degree of mobility. In contrast to in-plane confinement by wells and solely longitudinal mobility in wires, the extreme case of an atomistic confinement potential of a 0D box with all length scales on the

[1] Characterised by the de-Broglie wavelength $\lambda = h/p = h/mv = h/\sqrt{2mE}$, the thermal representation of which reads $\lambda_{\text{dB}}(T) = \sqrt{2\pi\hbar^2/(mk_\text{B}T)}$, using the relationship $E = \pi k_\text{B} T = p^2/2m$.

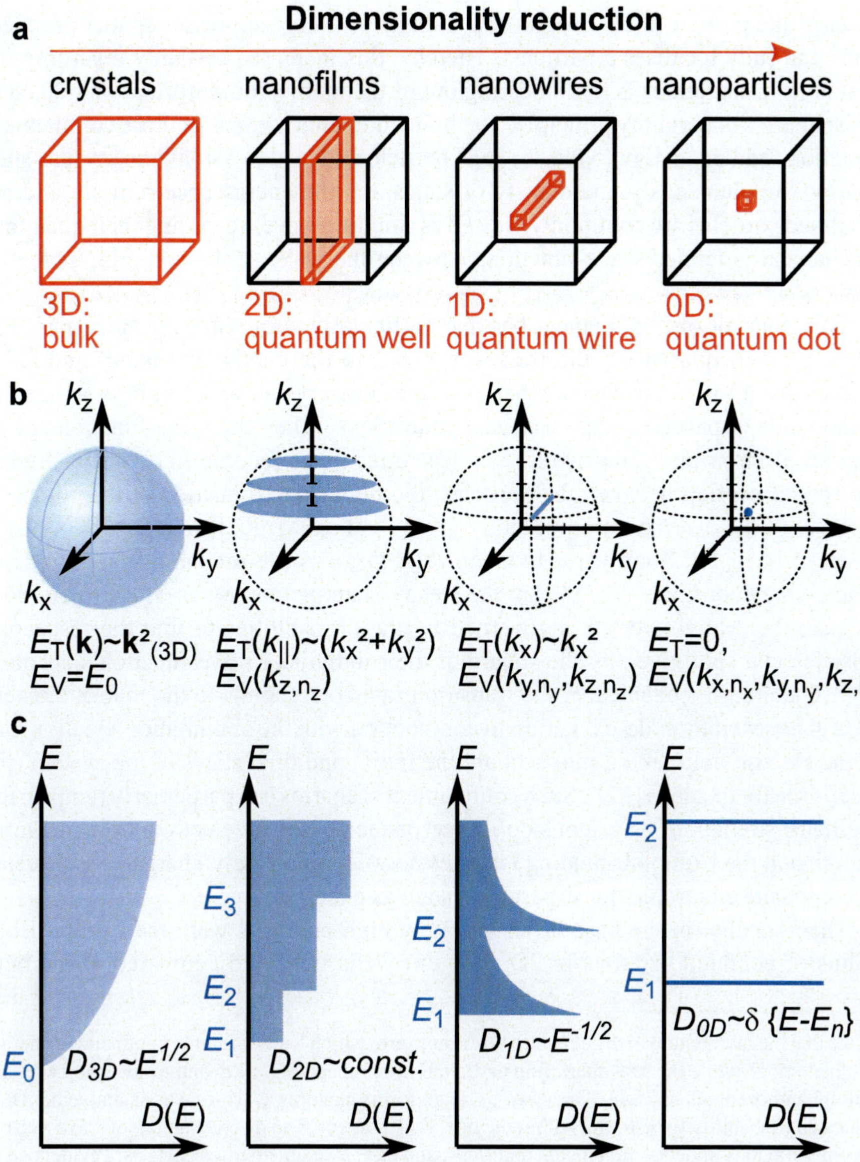

Fig. 6.1 **a** Sketch of the dimensionality reduction from bulk to zero-dimensional (from left to right). Red objects within the black box indicate the lower-dimensional structures, which confine particles in one, two or even three directions, resulting in quantum wells, wires and dots, respectively. The latter one is the equivalent of an atom-like entity often referred to as *artificial atom*, or *quantum box*. **b** Sketch of an electronic particle's phase-space states for the different confinement scenarios indicating the reduction of available states to a selection of allowed confinement states in the (3D) k space. Discretisation of k vectors is indicated. For simplicity, only the lowest k vector state for the respective confinement direction is sketched for wires and dots. Note that a modification of translational energy E_T and potential energy E_V from free (massive) particle ($E_\mathrm{kinetic} = \hbar^2 k^2/2m^*$) to fully confined particle occurs. **c** Schematic diagrams of the energy-dependent density of states $D_{3/2/1/0\mathrm{D}}(E)$ for the respective confinement situation from **a**, **b**. The proportionality is displayed. Thus, it can be easily evidenced that tailoring the dimensionality of a structure can have a serious impact on a system's energetics

order of the particle's (de-Broglie) wavelength entirely suppresses spatial propagation[2] and fully localises a particle.[3] Thereby, this increased certainty regarding the spatial information leads to a smearing out of the momentum information related to Heisenberg's uncertainty principle. The finite momenta in every of those confinement directions add up energy to the "boxed" particle and lead to a non-zero ground-state energy. Simultaneously, the density of states and the energy spectrum are altered. Confined particles are commonly viewed as standing waves regarding their quantum-mechanically-defined wave-functions between the walls of the confining potential barriers.

This concept of confinement became particularly successful for electronic particles, such as quasi-free[4] electrons (excited into the conduction band) and holes (defect electrons in the valence band) in the host crystal, and similarly applies to compounds of particles (e.g. excitons, polaritons). When the wave-functions of an excited electron and hole overlap, optical transitions become likely.[5] The higher the spatial overlap between the particles, the higher the transition probabilities. A measure of the overlap is conveniently given by the magnitude of a particle's wave-function $|\psi(\mathbf{r}, t)|$. Such wave-function $\psi(\mathbf{r}, t)$ gives the probability amplitude at space–time coordinate $(\mathbf{r}, t)$ (or in the steady-state picture $\varphi(\mathbf{r})$). Thus, to be more precise, the spatial overlap is governed by the probabilities to find the respective particles at a spot given by the square of their individual wave-function's absolute $|\psi|^2$. Quantum-mechanically, the transition rate corresponds to the matrix element for a given excitation/de-excitation in conjunction with the annihilation/creation part of the electric field taking into account the initial and final states of the system (for details see for instance [17]). Such confinement scenarios can particularly ramp up the oscillator strength of a system's optical resonance based on electron–hole recombination, and the Coulomb-binding energies between oppositely-charged constituents of composite electronic quasi-particles, such as excitons.

Theoretically, in the case of an infinitely high potential well, the exciton Bohr radius of confined excitons in 2D quantum wells is halved compared to the bulk

---

[2]Note that the momentum in the quantum box is not zero, which leads to a finite confinement energy. In other words, while the self-interfering particle is not propagating, its quantum-mechanical phase is finite and becomes discrete. The wave-package forms standing wave-functions and exhibits an enhanced probability to be found within the box. Amendment: The discrete nature of wave-vectors becomes obvious under confinement conditions, and else, available particle states are evidenced as quasi-continuous in macroscopic structures.

[3]The motion of the wave-package is restricted by the potential barriers of the confinement structure. Amendment: It means that the particle wave-function solutions must fulfil the (resonance) conditions imposed by the boundaries of the potential box. Only those wave-vectors which result in constructive interference forming standing waves prevail inside the box.

[4]The expression "quasi" takes into account the effect of a periodic potential landscape in the lattice of a crystal structure on propagating charge carriers.

[5]This is characterised by the dipole transition-matrix element $M \propto \mathbf{d}$, the emitter's dipole moment. See Fermi's golden rule, e.g. in [16, 17]. For an exciton, the dipole length is approximated by the exciton Bohr radius of the bound electron–hole pair at given main quantum number n. For an atom, the dipole moment is caused by the position change of electrons in the atom for a given transition $|b\rangle \rightarrow |a\rangle$ (ground to excited state) corresponding to the optical frequency ω_{ab}.

(3D) value. Accordingly, the binding energy should be raised by a factor of four compared to bulk. However, practical quantum wells have a finite thickness and finite potential height, which prevent them to reach that ideal value, as discussed in numerous studies from the 1980s on quantum-well systems [18–20]. Note that for practical quantum well systems, the thickness L plays an important role: a trade-off has to be found between an increased oscillator strength through a decrease of L and an undesired increased penetration of the wave-function into the barrier materials for finite barrier heights, as well as a rise of energy levels closer to the unbound energies of the surrounding materials with decreasing L (facilitating leakage of particles out of the potential well region). Indeed, monolayer 2D semiconductors of the TMDC family with their in-plane excitons strongly confined to the layer of transition metals offer a model system as natural nearly-ideal quantum wells.

6.2.2 Density of States

For quantum structures, dimensionality plays an important role, as it influences the distribution of energies of those systems. This distribution is described for any low-dimensional structure by its respective density of states (DOS), sketched in Fig. 6.1. In a periodic system, the DOS can be mathematically derived using periodic boundary conditions and the reciprocal relationship between real space and phase space (i.e. momentum space or, short, k space). To obtain this distribution for a given dimensionality d (considering integer positive numbers), one seeks to determine how many particle states around an energy E exist within the energy interval δE per k-space volume $(2\pi)^d/V_d$ per real-space volume $V_d = L^d$ in a (box-like) system of length L for each available direction. The typically-derived expression for the DOS, $D_d(E)\delta E$, is a characteristic for the dimensionality of the system and has a peculiar trend as a function of energy, which in a text-book manner can be described for the 4 distinct cases at the example of electronic states as follows.

The cubic 3D electronic system has an energy distribution (DOS) following a square-root-of-E proportionality (beginning at the ground state E_0, see first column of Fig. 6.1). The film-like 2D system exhibits a stairways-like DOS with a constant value added for each confinement energy level of the system[6] (second column of Fig. 6.1). The wire-like 1D system features spikes at every quantisation energy with inverse-square-root proportionality, i.e. $E^{-1/2}$, leading to a decrease of energy states

[6]Amendment: At the example of the quantum well, the determination of the DOS $D(E)$ can be summarised as follows. Taking into account spin (an energy state can host two fermions of opposite spin), a factor of two will be carried in $D(E)$. With the increment of dk in 2D k space, $N_{2D}(k)\,\mathrm{d}k = 2\pi k\,\mathrm{d}k$ states situated on the circumference of a circle with radius k add up at $E(k)$ per energy increment dE. Then, the DOS can be obtained through $D_{2D}(E)\,\mathrm{d}E = 2(N_{2D}(k)\delta k/\delta E)/(2\pi/L)^2 \cdot (1/L^2)\,\mathrm{d}E = 2(2\pi k)/(2\pi)^2 \cdot (\delta k/\delta E)\,\mathrm{d}E$. Due to $\mathrm{d}E/\mathrm{d}k = \hbar^2 k/m$, one obtains $D_{2D}(E)\,\mathrm{d}E = 2m/(2\pi\hbar^2)\,\mathrm{d}E$. Thus, $D_{2D}(E)$ shows no dependence on E. In the 3D and 1D case, $N_{3D}(k)\,\mathrm{d}k = 4\pi k^2\,\mathrm{d}k$ (surface of a sphere with radius k) and $N_{1D}(k)\,\mathrm{d}k = 2\,\mathrm{d}k$ (2 for both propagation directions on a line), respectively.

with increasing energy (third column of Fig. 6.1). The quantum box of a 0D system shows the characteristic signatures of atoms with completely discrete delta peaks, where states only exist at the defined energies of the quantum dot (last column of Fig. 6.1). Accordingly, the $D(E) \propto \delta(E - E_i)$, where i is the quantum number of the energy level.

While in three dimensions the DOS has its minimum at the bottom of the band, the feature at the bottom of the band is stronger in fewer dimensions. This influences heavily optical properties, such as absorption and the inverse process stimulated emission. Because of the higher density of states at the bottom of the band, low-dimensional structures are strongly favoured for optoelectronic devices.

6.2.3 Discrete Energies

The reduction of a structure's dimension down to the strong-confinement regime leads to quantisation effects that give rise to discrete energies for the respective direction. In the remaining directions without boundaries, particle propagation is not hindered and thus the free-particle kinetic energy term with the respective wave-vector for such propagation is preserved. In the case of a 0D system, which corresponds to a strong 3D confinement, atomistic levels are the consequence, similar to an atom's electron situated in the (effective) Coulomb potential of its positively-charged atom core. In fact, the distribution of energy levels strongly depends on the potential shape. For simplicity, the principle is discussed in common textbooks using rectangular potentials, with additional examples often providing triangular or parabolic wells for comparison. Another important factor regarding the energies in a confinement potential is the height of the walls, its potential barriers.

For every confinement direction, the wave-function must strictly fulfil certain requirements imposed by *fixed boundary conditions* (fbc). The solution of a potential-well Schrödinger equation consists of a sine behaviour (standing wave) for the restricted direction, e.g.

$$\psi \propto \sin(k_x x) \tag{6.1}$$

for confinement in x direction. Consequently, momenta become discrete, i.e.

$$k_x = n_x \pi / L_x, \text{ (fbc)}, \tag{6.2}$$

where L_x is the width of the quantum well in x and $n_x = 1, 2, \ldots$ are integer quantum numbers of the system.

In contrast, for those directions with negligible confinement and surface effects—which holds true for macroscopic bodies—*periodic boundary conditions* (pbc) are applied, mimicking a ring-like system where the tail ends up in the head of the ring, thereby leading to a closed entity resembling an infinite system. Here,

$$k_y = n_y (2\pi / L_y), \text{ (pbc)} \tag{6.3}$$

for instance in y direction, with $n_y = 0, \pm 1, \pm 2, \ldots$. The solution of the wave-function which fulfils these requirements are of the form

$$\psi \propto \exp(ik_y y). \tag{6.4}$$

This makes clear that the possible energies of a massive-particle system (e.g. electrons) range between fully free in 3D in the form of

$$E(\mathbf{k}) = \frac{\hbar^2 \mathbf{k}^2}{2m}, \text{ (free massive particle dispersion)}, \tag{6.5}$$

where m is the particle mass and $\mathbf{k} = \mathbf{k}_x + \mathbf{k}_y + \mathbf{k}_z$ the overall momentum in bulk, to partially free such as in wells according to

$$E_{n_z}(\mathbf{k}_{||}) = \frac{\hbar^2 \pi^2}{2mL_z^2} n_z^2 + \frac{\hbar^2 \mathbf{k}_{||}^2}{2m} \text{ (1D confined 2D system)}, \tag{6.6}$$

and to the case of full quantisation in 0D represented by

$$E_{n_x,n_y,n_z} = \frac{\hbar^2 \pi^2}{2mL_x^2} n_x^2 + \frac{\hbar^2 \pi^2}{2mL_y^2} n_y^2 + \frac{\hbar^2 \pi^2}{2mL_z^2} n_z^2 \text{ (3D confined 0D system)}. \tag{6.7}$$

Here, the terms with quantum numbers $n_{x/y/z}$ denote the confinement energies for the respective directions x, y, z, whereas for the particle Hamiltonian (with $E_{\text{tot}} = E_{\text{kin}}(\mathbf{k}) + V(\mathbf{r})$) in this steady-state consideration (i.e. without time-dependence) the potential $V(\mathbf{r}, t) = 0$ relative to the bulk ground-state in the well region.

For a schematic overview on the quantum-well confinement of electronic particles in semiconductor heterostructures see Fig. 6.2.

While size reduction itself enables the modification of energy states in systems, e.g. for charge carriers, quasi-particles or light, one can further play with energy levels by means of periodic arrangement.

Amazingly, artificial atom-like structures or quantum wells are not limited to a single entity. Indeed, coupled quantum wells can provide molecule-like hybrid states (besides increasing the number of confined particles in the total potential-wells structure with regard to a single well).

Going one step further, superlattices of wells or dots (e.g. deterministically achieved by epitaxy or naturally formed by moiré patterns) lift the degeneracy of energy states contributed from each quantum box completely by giving rise to mini-band formation (due to the periodic arrangement of these entities). This is analogous to band structure formation in crystals, in which the elements of a crystal are periodically arranged in space, as repetitions of a unit cell. Subbands in the conduction or valence band are for instance exploited in quantum-cascade lasers, unipolar devices that totally rely on intersubband transitions and cascaded (stimulated) relaxation processes for (coherent) light extraction.

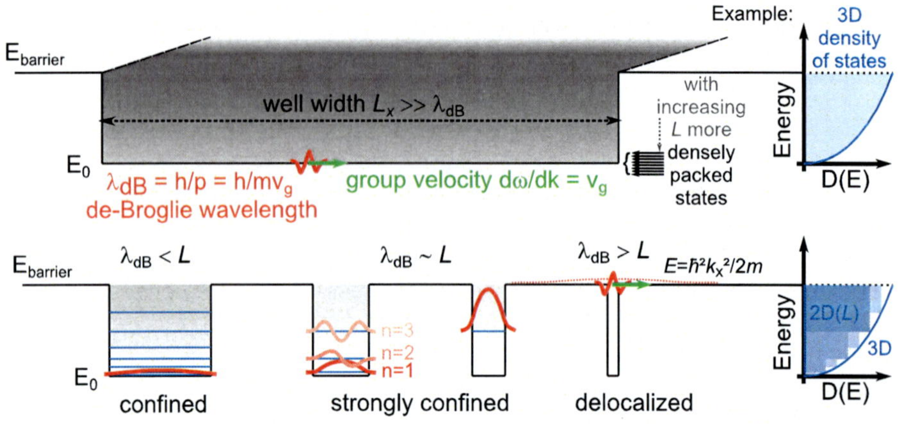

Fig. 6.2 Sketch of an energy potential well with relative ground state E_0 and potential barrier imposed by the energetics of the environment (barrier) E_{barrier}. In the top row, the situation is schematically drawn for a large well with negligible confinement effect on a massive electronic particle with quantum-mechanical wavelength λ_{dB}. The quantum particle's wave-function within the well does not self-interfere (relevant length scale is the coherence length) and behaves like an isolated wave-package with a group velocity and kinetic energy of a classical particle. The available energy states become quasi-continuous with increasing well width L (here in direction x), i.e. they are so densely packed that their discrete character becomes negligible. The corresponding density of states for (quasi-)3D is sketched as an example to the right of the figure. In the bottom row, the effect of size reduction is sketched for the confinement structure. With decreasing well width, the possible quantum states spread further apart energetically and the number of confined modes reduces. This is also represented by the schematic density of states for the three well width examples to the right of the figure. Strong confinement is achieved for potential well widths of the order of λ_{dB}. Too small features do not exhibit any bound states and the particle remains delocalised. This simple 1D representation can be easily used to imagine the quantisation effects for 2D and 3D confinement as well (also see Fig. 6.1). Discrete energy states have been at the core of various quantum devices and have delivered a strong impact on device functionalities in optoelectronics

Discrete energies or bands are not restricted to electronic systems, as photonic resonator modes in microresonators and energy *stop bands* in photonic crystals show. A 1D representative of an optical crystal is a distributed Bragg reflector, from which very high reflectivities can be obtained by a sequence of quarter-wavelength layers (a and b) with alternating refractive index[7] $n_{a/b}$. By sandwiching an optical half-

[7]Conveniently, this magnitude (i.e. the refractive index of a material) can be coarsely regarded as inverse to the electronic band gap: the higher the gap, the lower the refractive index. For a more accurate picture, the interested reader is referred to the Kramers–Kronig relations, see e.g. [21]. Amendment: The higher the refractive index contrast between the alternating layers (materials) of a Bragg mirror (photonic crystal), the better the overall reflectivity of the photonic stop band becomes. Note that the fundamental concept applies: Where there is no energy state for a particle, there can be no penetration of that particle into that specific region. More precisely expressed: If a physical environment lacks particle-specific energy states (modes) in a certain spectral region, e.g. by means of periodic arrangement of potential landscapes, such particle with energies within that range cannot be transmitted through that environment (unless through barrier tunnelling) and will be fully reflected. Nowadays, even electronic or photonic structures are investigated that can have

wavelength cavity between two DBRs, a quantum-well-like Fabry–Pérot resonator with finite ground-state energy for the cavity photon can be obtained. Note that the photonic DOS in 3D (vacuum) is $\rho \propto \omega^2$ due to the linear dispersion of photons in vacuum. In a Fabry–Pérot microcavity, the photon obtains an effective mass, kind of a *"rest mass"* $m_{eff} = E_0 n^2/c^2 \approx 10^{-5}\, m_0$, due to its approximately parabolic dispersion for small wave-vectors $k_{||} \approx 0$.

Further processing of such cavity into micropillars can form photonic quantum dots with atomistic energy levels for light, with which quantum coupling experiments on a single-emitter and empty-field level can be performed. Alternatively, planar systems can be etched to form waveguides in order to act as a photonic wire. Similarly, nanofibres or nanoneedles act as wires for light, and photonic-crystal fibres offer particularly high lateral confinement among the 1D photonic structures. Further slicing the photonic-crystal fibre into a half-wavelength thick photonic-crystal sheet can result in a very-high-quality optical resonator with properties of a 0D *photonic quantum box* (also referred to as *optical quantum dot*) based on strong lateral and out-of-plane confinement potentials. Similarly, if a planar Fabry–Pérot microcavity is etched into a circular (or square-shaped) micropillar shape with diameters of the order of few microns, the previously in-plane confined photonic modes experience strong lateral confinement by a cylindrically shaped cavity region with considerable refractive index contrast to the surrounding air/vacuum (cf. [6, 12]). In contrast, shallow confinement in photonic boxes can be achieved by buried sub-micron height modulations in monolithically grown planar microresonators [22–25]. In fact, one can also create photonic molecules and artificial lattices from all these optical quantum dots [26–31], which can even exhibit nontrivial topological properties (see for instance [32–35]). Such examples highlight the unique possibilities one obtains when reducing dimensionalities and modifying the potential landscape.

For the sake of compactness, Coulomb-binding phenomena, excitons, exciton complexes, phonons, polaritons and the like, their dispersions, the impact of potential landscapes on these quasi-particles, or other effects, such as tunneling phenomena, thermal and quantum-mechanical delocalisation effects, lifetimes, and so forth are not discussed in this brief introduction. On the one hand, this is omitted here due to the difficulty to compress the vast amount of contemporary knowledge and to find a beginning and an end; and on the other hand, due to the fact that excellent textbooks on semiconductors exist which describe all these fundamental aspects in various levels of details. Instead, a brief glimpse at the advantages that arise from low-dimensional structures is given, highlighting few selected examples in the following section.

---

topologically protected states which deliver very robust, scatter-free/dissipation-free propagation of their respective particles in such modes. Certain structures could in principle be even designed to provide only momentum states for unidirectional propagation (on their edge). However, since topological properties are not the subject of this chapter, such design principles and underlying theoretical concepts will not be detailed here.

6.3 Benefits and Applications

Typically, quantum structures are at the heart of many optoelectronic applications and quantum technologies of the first and second generation, as they enable non-classical device concepts, charge-carrier localisation and tailored energy transitions. A few examples are summarised below, mainly from the fields of laser physics and engineering, optoelectronic devices, as well as light–matter interactions.

Naturally, potential-well systems can be prepared for fermions and bosons equally, whereas the major difference comes from the occupancy of states and the corresponding nonclassical particle distribution functions. While *fermions* obeying the Fermi–Dirac (FD) distribution function can only occupy a quantum state once owing to the Pauli exclusion principle (exhibiting an anti-bunching effect in terms of statistics) and are distinguishable in the same energy state, e.g. by spin, *bosons* can occupy the same mode or energy state macroscopically (exhibiting a bunching effect). Their statistics differ fundamentally from classical particles—described by Maxwell–Boltzmann (MB) statistics—and that of non-integer spin particles (fermions). An overview on the three distribution functions (see Fig. 6.3) is provided for the sake of completeness (with $\mu = E_0$ the chemical potential, for fermions also known as Fermi energy E_F):

$$f_{\mathrm{classical}}(E) = \frac{1}{\exp\left(\frac{E-\mu}{k_\mathrm{B}T}\right)}, \text{ (MB)} \tag{6.8}$$

$$f_{\mathrm{fermion}}(E) = \frac{1}{\exp\left(\frac{E-\mu}{k_\mathrm{B}T}\right)+1}, \text{ (FD)} \tag{6.9}$$

$$f_{\mathrm{boson}}(E) = \frac{1}{\exp\left(\frac{E-\mu}{k_\mathrm{B}T}\right)-1}. \text{(BE)} \tag{6.10}$$

Thus, bosonic behaviour is characterised by the Bose–Einstein (BE) distribution function, which can give rise to an extreme aggregate state of bosonic matter referred to as Bose–Einstein condensate. In such a regime occurring after a spontaneous symmetry break (establishment of a single phase throughout the condensate), all condensed bosons are described by the same macroscopic wave-function and, thus, become indistinguishable.

The Bose–Einstein condensation effect is directly related to other intriguing many-particle phenomena, such as superfluidity (cf. [36, 37]) and superconductivity (condensation of Cooper pairs, composite bosons). In fact, bosons in the same mode with the same physical properties, such as energy, momentum, spin/polarisation and phase, are indistinguishable by nature. Below a critical temperature, the majority of Bose particles accumulates in the ground-state of the system through stimulated scattering processes,[8] when their wave packages strongly overlap with each other. Thereby, a quantum fluid with long-range order (spatial coherence) is established.

[8]The more particles are in the final state, the higher the final-state scattering rate becomes.

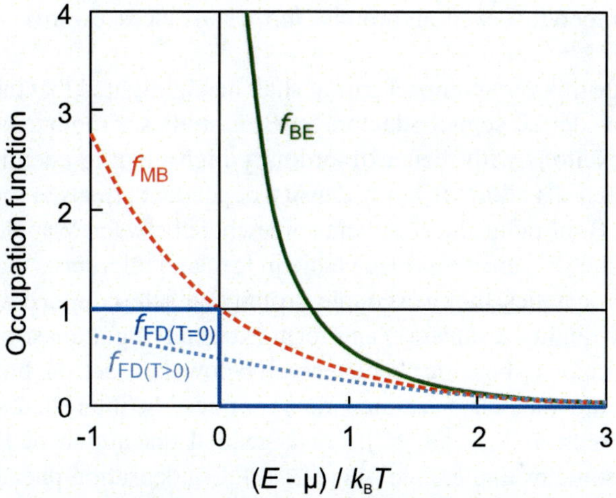

Fig. 6.3 Maxwell–Boltzmann (MB), Fermi–Dirac (FD) and Bose–Einstein (BE) distribution functions, representing the occupation functions for classical particles, fermions and bosons, respectively. They are plotted on a common scale vs. $(E - \mu)/k_B T$, that is with respect to the chemical potential μ as the reference energy. Drawn freely after [2], with the FD distribution at absolute zero added for comparison with the high-temperature behaviour

However, the particle density in the Bose cloud must be dilute enough so that three-particle interaction effects, which could lead to the formation of a solid instead of a BEC, can be neglected.

The condensation phenomenon, which is practically achievable at elevated temperatures for bosonic exciton–polaritons in optical microcavities, was in the past two decades exploited to demonstrate a novel type of coherent light source, labelled polariton laser. An overview on this subject is provided for instance in [15, 38, 39], with theoretical considerations provided e.g. in [40–42]. Numerous studies on optically-pumped polariton systems were performed to better understand the condensation behaviour for cavity–polaritons, their phase-transition criteria and the coherence signatures for such driven–dissipative condensates (cf. [43–46]), involving works with external fields. Nowadays, Bose fluids in solids are further studied with respect to utilisation in novel device concepts for optical circuitry and quantum information processing applications, as outlined in [47], and the utilisation of coherent polariton lasers is at the verge [48, 49].

Quantum structures such as photonic quantum boxes with their discrete energies are known to affect the condensation properties, for instance by sufficient decoupling of the condensate from uncondensed particles in the system, i.e. by reduced ground-state–reservoir interactions (as discussed in [50]). Also, quantum structures with strong confinement and thereby increased oscillator strength play an important role for light–matter interactions. In this context, semiconducting 2D materials[9] with

[9]The class of 2D materials is often also referred to as quantum materials.

their room-temperature excitons promise further advances towards practical polariton systems [51].

To conclude this very compact part, it shall be highlighted that there are two main reasons to consider 2D semiconductors for BEC studies at room temperature. Firstly, their robust excitons with their extraordinary high binding energies for inorganic crystalline materials allows to study density-dependent effects at room temperature. Their exciton Bohr radius locates them somewhere between Wannier–Mott [52, 53] and Frenkel-like [54] excitons[10] (see table in [55] and references therein). Secondly, in optical microcavities, they can couple strongly with the confined light field (even in relatively-low-quality resonators) and form extremely light quasi-particles—hybrid states with composite-bosonic character well below the electron–hole plasma (Mott) density. For fundamentals on semiconductor optics, the interested reader is referred to textbooks such as [21, 56, 57]. For a general discussion of BEC and topical overview on some examples, see [43, 58–62]. Condensation phenomena involving 2D-materials excitons are furthermore investigated without cavities at cryogenic temperatures, e.g. for bilayer systems [63].

6.3.1 Charge-Carrier Localisation and Tailored Transitions

Historically, the introduction of confinement schemes has both step-wise (for each major design change of active media) and continuously (due to maturity of the underlying technology) improved the performance of optical devices such as laser diodes. Beginning with bare pn-junctions, laser thresholds and operation temperatures benefited strongly from the introduction of double heterostructures (cf. [64, 65]) owing to an improved overlap between charge carriers and optical fields. Currently, quantum-well structures are indispensable elements of semiconductor gain media, whereas quantum dots open new pathways to ultra-efficient nanolasers [12, 66, 67], which are attractive for "green photonics" (e.g. [68]), and more uniquely for single-photon sources and cQED experiments (cf. [12, 69–73]). Beyond that, they find application in photodetectors, nanoelectronics and nonlinear materials. Ultimately, novel concepts for quantum technological devices fully rely on the control of electronic and optical properties of/by quantum structures.

In semiconductor lasers, the modification of the DOS yields noticeable improvements regarding the onset of lasing at reduced charge-carrier densities, as the number of available states right at the band-gap energy for the stimulated emission process is governed by the dimensionality. While bulk semiconductors with square-root DOS have exactly a minimum of states at the lowest transition energy between bands, quantum wells with their step function directly provide a constant number of states at a well-defined ground-state transition energy. Quantum dots further localise charge

---

[10]For details on these two exciton types and the differences between Wannier–Mott and Frenkel excitons, i.e. delocalised crystal lattice and localised molecule site excitons, respectively, see for instance [21].

carriers and provide excitons with even higher oscillator strength and an atomistic energy spectrum. These artificial atoms with high quantum yield and very narrow emission lines provide a further improvement of the lasing process in optical microcavities.

Also, superlattices, i.e. artificial electronic crystals, have had their impact on laser technology. In quantum-cascade lasers (e.g. [4]), minibands formed by periodically-arranged multiple quantum wells allow for efficient and well-defined intersubband transitions of charge carriers. Thereby, long-wavelength regions between the IR and THz can be addressed with laser transitions else not available (not efficient) using direct gap transitions. Sequences of barriers and wells not only act as the active region, but are also needed for the design of particle drainage and injector regions, which can be arranged around the active region for the recycling of transmitted and relaxed carriers to form a cascaded device. However, new challenges arise when aiming at room temperature operation due to the similarity of transition energies to $k_B T$ (thermal energy).

Quantisation effects can also be used for band-gap engineering, as the width of a quantum well directly influences its energy levels. Similarly, a combination of materials can be used to obtain a W-like band alignment, in which a well for holes is sandwiched by a material posing a well for electrons within the outer barriers. Since the double-well's non-zero wave-function in the conduction band significantly overlaps with the hole wave-function in the sandwiched material sheet, optical transitions from such a type-II heterostructure can act as laser transitions. Type-II active media harness the unique tunability of the optical energy gaps by the quantum-structure's design parameters, which give wider access to difficult-to-reach wavelength ranges in the infrared [74, 75] (under investigation for instance at the author's host department in Marburg for VECSEL chips operating close to telecom wavelengths), whereas transitions in type-I heterostructures heavily rely on the employed material system and its direct gap transition. Indeed, the gain dynamics of these different types of gain media are different, as a comparison between type-II and type-I VECSEL chips shows [76]. Moreover, type-II structures are designed to reduce parasitic charge-carrier loss channels in the gain region such as imposed by Auger processes, in order to improve the laser efficiency in those wavelength ranges compared to type-I structures based on suitable material systems.

6.3.2 Impact on Optoelectronics and Nanophotonics

A good example how photonic devices have improved due to the properties of quantum structures in their active region is given by the use of semiconductor quantum dots. For short pulse generation, the fact that charge-carrier lifetimes are generally lower than in quantum-well structures motivated quantum-dot devices development. This is because the shorter lifetimes, both in gain structures as well as in semiconductor saturable-absorber mirrors, promise higher repetition rates (for optical transmission schemes) and shorter pulses from mode-locked lasers. This is for instance discussed in [77].

Moreover, inhomogeneously broadened emission spectra of semiconductor quantum dots grown in the Stranski–Krastanov epitaxy mode enable wider wavelength tunability and promise shorter pulses as a consequence of an increased gain bandwidth. In addition, a stronger temperature dependence of emission helps variation of (sub-ps) pulse durations, for instance in self-mode-locked (SML) semiconductor disk lasers (SDLs) [78]. Although quantum-dot lasers cannot provide as high charge-carrier densities as in their quantum-well counterparts, record high output powers from quantum-dot SDLs have been demonstrated at 1 and 1.1 μm in excess of 8 and 7 W, respectively, for each single-chip laser device [79, 80]. To compensate for the reduced densities in such quantum-dots gain structures, the active region comprises numerous quantum-dot layer stacks distributed in the field maxima positions of these chips' (half-)microcavity. Further details about SDL achievements can be found for instance in [81, 82].

Besides conventional semiconductor lasers (cf. [83, 84]), nanolasers have strongly benefited from the development of quantum-dot systems incorporated into high-quality optical microresonators [85], for instance discussed in [12]. Conveniently, their exciton resonances and the coupling situation can be tuned in external fields for light–matter interaction studies [86, 87]. Particularly, single-quantum-dot lasers in the regime of a high cavity-mediated Purcell effect [88] promise ultralow thresholds (cf. [12, 66, 67]).

On the electronics side, nanotransistors based on single quantum dots,[11] such as single-electron transistors that employ resonant-tunneling effects and Coulomb blockade to control a one-electron flow and storage, could pave the way for higher bit densities and lower power consumption (see for instance [89, 90]).

On the quantum optics side, the nonclassical properties of individual dots that act as artificial atoms in solids provide a reliable source of single photons, even electrically-driven (cf. [91–93], and reviewed for instance in [94]). Such sources are highly attractive for quantum information and communication schemes, as their antibunching of the photon statistics and indistinguishability can be exploited in quantum cryptography (see for instance [73, 95–98]), as the non-cloning theorem and the quantum nature of photons promise detectable eavesdropping efforts (by a third party, referred to as person Eve) on the quantum channel established between two parties (commonly referred to as Alice and Bob in the literature).

Usually, single-photon sources are excited using commercially-available mode-locked Ti:sapphire lasers. However, as they have limited (upper) repetition rates of typically 80 MHz, the achievable single-photon flux is restricted by the excitation scheme. While electrical pumping is desired for practical applications, repetition rates are usually well below that of their optical counterparts. Without elaborating on the advantages and disadvantages of ultrafast Ti:sapphire lasers, it is worth noting that,

[11] Semiconductor quantum dots are sometimes also referred to as nanoislands, which typically form on a wetting layer, when epitaxially grown, or can be the result of nanopatterning into a predefined substrate area.

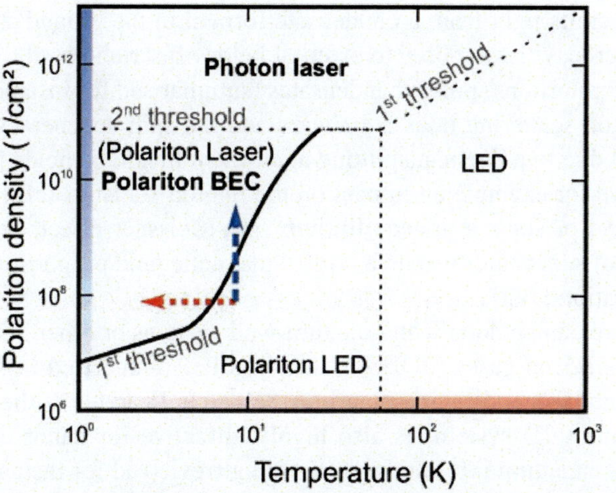

Fig. 6.4 Sketch of a polariton system's phase diagram, at the example of a GaAs-based microcavity device. Here polariton condensation, a BEC-like effect in a quantum-well microcavity structure in the strong light–matter coupling regime, is indicated for low temperatures and intermediate polariton densities when crossing the thick solid phase line in the directions indicated by red (cooling) and blue arrows (density increase). In fact, the dilute Bose gas of neutral excitons is only provided well below the Mott transition for that material system, where Coulomb binding is not spoiled by high charge-carrier densities and interparticle interactions remain little. At too high e–h densities, conventional lasing takes place, indicated by the horizontal dashed line. Owing to their light effective mass, polaritons promise observation of superfluid behaviour and condensation at elevated temperatures up to room temperature. Nevertheless, for a given material platform, at too high temperatures the excitons break up and the structure operates in a weak light–matter coupling regime, indicated by the vertical dashed line. For high temperatures and densities, conventional lasing occurs (loosely dotted line), corresponding to VCSEL operation. Drawn freely after [41] with additional markings and labels

alternatively, a quantum-dot-based single-photon source has been recently triggered by a mode-locked VECSEL, exhibiting a record-high single-photon flux of 143 MHz [99].

The use of quantum-well microcavity systems has not only enabled access to Bose condensation studies at elevated temperatures owing to the light effective mass of polaritons—hybrid light–matter quasi-particles [39, 47, 100, 101]. It has also brought up a new scheme of coherent light generation that does not rely on population inversion in the classical sense to obtain stimulated emission of radiation as in conventional lasers, but on stimulated scattering of bosons into a macroscopically-occupied ground state, in which coherence is established due to the indistinguishability of particles when described by the same macroscopic wave-function [40]. With the phase transition to occur in the excitonic regime well below a Mott transition in the semiconductor system, i.e. the transition from an insulator to a charge-carrier plasma state, the polariton condensation threshold naturally lies below that of conventional lasing [41, 102–104] (cf. Fig. 6.4). This motivates the development and study of a coherent light source referred to as polariton laser [40, 41, 48, 50, 105–

118], which emits light from a condensate formed in the ground state. In addition, polariton microcavities exposed to external fields offer rich possibilities to study the properties of polariton (spinor) condensates (summarised for instance in [119]), e.g. when polariton scattering rates are altered and the spin degeneracy of the ground-state is lifted due to a Zeeman splitting at elevated magnetic fields [112, 120, 121]. This can have for example an impact on the photon statistics (cf. Fig. 3.10), polarisation degree, or show a nonequilibrium spin-Meissner effect, understood as the suppression of superfluidity up to a critical magnetic field proportional to the polariton concentration [122].

Much more can be done with quantum-well systems or other 2D configurations, such as 2D electron gases (2DEGs), than can be summarised in short here using examples from laser and cavity–polaritons research. In addition, the use of naturally formed ultrathin 2D systems is also highly attractive for nanoelectronic applications, and for quantum information processing (reviewed for instance in [123]). To overcome limitations imposed by continuous miniaturisation efforts to keep up with Moore's law, new concepts are envisioned to reach beyond that point where mere shrinking of the channel size results in higher transistor densities and computational performance. 2D-material field-effect transistors [124, 125] promise such a path, but also the achievement of flexible nanoelectronics [126, 127] and novel concepts such as neuromorphic transistors for brain-inspired cognitive systems [128, 129], to name but a few.

Photodetectors and sensors based on graphene and graphene-related-materials are already heavily explored (see for instance [130]). Recently, such detectors have also been functionalised with colloidal quantum dots (cQDs, typically core–shell nanoparticles) that are deposited on the gate-tunable graphene channel of its field-effect transistor. While graphene absorbs little but over a wide range of wavelengths, quantum dots have a very narrow optical resonance related to 0D excitons with high oscillator strength. Due to the aforementioned size-dependence of the confinement energy (6.7), their optical band gap is easily tunable over a wide range for a fixed material system. However, cQDs also suffer from broadening of emission lines due to surface states and electronic defects, and these nanoclusters commonly need ligands for stabilisation, in order to prevent bunching in solution (not detailed here). Such hybrid scheme uses the spectrally tunable and usually strong photon absorption of cQDs in combination with charge- or energy-transfer processes to the underlying graphene layer for high-mobility photocurrent extraction, as studied for example in [131] with regard to gate-tunable Förster transfer (cf. Fig. 2.8).

These selected topics of quantum-structure and quantum-materials applications show some use of prominent quantum systems and their typical advantages. Certainly, towards the establishment of chip-integrated photonics for next generation high-speed optical computing and communications, photonic circuitries will strongly benefit from quantum-physical design principles and structure concepts realised on the nanoscale. Ultimately, based on quantum structures and devices, all-optical computers and quantum computers are envisioned, as discussed in the quantum-technology literature [8–10, 123].

References

1. C. Weisbuch, B. Vinter, *Quantum Semiconductor Structures: Fundamentals and Applications* (Academic, Cambridge, 1991)
2. J.H. Davies, *The Physics of Low-dimensional Semiconductors: An Introduction* (Cambridge University Press, Cambridge, 1998)
3. Y. Yamamoto, A. Imamoglu, *Mesoscopic Quantum Optics* (Wiley, Hoboken, 1999)
4. M. Razeghi, *Technology of Quantum Devices* (Springer, New York, 2010)
5. H. Lüth, *Quantum Physics in the Nanoworld: Schrödinger's Cat and the Dwarfs*, 2nd edn. Graduate Texts in Physics (Springer International Publishing, Berlin, 2015)
6. K.J. Vahala, Optical microcavities. Nature **424**, 839–846 (2003)
7. X. Guo, Y. Ying, L. Tong, Photonic nanowires: from subwavelength waveguides to optical sensors. Acc. Chem. Res. **47**, 656–666 (2014)
8. J.L. O'Brien, A. Furusawa, J. Vučković, Photonic quantum technologies. Nat. Photonics **3**, 687–695 (2009)
9. W.P. Schleich, K.S. Ranade, C. Anton, M. Arndt, M. Aspelmeyer, M. Bayer, G. Berg, T. Calarco, H. Fuchs, E. Giacobino, M. Grassl, P. Hänggi, W.M. Heckl, I.-V. Hertel, S. Huelga, F. Jelezko, B. Keimer, J.P. Kotthaus, G. Leuchs, N. Lütkenhaus, U. Maurer, T. Pfau, M.B. Plenio, E.M. Rasel, O. Renn, C. Silberhorn, J. Schiedmayer, D. Schmitt-Landsiedel, K. Schönhammer, A. Ustinov, P. Walther, H. Weinfurter, E. Welzl, R. Wiesendanger, S. Wolf, A. Zeilinger, P. Zoller, Quantum technology: from research to application. Appl. Phys. B **122**, 130 (2016)
10. A. Acín, I. Bloch, H. Buhrman, T. Calarco, C. Eichler, J. Eisert, D. Esteve, N. Gisin, S.J. Glaser, F. Jelezko, S. Kuhr, M. Lewenstein, M.F. Riedel, P.O. Schmidt, R. Thew, A. Wallraff, I. Walmsley, F.K. Wilhelm, The quantum technologies roadmap: a European community view. New J. Phys. **20**, 080201 (2018)
11. B. Deveaud, *The Physics of Semiconductor Microcavities* (Wiley-VCH Verlag, Weinheim, 2007)
12. S. Reitzenstein, A. Forchel, Quantum dot micropillars. J. Phys. D: Appl. Phys. **43**(3), 033001 (2010)
13. D. Sanvitto, V. Timofeev, *Exciton Polaritons in Microcavities: New Frontiers* (Springer, Berlin, 2012)
14. A.V. Kavokin, J.J. Baumberg, G. Malpuech, F.P. Laussy, *Microcavities*, vol. 1 (Oxford University Press, Oxford, 2017)
15. A. Rahimi-Iman, *Polariton Physics: From Dynamic Bose–Einstein Condensates in Strongly-Coupled Light–Matter Systems to Polariton Lasers*. Springer Series in Optical Sciences, vol. 229 (Springer International Publishing, Cham, 2020)
16. A.V. Kavokin, J.J. Baumberg, G. Malpuech, F.P. Laussy, *Microcavities* (Oxford University Press, Oxford, 2007)
17. M. Orszag, *Quantum Optics: Including Noise Reduction, Trapped Ions, Quantum Trajectories, and Decoherence*, 3rd edn. (Springer International Publishing, Berlin, 2016)
18. R.L. Greene, K.K. Bajaj, D.E. Phelps, Energy levels of Wannier excitons in GaAs-$Ga_{1-x}Al_xAs$ quantum-well structures. Phys. Rev. B **29**, 1807 (1984)
19. C. Priester, G. Allan, M. Lannoo, Wannier excitons in GaAs-$Ga_{1-x}Al_xAs$ quantum-well structures: influence of the effective-mass mismatch. Phys. Rev. B **30**, 7302 (1984)
20. R.P. Leavitt, J.W. Little, Simple method for calculating exciton binding energies in quantum-confined semiconductor structures. Phys. Rev. B **42**, 11774 (1990)
21. C.F. Klingshirn, *Semiconductor Optics* (Springer, Berlin, 2012)
22. O. El Daïf, A. Baas, T. Guillet, J.-P. Brantut, R.I. Kaitouni, J.L. Staehli, F. Morier-Genoud, B. Deveaud, Polariton quantum boxes in semiconductor microcavities. Appl. Phys. Lett. **88**, 061105–3 (2006)
23. R.I. Kaitouni, O. El Daïf, A. Baas, M. Richard, T. Paraiso, P. Lugan, T. Guillet, F. Morier-Genoud, J.D. Ganière, J.L. Staehli, V. Savona, B. Deveaud, Engineering the spatial confinement of exciton polaritons in semiconductors. Phys. Rev. B **74**, 155311 (2006)

24. A. Rahimi-Iman, C. Schneider, J. Fischer, S. Holzinger, M. Amthor, S. Höfling, S. Reitzenstein, L. Worschech, M. Kamp, A. Forchel, Zeeman splitting and diamagnetic shift of spatially confined quantum-well exciton polaritons in an external magnetic field. Phys. Rev. B **84**, 165325 (2011)
25. K. Winkler, C. Schneider, J. Fischer, A. Rahimi-Iman, M. Amthor, A. Forchel, S. Reitzenstein, S. Höfling, M. Kamp, Electroluminescence from spatially confined exciton polaritons in a textured microcavity. Appl. Phys. Lett. **102**, 041101 (2013)
26. M. Bayer, T. Gutbrod, J.P. Reithmaier, A. Forchel, T.L. Reinecke, P.A. Knipp, A.A. Dremin, V.D. Kulakovskii, Optical modes in photonic molecules. Phys. Rev. Lett. **81**, 2582–2585 (1998)
27. M. Bayer, T. Gutbrod, A. Forchel, T.L. Reinecke, P.A. Knipp, R. Werner, J.P. Reithmaier, Optical demonstration of a crystal band structure formation. Phys. Rev. Lett. **83**, 5374–5377 (1999)
28. C.W. Lai, N.Y. Kim, S. Utsunomiya, G. Roumpos, H. Deng, M.D. Fraser, T. Byrnes, P. Recher, N. Kumada, T. Fujisawa, Y. Yamamoto, Coherent zero-state and π-state in an exciton-polariton condensate array. Nature **450**, 529–532 (2007)
29. N.Y. Kim, K. Kusudo, C. Wu, N. Masumoto, A. Löffler, S. Höfling, N. Kumada, L. Worschech, A. Forchel, Y. Yamamoto, Dynamical d-wave condensation of exciton-polaritons in a two-dimensional square-lattice potential. Nat. Phys. **7**, 681–686 (2011)
30. N.Y. Kim, K. Kusudo, A. Löffler, S. Höfling, A. Forchel, Y. Yamamoto, Exciton-polariton condensates near the Dirac point in a triangular lattice. New J. Phys. **15**, 035032 (2013)
31. K. Winkler, J. Fischer, A. Schade, M. Amthor, R. Dall, J. Geßler, M. Emmerling, E.A. Ostrovskaya, M. Kamp, C. Schneider, S. Höfling, A polariton condensate in a photonic crystal potential landscape. New J. Phys. **17**, 023001 (2015)
32. M.A. Bandres, S. Wittek, G. Harari, M. Parto, J. Ren, M. Segev, D.N. Christodoulides, M. Khajavikhan, Topological insulator laser: experiments. Science 359, eaar4005 (2018)
33. G. Harari, M.A. Bandres, Y. Lumer, M.C. Rechtsman, Y.D. Chong, M. Khajavikhan, D.N. Christodoulides, M. Segev, Topological insulator laser: theory. Science **359**, eaar4003 (2018)
34. R. Ge, W. Broer, T.C.H. Liew, Floquet topological polaritons in semiconductor microcavities. Phys. Rev. B **97**, 195305 (2018)
35. S. Klembt, T.H. Harder, O.A. Egorov, K. Winkler, R. Ge, M.A. Bandres, M. Emmerling, L. Worschech, T.C.H. Liew, M. Segev, C. Schneider, S. Höfling, Exciton-polariton topological insulator. Nature **562**, 552–556 (2018)
36. N.N. Bogoliubov, On the theory of superfluidity. J. Phys. USSR **11**, 23 (1947)
37. O. Penrose, L. Onsager, Bose-Einstein condensation and liquid helium. Phys. Rev. **104**, 576 (1956)
38. V. Savona, D. Sarchi, Bose-Einstein condensation of microcavity polaritons. Phys. Status Solidi B **242**, 2290–2301 (2005)
39. H. Deng, H. Haug, Y. Yamamoto, Exciton-polariton Bose-Einstein condensation. Rev. Mod. Phys. **82**, 1489 (2010)
40. A. Imamoglu, R.J. Ram, S. Pau, Y. Yamamoto, Nonequilibrium condensates and lasers without inversion: exciton-polariton lasers. Phys. Rev. A **53**, 4250 (1996)
41. A. Kavokin, G. Malpuech, F.P. Laussy, Polariton laser and polariton superfluidity in microcavities. Phys. Lett. A **306**, 187–199 (2003)
42. D. Solnyshkov, E. Petrolati, A. Di Carlo, G. Malpuech, Theory of an electrically injected bulk polariton laser. Appl. Phys. Lett. **94**, 011110–3 (2009)
43. D. Snoke, Spontaneous Bose coherence of excitons and polaritons. Science **298**(5597), 1368–1372 (2002)
44. D. Sarchi, V. Savona, Long-range order in the Bose-Einstein condensation of polaritons. Phys. Rev. B **75**, 115326 (2007)
45. D.D. Solnyshkov, H. Tercas, K. Dini, G. Malpuech, Hybrid Boltzmann-Gross-Pitaevskii theory of Bose-Einstein condensation and superfluidity in open driven-dissipative systems. Phys. Rev. A: At. Mol. Opt. Phys. **89**, 033626 (2014)

46. A. Rahimi-Iman, *Optically-Excited Polariton Condensates*. Springer Series in Optical Sciences, vol. 229 (Springer International Publishing, Cham, 2020), pp. 195–240
47. D. Sanvitto, S. Kéna-Cohen, The road towards polaritonic device. Nat. Mater. **15**(10), 1061–1073 (2016)
48. M.D. Fraser, S. Höfling, Y. Yamamoto, Physics and applications of exciton-polariton lasers. Nat. Mater. **15**, 1049 (2016)
49. M.D. Fraser, Coherent exciton-polariton devices. Semicond. Sci. Technol. **32**, 093003 (2017)
50. S. Kim, B. Zhang, Z. Wang, J. Fischer, S. Brodbeck, M. Kamp, C. Schneider, S. Höfling, H. Deng, Coherent polariton laser. Phys. Rev. X **6**, 011026 (2016)
51. C. Schneider, M.M. Glazov, T. Korn, S. Höfling, B. Urbaszek, Two-dimensional semiconductors in the regime of strong light-matter coupling. Nat. Commun. **9**, 2695 (2018)
52. G.H. Wannier, The structure of electronic excitation levels in insulating crystals. Phys. Rev. **52**, 191 (1937)
53. N.F. Mott, Conduction in polar crystals. II. The conduction band and ultra-violet absorption of alkali-halide crystals. Trans. Faraday Soc. **34**, 500 (1938)
54. J. Frenkel, *Wave Mechanics. Elementary Theory* (Clarendon Press, Oxford, 1932)
55. K.F. Mak, J. Shan, Photonics and optoelectronics of 2D semiconductor transition metal dichalcogenides. Nat. Photonics **10**, 216–226 (2016)
56. N. Peyghambarian, S.W. Koch, A. Mysyrowicz, *Introduction to Semiconductor Optics* (Prentice Hall, Upper Saddle River, 1993)
57. H. Haug, S.W. Koch, *Quantum Theory of the Optical and Electronic Properties of Semiconductors* (World Scientific, Singapore, 1994)
58. A. Griffin, D.W. Snoke, S. Stringari, *Bose-Einstein Condensation* (Cambridge University Press, Cambridge, 1995)
59. S.A. Moskalenko, D.W. Snoke, *Bose-Einstein Condensation of Excitons and Biexcitons: And Coherent Nonlinear Optics with Excitons* (Cambridge University Press, Cambridge, 2000)
60. A.J. Leggett, Bose-Einstein condensation in the alkali gases: some fundamental concepts. Rev. Mod. Phys. **73**, 307–356 (2001)
61. L.P. Pitaevskii, S. Stringari, *Bose-Einstein Condensation* (Oxford University Press, Oxford, 2003)
62. A. Posazhennikova, Colloquium: weakly interacting, dilute Bose gases in 2D. Rev. Mod. Phys. **78**(4), 1111–1134 (2006)
63. Z. Wang, D.A. Rhodes, K. Watanabe, T. Taniguchi, J.C. Hone, J. Shan, K.F. Mak, Evidence of high-temperature exciton condensation in two-dimensional atomic double layers. Nature **574**, 76–80 (2019)
64. H. Kroemer, Nobel lecture: quasielectric fields and band offsets: teaching electrons new tricks. Rev. Mod. Phys. **73**, 783–793 (2001)
65. Z.I. Alferov, Nobel lecture: the double heterostructure concept and its applications in physics, electronics, and technology. Rev. Mod. Phys. **73**, 767–782 (2001)
66. S. Noda, Seeking the ultimate nanolaser. Science **314**, 260 (2006)
67. S. Strauf, F. Jahnke, Single quantum dot nanolaser. Laser Photonics Rev. **5**, 607–633 (2011)
68. G. Eisenstein, D. Bimberg, *Green Photonics and Electronics* (Springer, Cham, 2017)
69. Y. Yamamoto, F. Tassone, H. Cao, *Semiconductor Cavity Quantum Electrodynamics* (Springer, Berlin, 2000)
70. J.-M. Gérard, Solid-state cavity-quantum electrodynamics with self-assembled quantum dots, in *Single Quantum Dots: Fundamentals, Applications, and New Concepts* (Springer, Berlin, 2003), pp. 269–314
71. P. Michler (ed.), *Single Quantum Dots: Fundamentals, Applications and New Concepts* (Springer, Berlin, 2003)
72. P. Michler (ed.), *Single Semiconductor Quantum Dots* (Springer, Berlin, 2009)
73. S. Rodt, S. Reitzenstein, T. Heindel, Deterministically fabricated solid-state quantum-light sources. J. Phys. Condens. Matter **32**, 153003 (2020)

74. C. Möller, C. Fuchs, C. Berger, F. Zhang, A. Rahimi-Iman, M. Koch, A.R. Perez, S.W. Koch, J. Hader, J.V. Moloney, W. Stolz, The development and fundamental analysis of type-II VECSELs at 1.2 μm (Conference Presentation), in *Vertical External Cavity Surface Emitting Lasers (VECSELs) VII*, vol. 10087 (International Society for Optics and Photonics, 2017), p. 100870L
75. C. Möller, F. Zhang, C. Fuchs, C. Berger, A. Rehn, A.R. Perez, A. Rahimi-Iman, J. Hader, M. Koch, J.V. Moloney, S.W. Koch, W. Stolz, Fundamental transverse mode operation of a type-II vertical-external-cavity surface-emitting laser at 1.2 μm. Electron. Lett. **53**, 93–94 (2017)
76. C. Lammers, M. Stein, C. Berger, C. Möller, C. Fuchs, A. Ruiz Perez, A. Rahimi-Iman, J. Hader, J.V. Moloney, W. Stolz, S.W. Koch, M. Koch, Gain spectroscopy of a type-II VECSEL chip. Appl. Phys. Lett. **109**, 232107 (2016)
77. E.U. Rafailov, M.A. Cataluna, W. Sibbett, Mode-locked quantum-dot lasers. Nat. Photonics **1**, 395–401 (2007)
78. M. Gaafar, D.A. Nakdali, C. Möller, K.A. Fedorova, M. Wichmann, M.K. Shakfa, F. Zhang, A. Rahimi-Iman, E.U. Rafailov, M. Koch, Self-mode-locked quantum-dot vertical-external-cavity surface-emitting laser. Opt. Lett. **39**, 4623–4626 (2014)
79. D. Al Nakdali, M.K. Shakfa, M. Gaafar, M. Butkus, K.A. Fedorova, M. Zulonas, M. Wichmann, F. Zhang, B. Heinen, A. Rahimi-Iman, W. Stolz, E.U. Rafailov, M. Koch, High-power quantum-dot vertical-external-cavity surface-emitting laser exceeding 8 W. IEEE Photonics Technol. Lett. **26**, 1561–1564 (2014)
80. D. Al Nakdali, M. Gaafar, M.K. Shakfa, F. Zhang, M. Vaupel, K.A. Fedorova, A. Rahimi-Iman, E.U. Rafailov, M. Koch, High-power operation of quantum-dot semiconductor disk laser at 1180 nm. IEEE Photonics Technol. Lett. **27**, 1128–1131 (2015)
81. A. Rahimi-Iman, Recent advances in VECSELs. J. Opt. **18**, 093003 (2016)
82. M.A. Gaafar, A. Rahimi-Iman, K.A. Fedorova, W. Stolz, E.U. Rafailov, M. Koch, Mode-locked semiconductor disk lasers. Adv. Opt. Photonics **8**, 370–400 (2016)
83. G.P. Agrawal, N.K. Dutta, *Semiconductor Lasers* (Springer US, New York, 1993)
84. T. Numai, *Fundamentals of Semiconductor Lasers* (Springer Japan, Tokyo, 2015)
85. S. Reitzenstein, T. Heindel, C. Kistner, A. Rahimi-Iman, C. Schneider, S. Höfling, A. Forchel, Low threshold electrically pumped quantum dot-micropillar lasers. Appl. Phys. Lett. **93**(6), 061104 (2008)
86. C. Kistner, T. Heindel, C. Schneider, A. Rahimi-Iman, S. Reitzenstein, S. Höfling, A. Forchel, Demonstration of strong coupling via electro-optical tuning in high-quality QD-micropillar systems. Opt. Express **16**, 15006–15012 (2008)
87. S. Reitzenstein, S. Münch, P. Franeck, A. Rahimi-Iman, A. Löffler, S. Höfling, L. Worschech, A. Forchel, Control of the strong light-matter interaction between an elongated $In_{0.3}Ga_{0.7}As$ quantum dot and a micropillar cavity using external magnetic fields. Phys. Rev. Lett. **103**, 127401 (2009)
88. E.M. Purcell, Spontaneous emission probabilities at radio frequencies. Phys. Rev. **69**, 681 (1946)
89. L.-J. Wang, G. Cao, T. Tu, H.-O. Li, C. Zhou, X.-J. Hao, Z. Su, G.-C. Guo, H.-W. Jiang, G.-P. Guo, A graphene quantum dot with a single electron transistor as an integrated charge sensor. Appl. Phys. Lett. **97**, 262113 (2010)
90. R. Waser, *Nanoelectronics and Information Technology: Advanced Electronic Materials and Novel Devices* (Wiley, New York, 2012)
91. D.J.P. Ellis, A.J. Bennett, S.J. Dewhurst, C.A. Nicoll, D.A. Ritchie, A.J. Shields, Cavity-enhanced radiative emission rate in a single-photon-emitting diode operating at 0.5GHz. New J. Phys. **10**, 043035 (2008)
92. J. Claudon, J. Bleuse, N.S. Malik, M. Bazin, P. Jaffrennou, N. Gregersen, C. Sauvan, P. Lalanne, J.-M. Gérard, A highly efficient single-photon source based on a quantum dot in a photonic nanowire. Nat. Photonics **4**, 174–177 (2010)
93. T. Heindel, C. Schneider, M. Lermer, S.H. Kwon, T. Braun, S. Reitzenstein, S. Höfling, M. Kamp, A. Forchel, Electrically driven quantum dot-micropillar single photon source with 34% overall efficiency. Appl. Phys. Lett. **96**, 011107 (2010)

94. A. Boretti, L. Rosa, A. Mackie, S. Castelletto, Electrically driven quantum light sources. Adv. Opt. Mater. **3**, 1012–1033 (2015)
95. M. Pelton, C. Santori, J. Vučković, B. Zhang, G.S. Solomon, J. Plant, Y. Yamamoto, Efficient source of single photons: a single quantum dot in a micropost microcavity. Phys. Rev. Lett. **89**, 233602 (2002)
96. C. Santori, D. Fattal, J. Vučković, G.S. Solomon, Y. Yamamoto, Indistinguishable photons from a single-photon device. Nature **419**, 594–597 (2002)
97. A.J. Bennett, R.B. Patel, A.J. Shields, K. Cooper, P. Atkinson, C.A. Nicoll, D.A. Ritchie, Indistinguishable photons from a diode. Appl. Phys. Lett. **92**(19), 193503 (2008)
98. P. Senellart, G. Solomon, A. White, High-performance semiconductor quantum-dot single-photon sources. Nat. Nanotechnol. **12**, 1026–1039 (2017)
99. A. Schlehahn, M. Gaafar, M. Vaupel, M. Gschrey, P. Schnauber, J.-H. Schulze, S. Rodt, A. Strittmatter, W. Stolz, A. Rahimi-Iman, T. Heindel, M. Koch, S. Reitzenstein, Single-photon emission at a rate of 143 MHz from a deterministic quantum-dot microlens triggered by a mode-locked vertical-external-cavity surface-emitting laser. Appl. Phys. Lett. **107**, 041105 (2015). https://doi.org/10.1063/1.4927429
100. C. Weisbuch, M. Nishioka, A. Ishikawa, Y. Arakawa, Observation of the coupled exciton-photon mode splitting in a semiconductor quantum microcavity. Phys. Rev. Lett. **69**, 3314 (1992)
101. A. Rahimi-Iman, *Towards Polariton Condensates and Devices*. Springer Series in Optical Sciences, vol. 229 (Springer International Publishing, Cham, 2020), pp. 1–32
102. H. Deng, G. Weihs, C. Santori, J. Bloch, Y. Yamamoto, Condensation of semiconductor microcavity exciton polaritons. Science **298**(5591), 199–202 (2002)
103. H. Deng, G. Weihs, D. Snoke, J. Bloch, Y. Yamamoto, Polariton lasing vs. photon lasing in a semiconductor microcavity. Proc. Natl. Acad. Sci. **100**(26), 15318–15323 (2003)
104. J.-S. Tempel, F. Veit, M. Assmann, L.E. Kreilkamp, A. Rahimi-Iman, A. Löffler, S. Höfling, S. Reitzenstein, L. Worschech, A. Forchel, M. Bayer, Characterization of two-threshold behavior of the emission from a GaAs microcavity. Phys. Rev. B **85**, 075318 (2012)
105. L.S. Dang, D. Heger, R. André, F. Bœuf, R. Romestain, Stimulation of polariton photoluminescence in semiconductor microcavity. Phys. Rev. Lett. **81**, 3920 (1998)
106. G. Malpuech, A. Kavokin, A. Di Carlo, J.J. Baumberg, Polariton lasing by exciton-electron scattering in semiconductor microcavities. Phys. Rev. B **65**, 153310 (2002)
107. S. Christopoulos, G.B.H. von Högersthal, A.J.D. Grundy, P.G. Lagoudakis, A.V. Kavokin, J.J. Baumberg, G. Christmann, R. Butté, E. Feltin, J.-F. Carlin, N. Grandjean, Room-temperature polariton lasing in semiconductor microcavities. Phys. Rev. Lett. **98**, 126405 (2007)
108. D. Bajoni, P. Senellart, E. Wertz, I. Sagnes, A. Miard, A. Lemaître, J. Bloch, Polariton laser using single micropillar GaAs-GaAlAs semiconductor cavities. Phys. Rev. Lett. **100**, 047401 (2008)
109. J.J. Baumberg, A.V. Kavokin, S. Christopoulos, A.J.D. Grundy, R. Butté, G. Christmann, D.D. Solnyshkov, G. Malpuech, G. Baldassarri Höger von Högersthal, E. Feltin, J.-F. Carlin, N. Grandjean, Spontaneous polarization buildup in a room-temperature polariton laser. Phys. Rev. Lett. **101**, 136409 (2008)
110. D. Bajoni, Polariton lasers. Hybrid light-matter lasers without inversion. J. Phys. D: Appl. Phys. **45**(31), 313001 (2012)
111. A. Rahimi-Iman, *Nichtlineare Effekte in III/V Quantenfilm-Mikroresonatoren: Von dynamischer Bose–Einstein-Kondensation hin zum elektrisch betriebenen Polariton-Laser* (Cuvillier Verlag, Göttingen, 2013)
112. C. Schneider, A. Rahimi-Iman, N.Y. Kim, J. Fischer, I.G. Savenko, M. Amthor, M. Lermer, A. Wolf, L. Worschech, V.D. Kulakovskii, I.A. Shelykh, M. Kamp, S. Reitzenstein, A. Forchel, Y. Yamamoto, S. Höfling, An electrically pumped polariton laser. Nature **497**, 348–352 (2013)
113. A. Rahimi-Iman, C. Schneider, S. Höfling, Electrically driven polariton lasing. Opt. Photonics News **24**, 30–30 (2013)
114. P. Bhattacharya, B. Xiao, A. Das, S. Bhowmick, J. Heo, Solid state electrically injected exciton-polariton laser. Phys. Rev. Lett. **110**, 206403 (2013)

115. P. Bhattacharya, T. Frost, S. Deshpande, M.Z. Baten, A. Hazari, A. Das, Room temperature electrically injected polariton laser. Phys. Rev. Lett. **112**, 236802 (2014)
116. C.P. Dietrich, A. Steude, L. Tropf, M. Schubert, N.M. Kronenberg, K. Ostermann, S. Höfling, M.C. Gather, An exciton-polariton laser based on biologically produced fluorescent protein. Sci. Adv. **2**, e1600666 (2016)
117. S.A. Moskalenko, I.M. Tiginyanu, Exciton-polariton laser. Low Temp. Phys. **42**, 330–339 (2016)
118. A. Rahimi-Iman, *The Concept of Polariton Lasing*. Springer Series in Optical Sciences, vol. 229 (Springer International Publishing, Cham, 2020), pp. 87–118
119. A. Rahimi-Iman, *Polaritons in External Fields*. Springer Series in Optical Sciences, vol. 229 (Springer International Publishing, Cham, 2020), pp. 241–262
120. J. Fischer, S. Brodbeck, A.V. Chernenko, I. Lederer, A. Rahimi-Iman, M. Amthor, V.D. Kulakovskii, L. Worschech, M. Kamp, M. Durnev, C. Schneider, A.V. Kavokin, S. Höfling, Anomalies of a nonequilibrium spinor polariton condensate in a magnetic field. Phys. Rev. Lett. **112**, 093902 (2014)
121. A.V. Chernenko, A. Rahimi-Iman, J. Fischer, M. Amthor, C. Schneider, S. Reitzenstein, A. Forchel, S. Hoefling, Polariton condensate coherence in planar microcavities in a magnetic field. Semiconductors **50**, 1609–1613 (2016). https://doi.org/10.1134/S1063782616120058
122. Y.G. Rubo, A. Kavokin, I. Shelykh, Suppression of superfluidity of exciton-polaritons by magnetic field. Phys. Lett. A **358**, 227–230 (2006)
123. X. Liu, M.C. Hersam, 2D materials for quantum information science. Nat. Rev. Mater. **4**, 669–684 (2019)
124. B. Radisavljevic, A. Radenovic, J. Brivio, V. Giacometti, A. Kis, Single-layer MoS_2 transistors. Nat. Nanotechnol. **6**, 147–150 (2011). https://doi.org/10.1038/nnano.2010.279
125. Q.H. Wang, K. Kalantar-Zadeh, A. Kis, J.N. Coleman, M.S. Strano, Electronics and optoelectronics of two-dimensional transition metal dichalcogenides. Nat. Nanotechnol. **7**, 699–712 (2012)
126. K.S. Kim, Y. Zhao, H. Jang, S.Y. Lee, J.M. Kim, K.S. Kim, J.-H. Ahn, P. Kim, J.-Y. Choi, B.H. Hong, Large-scale pattern growth of graphene films for stretchable transparent electrodes. Nature **457**, 706–710 (2009)
127. D. Akinwande, N. Petrone, J. Hone, Two-dimensional flexible nanoelectronics. Nat. Commun. **5**, 5678 (2014)
128. C.J. Wan, Y.H. Liu, P. Feng, W. Wang, L.Q. Zhu, Z.P. Liu, Y. Shi, Q. Wan, Flexible metal oxide/graphene oxide hybrid neuromorphic transistors on flexible conducting graphene substrates. Adv. Mater. **28**, 5878–5885 (2016)
129. C.J. Wan, L.Q. Zhu, Y.H. Liu, P. Feng, Z.P. Liu, H.L. Cao, P. Xiao, Y. Shi, Q. Wan, Proton-conducting graphene oxide-coupled neuron transistors for brain-inspired cognitive systems. Adv. Mater. **28**, 3557–3563 (2016)
130. F.H.L. Koppens, T. Mueller, P. Avouris, A.C. Ferrari, M.S. Vitiello, M. Polini, Photodetectors based on graphene, other two-dimensional materials and hybrid systems. Nat. Nanotechnol. **9**, 780–793 (2014)
131. R. Li, L.M. Schneider, W. Heimbrodt, H. Wu, M. Koch, A. Rahimi-Iman, Gate tuning of Förster resonance energy transfer in a graphene - quantum dot FET photo-detector. Sci. Rep. **6**, 28224 (2016). https://doi.org/10.1038/srep28224

Chapter 7
Structuring Possibilities

Abstract Recent developments in the field of semiconductor physics and material sciences strongly rely on high-quality materials synthesis, such as epitaxial growth and deposition, and ultra-precise processing techniques, such as etching and milling, that result in defined mesoscale or nanoscale structures for various studies and applications. Typically, experiments rely on the availability of suitable structures, and advances in the domain of quantum technologies, laser sciences and nanomaterial systems would be hardly imaginable without the many key tools developed over the past decades. In this chapter, some important structuring possibilities that enabled the work with 2D materials, quantum structures and nanoparticles in the author's Habilitation project are briefly summarised. The arbitrary and narrow selection of examples in the different domains addressed here may merely serve explanatory purposes and provide an overview on possible techniques used in the overall presented work.

7.1 Epitaxy

Epitaxy refers to the concept of ordered deposition of crystalline materials on top of crystalline materials. In the simplest case, layered growth of the bare substrate material can be performed on a crystalline substrate. Typically, for good growth, a buffer layer is grown on top of the substrate to provide a smooth, as well as defect- and strain-free, crystal surface—before the actual structure growth is carried out.

The most prominent use of epitaxy methods is represented by heterostructure growth. Multilayered structures and thin-layer heterostructures are commonly grown by molecular beam epitaxy (MBE) [1, 2] and metal-organic vapour-phase epitaxy (MOVPE) [3], but there are indeed also other (derivatives of these) techniques not detailed here for the sake of compactness. The map of semiconductors, which plots the common semiconductors' energies as a function of their lattice parameters, is well known in this domain. This bears importance for every epitaxial growth process due to the strong impact of lattice mismatch on growth quality, as strain and defects can be drastically reduced when materials with similar lattice constant are grown on

A. Rahimi-Iman, *Semiconductor Photonics of Nanomaterials and Quantum Structures*, Springer Series in Solid-State Sciences 196,
https://doi.org/10.1007/978-3-030-69352-7_7

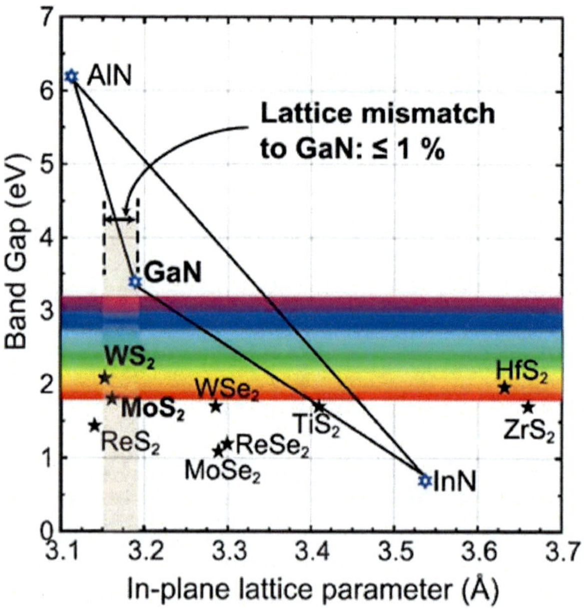

Fig. 7.1 Band gap as a function of the in-plane lattice parameter. The common semiconductor materials in this range are the wurtzite-like nitrides. Remarkably, some sulfur-based TMDCs such as WS_2 and Mo_2 have a very similar lattice parameter among themselves and are almost matched to GaN. Other transition-metal diselenides feature a wider lattice, e.g. TiS_2 is in principle lattice-matched to a GaInN. Similarly, the transition-metal diselenides lie closely together. For a more elaborate depiction of the alloy lines between GaN, AlN and InN, and the relation/comparison to other typical semiconductor systems (many of them zinc-blende type and with larger lattice parameter), the interested reader is referred to semiconductor physics textbooks. The insulating hBN with band gap of about 5–6 eV and a lattice constant of about 0.25 nm does not fit into this chart's range. The rainbow-coloured box indicates the visible spectral region.

top of each other. Of course, techniques have been developed to counter the influence of lattice mismatch, for instance by introducing strain compensating layers.

Nowadays, with the revival of studies on 2D materials, which may one day become part of integrated circuits or semiconductor optoelectronic devices, they are eagerly integrated into the existing epitaxy map of conventional semiconductors. This is pursued with eyes towards van-der-Waals (vdW) epitaxy, i.e. the controlled growth of vertical stacks based on layered materials. For such a map including TMDCs[1] (Fig. 7.1), see for instance [4], in which TMDCs are discussed as a substrate material for crystalline GaN growth.

As a general rule, one can say that for optoelectronic applications very high-quality materials are usually required, because charge carriers recombine nonradiatively at defects—unless the defects are rather a feature than a bug. These quality-crystalline

[1] Acronym for ***t****ransition-****m****etal* ***d****ichalcogenides.*

materials may be incorporated layer-wise as quantum wells or differently arranged heterostructures inside a device's semiconductor structure, or even be deposited to form nano-sized islands in a self-organised fashion.

Samples grown by epitaxy are widely used in science and industry for the production of various electronic, optical and optoelectronic devices. Epitaxy has become indispensable in research activities on nanostructured systems and quantum-physical devices. Naturally, different methods are applied to serve different purposes, which are briefly summarised using two common examples. More information is naturally found in the expert literature on these subjects (in the field of semiconductor technology) and it is the aim of this section to merely provide a basic background.

7.1.1 Molecular Beam Epitaxy

Molecular-beam epitaxy (MBE) provides a powerful tool for structure growth with enourmous precisions down to the sub-monolayer limit. It serves as the best example for well-controlled growth in an evacuated environment using a beam of molecules in which particles impinge on the substrate surface one after the other, with the beam being in the so-called Knudsen regime. This regime is characterised by the mean-free path of a particle in the beam before a scattering process occurs which must be larger than the chamber dimensions. Thus, ultra-high vacuum is necessary to provide predictable growth.

Materials are provided by Knudsen cells that serve as furnaces from which the molecular beam originates as a consequence of a heating process. While the temperature of the furnace regulates the flux, a shutter in front of the exit aperture allows to control exposure times. For radial symmetric growth, the substrate holder is rotating, while adequate heating of the substrate provides kinetic energy to the deposited molecules to form homogeneous films and enable surface reconstructions by means of diffusion and migration. With the help of reflected high-energy electrons diffraction (RHEED), monolayer growth can be monitored in situ.

MBE is widely known for its high-quality growth, lab-scale production and good material-interface control. However, at the same time, it is costly, lacks mass production capabilities and is relatively slow with one monolayer-per-second growth rate. A micrometer-scale structure could take hours to grow. Moreover, precise control and knowledge of the various growth parameters is needed to produce high-quality structures deterministically. Ideally, the apparatus is operated in a clean-room environment to facilitate high-quality growth.

7.1.2 Chemical Vapour Deposition

An alternative method for epitaxy is given by chemical vapour deposition (CVD). The concept of CVD differs drastically from MBE. Its layered growth results from

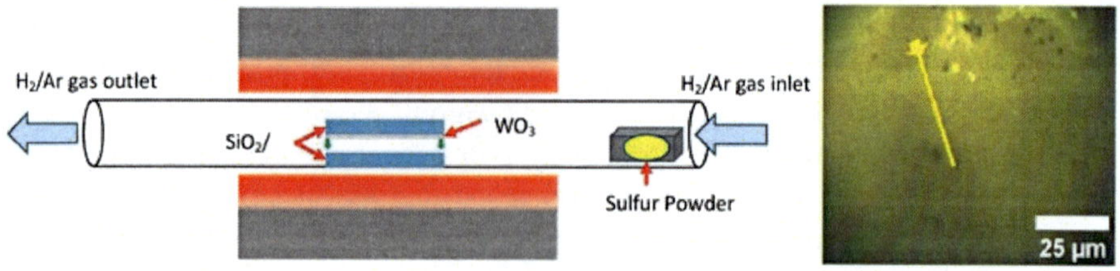

Fig. 7.2 Sketch of the CVD-growth process for controlled monolayer TMDC deposition in a hot furnace. The method described in [6] has been further refined to date and enables different growth modes, such as large area single-crystal growth, full monolayer polycrystalline coverage, selective or even site-controlled growth. The interested reader is referred to the growth-specific TMDC literature. Reproduced under the terms of the CC-BY 4.0 Licence (http://creativecommons.org/licenses/by/4.0/). [6] Copyright 2015 The Author(s) and Springer Nature. Right inset: CVD-grown WSe_2 monolayer flake obtained from this method (also see caption of Fig. 1.1). Adapted under the terms of the CC-BY 3.0 Licence (http://creativecommons.org/licenses/by/3.0/) from Lippert et al., 2D Mater., "Influence of the substrate material on the optical properties of tungsten diselenide monolayers", https://doi.org/10.1088/2053-1583/aa5b21 [7] Copyright 2017 IOP Publishing Ltd

the combination of different evapourated or gaseous precursor materials under atmospheric pressures and high temperatures (usually hundreds of degrees more than in the case of MBE) in a growth reactor.

Chemical reactions taking place on the surface then lead to material-specific growth on the rotated and heated substrate. In contrast to MBE, the chamber is full of material vapour and very high amounts of growth materials end up in an exhaust. However, the major advantage of CVD lies in the large-scale production possibilities in comparison to MBE.

Particularly, metal-organic CVD (MOCVD[2]) is very prominent in the domain of semiconductor devices growth. In fact, operation costs and quality issues are comparable to that of MBE according to textbooks, although it can be said that interface sharpness might not match those of MBE-grown heterostructures. However, it should be noted as a clear advantage that deterministic and repeatable production output can become scalable when using industry-scale shower head reactors. Still, as examples from the field of semiconductor-disk-laser research show [5], MOCVD is a very attractive production method for lab-scale activities and is known to deliver high-performance photonic devices.

In recent times, the need for large-area production in the field of monolayer 2D-materials synthesis has also grown. In this context, CVD techniques have provided access to large-scale pattern growth of graphene and TMDC films, for instance for stretchable transparent electrodes [8] or fundamental studies, such as on the growth scale and kinetics [6, 9]. A typical growth setup is schematically shown in Fig. 7.2a together with polycrystalline CVD-grown monolayer WSe_2 flakes of intermediate size (b) from the generation of samples studied in [7] in comparison with mechanically exfoliated monolayers.

[2] Also known as ***metal-organic vapor-phase epitaxy***, MOVPE.

Currently, more advanced growth-modes are applied to achieve location-specific growth and size control [10] or even lateral or vertical heterostructures [11–13]. Under certain growth conditions, even certain (high-)symmetry configurations of homobilayers can be favoured [14] based on how the grown monolayer is chemically terminated [15]. TEM high-resolution images and diffraction patterns could verify such targeted alignment (not shown or discussed here), as addressed for instance in the Supporting Information section of [14].

Some studies even indicated that, while the quality of CVD-grown large-area monolayers may not have been as good as that of mechanically exfoliated flakes, the optical properties on the same substrate did not differ much. However, growth-induced strain and defects can alter the charge-carrier lifetimes and the spectral features [7].

For example, CVD growth of WSe_2 monolayers in the study of [7] was achieved using a tungsten source carrier chip (5 nm WO_3[3] thin film on 90 nm SiO_2) and a sapphire substrate. The tungsten source chip was covered by the sapphire growth substrate in a face-to-face contact configuration. The sample was loaded into the center of a 2" diameter and 24" long quartz tube (comparable to the sketch in Fig. 7.2), and a ceramic boat with 1 g of selenium powder was located upstream in the quartz tube. After loading, the ambient gas of the tube was purged out by a mechanical pump. At a typical base pressure of 10 mTorr, the furnace was heated to 750°C at a specific ramping rate (13 min^{-1}) and the temperature held at 750°C for 4 min. Afterwards, the temperature was raised to 850°C at the same ramping rate. 20 sccm of Ar gas was introduced at 500°C during the temperature increase to reduce moisture inside of the tube, and the flow was ended at 500°C with decreasing temperatures. During the process, hydrogen gas was supplied to improve WO_3 reduction temperature upwards from 700°C to 600°C temperature downwards. During growth, a 1.6 Torr pressure was maintained in the furnace. After 20 min at 850°C, the furnace was cooled down to room temperature naturally. For the resulting optical properties of the obtained samples, see e.g. [7, 16].

Naturally, the outcome of a CVD growth process depends on various parameters and is based on profound long-term experience. However, as examples from the literature reveal [17], CVD-grown monolayers are not perfectly monocrystalline with large flakes exhibiting numerous grain boundaries (cf. [18] on imaging second-harmonic generation studies) and are full of defects, mainly chalcogen vacancies (cf. [19] on the control of point defects in TMDCs). Commonly, said from the author's experience, monolayers of large domain size are very rare, and full-monolayer coverage of substrates comes at the cost of polycrystallinity. Moreover, monolayers often exhibit bilayer and few-layer features on top, if the growth process is not optimised for (nearly-pure) monolayer yield, and in some cases even carry clusters of material spot-wise on the ultrathin sheets.

In contrast, preferentially-grown WS_2 homobilayers can be synthesised via a two-step low-pressure chemical-vapour deposition (LPCVD) process [14]. In general, monolayers obtained from the first growth step will provide seeding sites for another monolayer of the same material to grow on top during the second run, resulting in

[3]Tungsten trioxide.

a homobilayer. In the example here, these are composed of two WS_2 monolayers with distinct stacking-twist angles achieved by the choice of growth parameters. Firstly, a source substrate was prepared by depositing 5 nm WO_3 on a sapphire or SiO_2/Si substrate using physical vapour deposition (PVD). The source substrate was placed face-to-face atop another sapphire or SiO_2/Si substrate, which was the bilayer-growth substrate. During the first growth, WS_2 monolayers were grown on the growth substrate at 950°C. In the second growth step, a new source substrate was placed face-to-face onto the growth substrate, which carried the WS_2 monolayers formed during the first-step growth.

In such a two-step-growth scenario utilised in [14], the monolayers grown during the first step provide seeding sites for the material deposited during the second growth. In that example, the second growth step took place for WS_2 at 850°C, resulting in homobilayer growth of WS_2 with an AA' or AB stacking mode. Remarkably, these two different stacking modes were determined by adjusting the growth parameters of the WS_2 monolayers obtained during the initial growth step. The amount of sulfur can be adjusted during the first-step growth, which favours either a sulfur-rich environment or a WO_3-rich environment. It was found previously in [15] that the sulfur-rich environment results in sulfur-terminated WS_2 monolayers, whereas the WO_3-rich environment was found to give tungsten-terminated WS_2 monolayers. In the preferential two-step bilayer-growth performed for the study of [14], the former case triggered the AB stacking of the homobilayers, whereas the latter triggered the AA' stacking of the homobilayers.

7.2 Patterning and Assembly

In the following, common concepts for the production and assembly of nanostructured systems are briefly summarised. Thereby, "top–down" and "bottom–up" nanotechnological approaches are addressed which can be utilised to achieve various nanoscale components for optoelectronic and quantum technological devices. Typical scenarios are the achievement of confinement potentials or waveguides structures in semiconductor devices through patterning, as well as electrical contacting schemes through metal deposition on masked surfaces or nanoparticle growth through synthesis in solution and induced-/self-assembly on exposed substrate facets.

7.2.1 Lithography, Deposition and Etching

Translating patterns from a design template into a solid has wide use in the production and development of nanotechnological devices. To manipulate (VIS–IR) light on the length scale of its wavelength, microstructuring with sub-micrometer precision is required. This can be conveniently achieved using optical lithography for masking of a sample surface. Commonly, a photosensitive resist is exposed by a shadow mask and the defined pattern in the resist is developed for subsequent etching.

In contrast to optical confinement, electronic confinement takes place on the nanoscale. Thus, quantum structures inscribed top-down (instead of bottom-up by growth) can only fulfil their role if length scales comparable to the electronic particle's de-Broglie wavelength are patterned. Moreover, in order to achieve denser and denser transistor packaging on integrated circuits, transistor structures are further and further miniaturised, until the natural limitations imposed by quantum tunnelling are reached on the electronic side or the processing precision is exhausted on the lithography and etching side. While mass production of integrated-circuit chips requires a parallel, scalable process such as photolithography, it also faces resolution limitations imposed by the optical wavelength employed for lithography.

An alternative lithography technique uses processing precision is exhausted directed onto the sample for serial position-after-position development of a defined pattern into the resist. Its working principle is based on that of a scanning electron microscope, which rasters the resist with a focused electron beam. While the precision can reach well below the microscale down to the lower nanoscale, the speed of inscribing the pattern is very low. Thus, this time-costly method is mainly attractive for lab-scale research purposes and remains an indispensable tool for nanostructuring facilities.

Once the pattern is defined and developed, the residue is rinsed off to provide the mask. The exposed facet of the sample material can then for instance be etched or metallised. Different etching techniques result in different structuring qualities, as can be seen for the examples of micropillar etchings based on wet-chemical etching, plasma or reactive-ion etching discussed for instance in [20, 21]. It should be noted that etching rates can be different for different materials and crystal orientations, and that the aspect ratio, shape and amount of underetching strongly depend on the combination of etch, sample and exposure time. The interested reader is referred to textbooks on processing techniques, e.g. [22].

More straight forward is the metallisation process of a masked sample. Either a patterned resist is used, which is lifted off after metal deposition (e.g. by physical vapour deposition or sputtering), or by a shadow mask, which is placed on top of the sample and exposes the desired sample region to the deposition of metal. In both cases, removal of the mask leaves behind the metallised areas and the previously-protected non-metallised surfaces.

For high-precision contacting schemes, typically lithography steps are involved, which require homogeneous plane deposition of (polymer) resist material, development of the resist, setting free of facets for metallisation, and rinsing of the residues after the whole process. Indeed, this is the preferred method for device fabrication with high-precision contacts, and widely used for 2D-materials optoelectronic device preparation.

In (nowadays rather rare) situations, where no micron or sub-micron precision is needed, coarse masking with a shadow mask remains an option. A shadow mask can be produced for instance by laser cutting from a ten-micrometer thin sheet of aluminium foil. In [23], such shadow mask has been used to deposit a sequence of 20 nm titanium and 200 nm gold as films on top of each other to form electric contacts on graphene, to give an example (see Fig. 7.3). However, with this approach, electrode gaps on the order of 30 to 40 nm pose a lower limit.

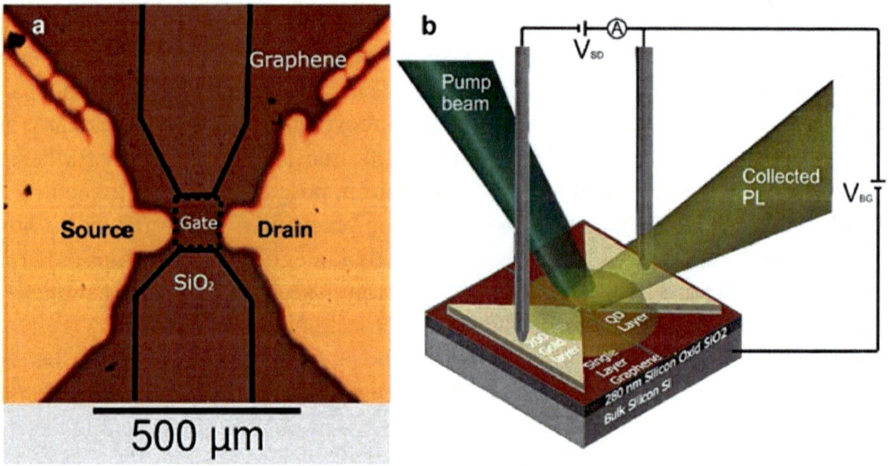

Fig. 7.3 **a** Graphene field-effect transistor (FET) achieved from processing steps involving laser cutting. In this example, the graphene channel was defined by laser cutting/ablation, resulting in an approximately $100\,\mu$m channel width. The source and drain contacts were achieved by laser cutting of an aluminum mask for physical-vapour deposition of metals on the target substrate. Remarkably, this yielded good enough phototransistor structures to perform the experiments with homemade cQD-decorated graphene FETs [23], sketched in (**b**). Reproduced under the terms of the CC-BY 4.0 Licence (http://creativecommons.org/licenses/by/4.0/). [23] Copyright 2016 The Author(s), published by Springer Nature

It shall be also briefly mentioned that etching is naturally not only used in connection with lithography but also independently. For instance, when a laser chip's top capping layer needs to be thinned down from half-wavelength to nearly quarter-wavelength thickness to enable an anti-resonant chip design based on a resonant chip structure, wet chemical etching with known etching rates for the target material is applied (see for instance works on mode-locked VECSELs [24]). Another example used selective etching to carve out a concave cavity from a fibre tip, removing predominantly the core and not the cladding, as successfully demonstrated for microcavity experiments by the author's partner group of W. Fang at the Zhejiang University [25] (see below and Fig. 7.7 at the end of this section).

7.2.2 Synthesis of Nanoparticles

The use of nanoparticles in the form of colloidal quantum dots (cQDs) or plasmonic nanoparticles in science is huge, and even industry has been eyeing them for different purposes [26]. A prominent example how cQDs can improve industrial products is given by the developments in high-quality display technology [27]. Quantum dots had become particularly attractive because of their size-tunable emission wavelength, saturated emission colours and near-unity luminance efficiency. Moreover, they are known to feature inherent photo- and thermal stability and excellent solution processability. In fact, light-emitting diodes based on cQDs had been demonstrated few years ago [28].

Currently, efforts are driven to achieve high-brightness, high-contrast, high-resolution television screens based on cQDs as emitters. Owing to the high colour purity and relatively easy production techniques, they became attractive for application in screens. For mass production, dispersed particles could be directly inkjet printed to form pixel-wise arranged active media of the device, to give a practical example. Based on improvements in the efficiency and operation lifetime of quantum-dot LEDs (QLEDs), ultra-thin and flexible, wide-colour-gamut, large-area, energy-saving, and cost-effective displays are expected from their commercialisation (for details, see [27]).

The only (obvious) drawback is that most useful core/shell cQDs are composed of heavy metals. This renders the hunt for ecologically-friendly siblings, which also perform well and are durable, a remaining challenge in order to enable the wider employment of these nanoparticles in applications. Despite their ecological drawbacks, cQDs such as CdSe/ZnS or similar heterostructure clusters (and increasingly perovskite cQDs [29, 30]) are still used in research to explore their advantages for optoelectronic device concepts.

Remarkably, the else famously-known as indirect-gap material silicon (an archetype lecture example for a bad emitter, although the role of defect states is completely ignored for simplicity) has become an attractive platform for optoelectronics in the form of nanoparticles, which have successfully ended up in photodetection and LED device schemes [31–36]. This is easily understood, as the strong spatial localisation leads to a smearing out of the energy states in the phase space due to Heisenberg's uncertainty relation $\Delta x \Delta k \geq 1$. Thereby, the harsh optical selection rules applying to bulk band structures are washed out due to the strong confinement effects in colloidal nanocrystals.

Recently, hybrid systems of 2D materials decorated by nanocrystals gained popularity for improved photodetection and sensing applications (see for instance [23, 37–42]). For instance, monolayer-based field-effect transistors (FET) can be decorated with cQDs in order to functionalise the active region for sensitive detection of visible to near-infrared light, as was done for a graphene FET in [23] for a study of gate-tunable Förster energy transfer from cQDs to graphene (Fig. 7.3, also see Fig. 2.8 in Chap. 2). From a dispersion of less than 6-nm-sized CdSe/ZnS cQDs with oleic acid (OA) surface ligands, which are commercially available, a small amount of dots was drop-casted on the graphene channel of the device and the solvent was evapourated. The ligand type determines whether the clusters can bind covalently or not. To limit the modification of graphene's electrical properties, the ligands' end group was chosen to be a methyl group. If necessary, for a given cQD dispersion, a ligand exchange can be performed in the solution by which, for instance, the length of the ligands and, thereby, the mean distance of cQDs to each other can be adjusted.

In another example, a hybrid structure of 2D layered GaTe with gold nanoparticles was studied and demonstrated to be useful for an ultrasensitive detection of aromatic molecules [42]. Gold nanoparticles were grown directly onto mechanically-

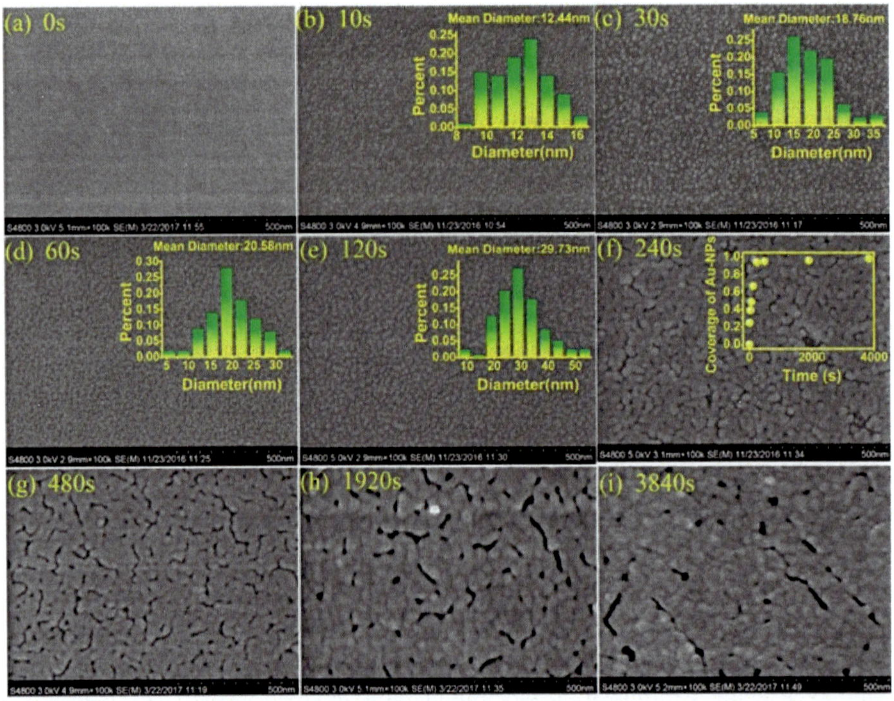

Fig. 7.4 "Golden-GaTe" towards ultra-sensitive detection of aromatic molecules and superior substrates for surface-enhanced Raman spectroscopy. Growth results of gold on the surface of few-layer GaTe films obtained from differently-timed immersion in $HAuCl_4$ solution, i.e. for **a–i** 0, 10, 30, 60, 120, 240, 480, 1920, and 3840 s, respectively [42]. Insets of **b–e**: Size distribution as well as the mean diameter of gold nanoparticles. Inset of **f**: Immersion-time-dependent coverage of Au nanoparticles on the GaTe films. (**a–i**) Reproduced with permission. [42] Copyright 2017 American Chemical Society

exfoliated few-layer 2D GaTe flakes (Fig. 7.4). Therefor, the as-prepared GaTe on SiO_2 substrate was immersed in a 0.2 mg/mL $HAuCl_4$ aqueous solution at controlled temperatures for different periods, before being cleaned and dried for experiments. While defects are often parasitic, in this case the density of defects in GaTe films was utilised for a facile decoration of GaTe surfaces with gold nanoparticles. The sizes and coverage ratios for these nanoparticles on GaTe could be tuned by varying the immersion time, providing a maximum coverage up to 98% with the help of the Ga vacancies in the layered GaTe. These golden-GaTe synthesis results were achieved in the author's partner group of H.Z. Wu at the Zhejiang University.

Recently, the enhancement of the performance of low-temperature-grown GaAs (LT-GaAs) THz antennae had been proposed by decorating their surface in the metal-antenna gap region with nanoparticles [43]. The 300-nm LT-GaAs on semi-insulating GaAs substrate, which was grown at a relatively low temperature of 300°C to control the defect types for tailored charge-carrier trapping and recombination rates, provided the necessary photoexcited charge carriers. These charge carriers can be

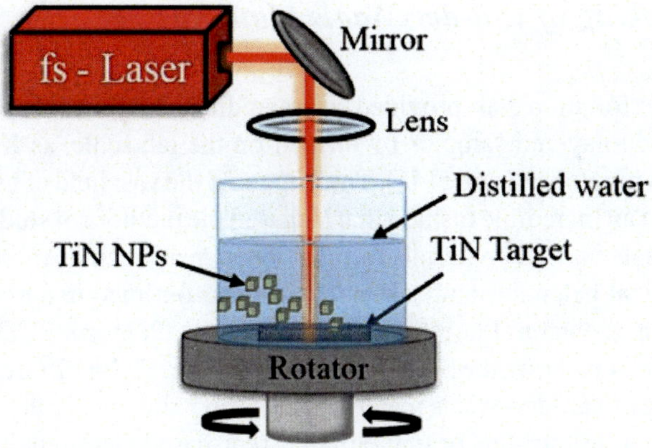

Fig. 7.5 Sketch of a simple production method for homemade plasmonic nanoparticles. From a laser source, 100-fs pulses are directed onto a TiN target in distilled water. By ablation, colloidal TiN nanoparticles disperse in the solution and can be characterised for instance with a so-called Nanosizer and Zetasizer [43]. Reproduced with permission. [43] Copyright 2017 Springer Nature

used for the ultrafast generation or the optically-gated detection of transient electric fields. Plasmonic nanoparticles (PNPs) atop had been only introduced to serve as scattering sites and light concentrators for better (optical) coupling into the underlying semiconductor in order to obtain a higher device performance. For such study, TiN PNPs, which provide a comparably broad spectral response in the visible and near-infrared region and a non-toxic composition, were prepared using two methods and, thereby, different particle size distributions.

It is interesting to show that for lab-scale experiments, nanoparticles can be prepared by relatively simple means, e.g. direct ultrasonication and pulsed-laser ablation in water using commercial nanopowders (TiN particles were sized narrowly around 20 nm in this example). The former technique resulted in an average PNP size of about 200 nm in solution with a wide distribution of sizes, while the latter provided particles in the range of 40 to 100 nm with average size about 60 nm. For laser ablation, a TiN target was produced by nanopowder compression and ablated by a 150-fs pulsed Ti:sapphire laser at the bottom of a glass vessel filled with double-distilled water (Fig. 7.5). After PNP production, a layer of self-prepared polydispersed TiN PNPs was deposited on the surface of the employed bow-tie antennae using the drop-casting method. Remarkably, this simple approach led to a noticeable—while not ground-breaking—improvement of an existing THz antenna's performance [43], making use of plasmonics and nanoantenna concepts. A similar experiment with THz antennae and nanoparticles was demonstrated shortly after with the similar aim and conclusions [44].

7.2.3 Stacking of van-der-Waals Materials

Mechanical exfoliation has provided an incredibly easy and hands-on technique to produce monolayered samples by oneself on the lab scale, as K. Novoselov has highlighted in his Nobel lecture [45]. This triggered the vast field of graphene research and beyond, in an incredible fashion that it caused the plethora of studies on graphene-related materials and 2D materials in the wider sense, such as monolayer TMDCs, and their vertical heterostructures [46–48]. As was outlined in a news feature in the journal Nature, the natural question which arises is, *"2D, or not 2D"* [49].

In the sense of entering the 3D world with 2D materials for different purposes, heterostructuring of van-der-Waals materials has grown drastically in significance—for instance for charge-transfer- or tunneling-current-based electronic or optoelectronic devices, which may need schemes for electronic coupling between layers or charge separation at interfaces.

Moreover, it has been understood that the optical quality of monolayer semiconductors and the energetics of their resonances [51–55] are strongly altered by the environment, which in the most fundamental case is encapsulation by insulating 2D material hexagonal boron nitride (hBN, also known as α-BN, labelled graphitic BN). But even buffer layers of hBN, which provide an atomically-smooth, chemically-inert and undoped interface as (hydrophobic) vdW substrate material, are known to improve optoelectronic properties such as the luminescence yield from common TMDCs, as investigated in [53] (see Fig. 2.9 for optical micrographs of various manually stacked vdW heterostructures using primarily hBN and WSe_2). Moreover, they can practically act as an insulator, as needed for instance in [50] to suppress the (unexpected) quenching of PL when TMDCs are deposited directly on transparent GaP. To achieve the goal of PL enhancement by employing a specially designed and processed photonic nanostructure in GaP, different mechanical exfoliation, transfer and stacking steps were needed, the overview of which is shown as a microscope image series in Fig. 7.6.

Recent experiments with high-quality tungsten-diselenide monolayers sandwiched between hBN flakes (Fig. 5.17a) even showed a considerable energy–momentum dispersion of the excitonic resonances in optical measurements [55]. Such measurements also demonstrated considerable helicity variations for emerging exciton complexes. The radiation patterns of these exciton complexes were separately characterised in a study of in-plane and out-of-plane exciton species in WSe_2 [56], and probed as a function of the excitation detuning as well as detected with a momentum-dependent valley-pseudo-spin texture [57]. Those experiments rely on high-quality WSe_2 bulk single crystals, which were grown in an excess selenium flux with a defect density of $5 \times 10^{10}\,\mathrm{cm}^{-2}$ (see [19]).

Encapsulated samples comprising monolayer WSe_2 and hBN with very high quality are produced for instance in the author's partner groups of J. Hone and K. Barmak at the Columbia University, with the help of S. Esdaille, D. A. Rhodes, and provided for joint experimental work. Therefor, 2D-materials flakes are initially exfoliated from bulk single crystals onto SiO_2 before stacking them sequentially

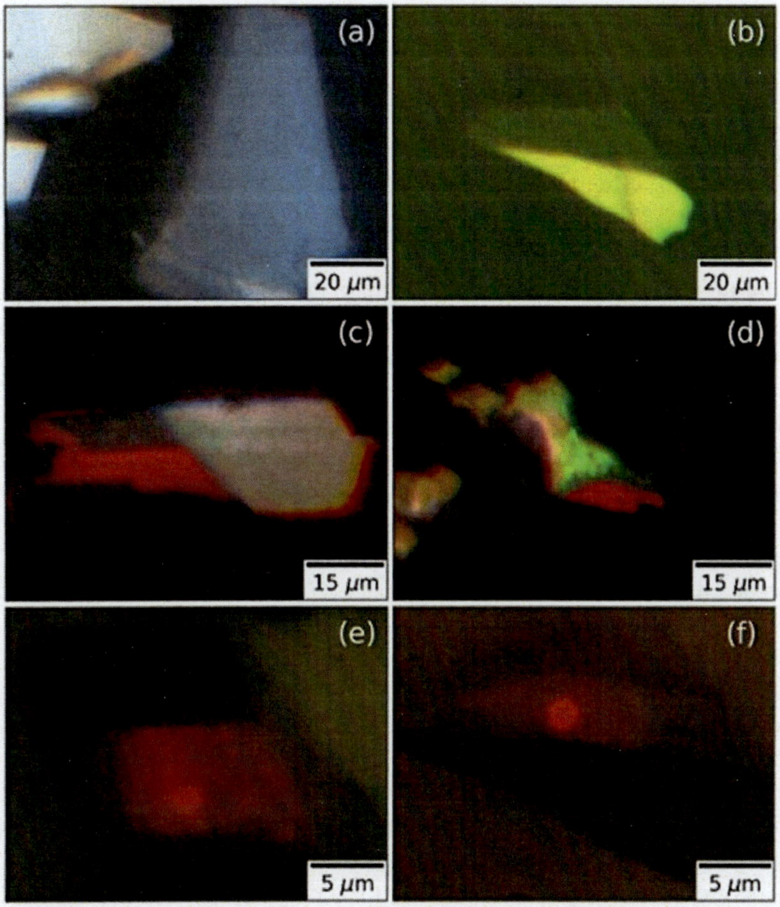

Fig. 7.6 Example of a stacking process in which the 2D-materials flakes are imaged under laser illumination at a wavelength of 532 nm (blocked in the detection path by proper filters), which excites pronounced sample fluorescence around 620 nm (red glow) only for the direct-gaped monolayer TMDCs. In the experiment discussed in [50], the flakes on specially prepared optical in-plane microresonators with tailored height-profile modulation of the substrate show strong emission from the cavity region, considerably stronger than for a simple reference hole in the same substrate material with same depth. Therefor, monolayer WS_2 (**c**, **d**) was placed on a multilayer thin hBN (**a**, **b**), which was before that transferred to a FIB-patterned GaP reference (**e**) or microcavity (**f**) structure. Note that laser light is generally hazardous and can permanently damage the eyes, if they are exposed to laser radiation. Working with lasers makes proper and accurate safety measures mandatory. One must not repeat the here described experiments using laser light coupled into a microscope without accurate and proper safety measures in compliance with laser safety norms, particularly when using magnifying optics. (**a–f**) Reproduced with permission. [50] Copyright 2019 American Chemical Society

with the temperature-controlled pick-and-lift technique to form the desired vertical structure. Here, an oxygen plasma treatment of the substrate surface for WSe_2 before exfoliation helps to reduce contamination of the flake.

Owing to the sufficient contrast under a light microscope, monolayers and thin hBN can be both well identified by optical contrast. Typically, a dry stacking technique is then used with polypropylene carbonate (PPC) on PDMS to pickup and stack h-BN/TMDC layers.

Interestingly, the preferred way of heterostructuring is based on consecutive picking-up of the identified flakes resulting in the desired sequence on the polymer-based viscoelastic stamp. For the above examples [55–57], firstly, a top layer of hBN was picked up at 48°C, then WSe_2, and finally the bottom layer of hBN. To re-smooth the PPC and ensure a clean wave-front after each hBN pick-up step, the PPC was briefly heated to 90°C. For the final transfer of the stack onto a clean substrate, the substrate was heated to 75°C at first, and once the stack had been put into contact, it was gradually heated to 120°C. Consecutively, the PPC/PDMS was lifted. To remove polymer residue, the substrate with stack on top was immersed in chloroform and rinsed with isopropyl alcohol (IPA). This procedure follows the method used by the group of J. Hone at the Columbia University [19].

While simple viscoelastic stamping of flakes can be performed with only one polymer component as a sticky film, the combination of two films allows one to modify the stickiness of the different polymer layers of the joint film by means of heating, owing to the different glass temperatures of each of the used polymer materials.

In contrast to exfoliated samples, CVD growth can directly deliver epitaxially stacked homo- and heterolayers of 2D materials. While current work is primarily focusing on both lateral and vertical heterostructure fabrication [12, 48], the deterministic growth of bilayers with different stacking angle is equally relevant, being motivated by recent observations of a superconducting phase [58, 59] and other interesting phenomena [60, 61]. In that specific superconductivity study on bilayer graphene by Cao et al., graphene from the same flake was merely picked up by mechanical break off and transferred with a given twist angle on top of its substrate-bound remainder, to give an example.

Bilayers of CVD-grown TMDCs with different symmetries such as AA' and AB-type homobilayers of WS_2 were readily obtained after a two-step growth process, as described in the previous section. Yet, a transfer step is typically applied to provide the samples on the desired substrate for optical studies [14], which is as an example briefly outlined here. By covering the sample with poly(methyl methacrylate) (PMMA) by a dropper and letting it dry, a subsequent dive in 30% KOH (aq) for a few minutes left behind a floating PMMA/bilayer stack separated from the growth substrate. The KOH residue was then removed from the PMMA/bilayer stack during a repeated cleaning procedure in deionized (DI) water. Afterwards, the cleaned PMMA/TMDC sheet was attached to the target substrate, which could be for instance SiO_2, sapphire or hBN, dried in ambient conditions for 1 hour and later backed at 90 °C for 1.5 min in order to enhance the bonding between the substrate and the bilayers. Following this transfer procedure, PMMA was removed using 50°C warm acetone for 30 min. Thereafter, the sample was freed from residues via an IPA rinsing, DI water rinsing and hotplate baking at 130°C for 5 min.

In fact, by stacking van-der-Waals materials arbitrarily using different sequences of material sheets, different twist angles and energy band gaps, a plethora of configurations become imaginable, and different effects on the electronic or optoelectronic properties can be obtained, as aforementioned studies and reviews indicate.

7.2.4 Laser Processing and Ion Beam Milling

In addition to etching-based structuring techniques, laser processing can provide a dry and similarly well-controllable method for microstructuring, as examples such as beam-machined free-standing THz metamaterials indicate [62]. For metamaterials preparation in metal foils by laser ablation, a shutter-controlled 10-Hz ns-pulsed Nd:YAG laser and two motorised stages were employed in Marburg previously. However, this comes at the cost of precision, as ablation of material by a laser beam is not only limited by the beam's focused spot size, but also by the heat transferred to the material. Additional improvements can be obtained by ultrafast pulsed lasers (cf. [24]), when samples are exposed to high peak powers within very short time scales that prevent a large footprint due to suppressed heat transfer, which takes place on longer time scales. Recently, pulsed CO_2-laser pulses had been also used to post-process selectively-etched fibre cores at their tips as a means of thermal treatment to obtain smooth concave micromirror shapes [25] (Fig. 7.7).

In fact, laser cutting can be also applied for experimental studies on large-area 2D materials. In one example, patterning of graphene and of aluminum shadow masks were performed for channel definition and electrical-contact deposition, respectively, in order to produce a field-effect-transistor device for photodetector studies involving light-absorbing nanoparticles [23] (Fig. 7.3).

In contrast to laser cutting, milling on the nanoscale becomes feasible using focused ion beams (FIBs). Typically, a FIB gun is found in scanning-electron microscopes used for sample preparation, e.g. prior to transmission-electron microscopy.

It has been recently shown that photonic structures of reasonable quality can be achieved using FIB for experiments combining patterned landscapes and 2D materials [50] (Fig. 7.8). In the work by Mey et al., FIB milling took place directly on the GaP sample inside the evacuated chamber of a FIB device loaded with a gallium gun, using a 30-KV ion beam with a beam current of 500 pA. Due to the integrated scanning-electron microscope in such apparatus, in situ imaging of the structure is usually possible.

Although cutting and milling techniques (both top–down techniques) can be very useful for prototyping, a major drawback originates from the serial writing of patterns into the target material, which is time consuming. The same drawback holds true for bottom-up laser writing (a form of 3D printing), which is an excellent tool for prototyping and which even opens up the possibility of optically inscribing sub-micron-precise structures into photoresist material utilising two-photon polymerisa-

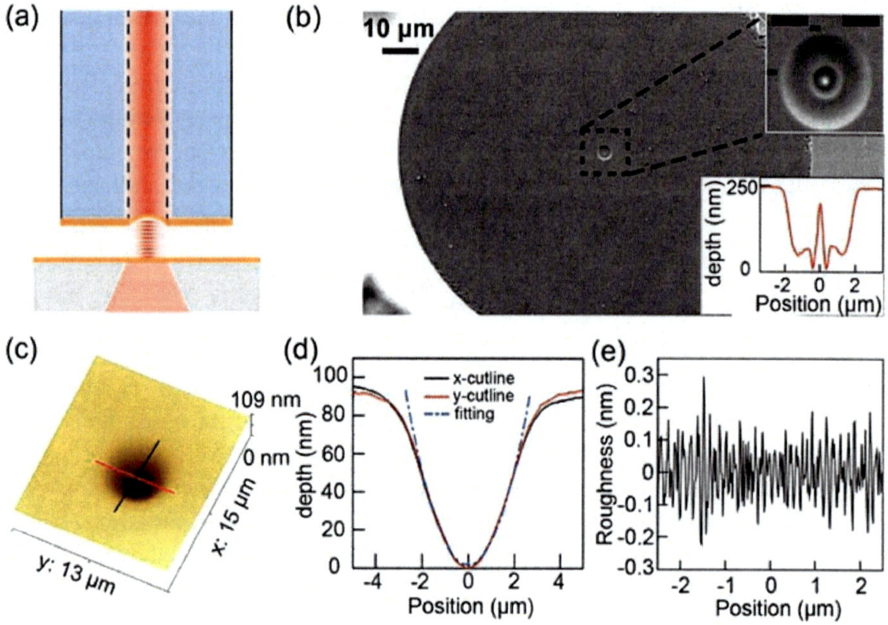

Fig. 7.7 **a** Sketch of a microcavity based on a planar dielectric mirror and a dielectric-mirror coated fibre tip with concave shape in the core region. **b** Chemical-etching result. The fibre tip's core was selectively etched and a nearly concave shape was produced with visible surface inhomogeneities. To improve smoothness of this dip, a laser pulse thermal treatment was applied, which resulted in a very low deviation from an ideal circle (**c**), as the cross-section (**d**) and the roughness plot (**e**) demonstrated. For details, see [25]. Reproduced with permission. [25] Copyright 2019 AIP Publishing

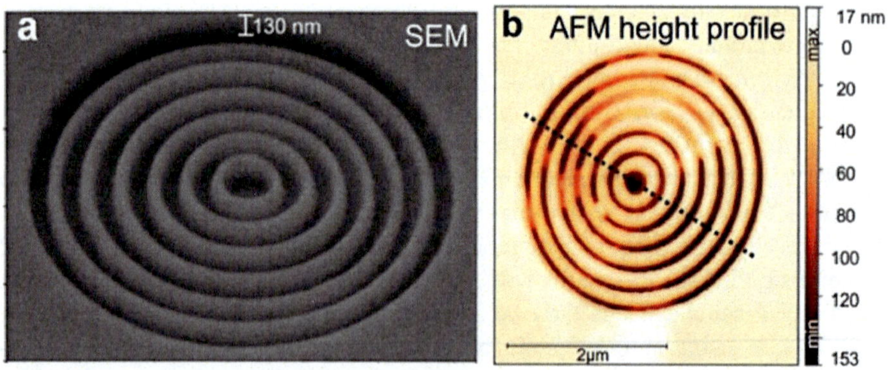

Fig. 7.8 Example of a FIB-milled photonic nanostructure with circular in-plane microcavity based on a Bragg grating, as studied for PL enhancement through improved in- and outcoupling of light [50] (also see Fig. 2.13). **a** SEM image. **b** False-colour AFM topography. (**a–b**) Adapted with permission. [50] Copyright 2019 American Chemical Society

tion, i.e. development of resist with a strong spatial selectivity of optical exposure due to nonlinear absorption. Additionally, milling aspect ratios, precision and speed strongly depend on the flux, focusing capabilities and the displacement precision, e.g. that of a computer-controlled translation stage or beam deflection components, and limits the use of these techniques with respect to sample processing. Nevertheless, these key tools are successfully employed in a wide array of research activities and enable various exciting studies, some of which are briefly reported in this work.

References

1. R. Dingle, W. Wiegmann, C.H. Henry, Quantum states of confined carriers in very thin $Al_xGa_{1-x}As$-GaAs-$Al_xGa_{1-x}As$ heterostructures. Phys. Rev. Lett. **33**, 827 (1974)
2. M.A. Herman, H. Sitter, *Molecular Beam Epitaxy: Fundamentals and Current Status* (Springer, Berlin Heidelberg, 1996)
3. G.B. Stringfellow, *Organometallic Vapor-Phase Epitaxy: Theory and Practice* (Elsevier Science, 2012)
4. P. Gupta, A.A. Rahman, S. Subramanian, S. Gupta, A. Thamizhavel, T. Orlova, S. Rouvimov, S. Vishwanath, V. Protasenko, M.R. Laskar, H.G. Xing, D. Jena, A. Bhattacharya, Layered transition metal dichalcogenides: promising near-lattice-matched substrates for GaN growth. Sci. Rep. **6**, 1–8 (2016). https://doi.org/10.1038/srep23708
5. M. Guina, A. Rantamäki, A. Härkönen, Optically pumped vecsels: review of technology and progress. J. Phys. D: Appl. Phys. **50**, 383001 (2017)
6. K.N. Kang, K. Godin, E.-H. Yang, The growth scale and kinetics of WS_2 monolayers under varying H_2 concentration. Sci. Rep. **5**, 13205 (2015). https://doi.org/10.1038/srep13205
7. S. Lippert, L.M. Schneider, D. Renaud, K.N. Kang, O. Ajayi, J. Kuhnert, M.-U. Halbich, O.M. Abdulmunem, X. Lin, K. Hassoon, S. Edalati-Boostan, Y.D. Kim, W. Heimbrodt, E.-H. Yang, J.C. Hone, A. Rahimi-Iman, Influence of the substrate material on the optical properties of tungsten diselenide monolayers. 2D Mater. **4**, 025045 (2017). https://doi.org/10.1088/2053-1583/aa5b21
8. K.S. Kim, Y. Zhao, H. Jang, S.Y. Lee, J.M. Kim, K.S. Kim, J.-H. Ahn, P. Kim, J.-Y. Choi, B.H. Hong, Large-scale pattern growth of graphene films for stretchable transparent electrodes. Nature **457**, 706–710 (2009)
9. X. Wang, Y. Gong, G. Shi, W.L. Chow, K. Keyshar, G. Ye, R. Vajtai, J. Lou, Z. Liu, E. Ringe, B.K. Tay, P.M. Ajayan, Chemical vapor deposition growth of crystalline monolayer $MoSe_2$. ACS Nano **8**, 5125–5131 (2014)
10. X. Wang, K. Kang, S. Chen, R. Du, E.-H. Yang, Location-specific growth and transfer of arrayed MoS_2 monolayers with controllable size. 2D Mater. **4**, 025093 (2017)
11. J. Kang, S. Tongay, J. Zhou, J. Li, J. Wu, Band offsets and heterostructures of two-dimensional semiconductors. Appl. Phys. Lett. **102**, 012111 (2013)
12. Y. Gong, J. Lin, X. Wang, G. Shi, S. Lei, Z. Lin, X. Zou, G. Ye, R. Vajtai, B.I. Yakobson, H. Terrones, M. Terrones, B.K. Tay, J. Lou, S.T. Pantelides, Z. Liu, W. Zhou, P.M. Ajayan, Vertical and in-plane heterostructures from WS_2/MoS_2 monolayers. Nat. Mater. **13**, 1135–1142 (2014)
13. D. Jariwala, A.R. Davoyan, J. Wong, H.A. Atwater, Van der Waals materials for atomically-thin photovoltaics: promise and outlook. ACS Photon. **4**, 2962–2970 (2017)
14. L.M. Schneider, J. Kuhnert, S. Schmitt, W. Heimbrodt, U. Huttner, L. Meckbach, T. Stroucken, S.W. Koch, S. Fu, X. Wang, K. Kang, E.-H. Yang, A. Rahimi-Iman, Spin-layer and spin-valley locking in CVD-grown AA'- and AB-stacked tungsten-disulfide bilayers. J. Phys. Chem. C **123**, 21813–21821 (2019). https://doi.org/10.1021/acs.jpcc.9b07213
15. D. Zhu, H. Shu, F. Jiang, D. Lv, V. Asokan, O. Omar, J. Yuan, Z. Zhang, C. Jin, Capture the growth kinetics of CVD growth of two-dimensional MoS_2. npj 2D Mater. Appl. **1**, 8 (2017)

16. L.M. Schneider, S. Lippert, J. Kuhnert, D. Renaud, K.N. Kang, O. Ajayi, M.-U. Halbich, O.M. Abdulmunem, X. Lin, K. Hassoon, S. Edalati-Boostan, Y.D. Kim, W. Heimbrodt, E.H. Yang, J.C. Hone, A. Rahimi-Iman, The impact of the substrate material on the optical properties of 2D WSe_2 monolayers. Semiconductors **52**, 565–571 (2018). https://doi.org/10.1134/S1063782618050275
17. D. Rhodes, S.H. Chae, R. Ribeiro-Palau, J. Hone, Disorder in van der Waals heterostructures of 2D materials. Nat. Mater. **18**, 541–549 (2019)
18. J.E. Zimmermann, B. Li, J. Hone, U. Höfer, G. Mette, Second-harmonic imaging microscopy for time-resolved investigations of transition metal dichalcogenides (2016). arxiv:1608.03434
19. D. Edelberg, D. Rhodes, A. Kerelsky, B. Kim, J. Wang, A. Zangiabadi, C. Kim, A. Abhinandan, J. Ardelean, M. Scully, D. Scullion, L. Embon, R. Zu, E.J.G. Santos, L. Balicas, C. Marianetti, K. Barmak, X. Zhu, J. Hone, A.N. Pasupathy, Approaching the intrinsic limit in transition metal diselenides via point defect control. Nano Lett. **19**, 4371–4379 (2019)
20. J.M. Gérard, D. Barrier, J.Y. Marzin, R. Kuszelewicz, L. Manin, E. Costard, V. Thierry Mieg, T. Rivera, Quantum boxes as active probes for photonic microstructures: the pillar microcavity case. Appl. Phys. Lett. **69**, 449 (1996)
21. S. Reitzenstein, A. Forchel, Quantum dot micropillars. J. Phys. D: Appl. Phys. **43**(3), 033001 (2010)
22. M. Sugawara, *Plasma Etching: Fundamentals and Applications* (Oxford Science Publications, 1998)
23. R. Li, L.M. Schneider, W. Heimbrodt, H. Wu, M. Koch, A. Rahimi-Iman, Gate tuning of Förster resonance energy transfer in a graphene - quantum dot FET photo-detector. Sci. Rep. **6**, 28224 (2016). https://doi.org/10.1038/srep28224
24. M.A. Gaafar, A. Rahimi-Iman, K.A. Fedorova, W. Stolz, E.U. Rafailov, M. Koch, Mode-locked semiconductor disk lasers. Adv. Opt. Photonics **8**, 370–400 (2016)
25. P. Qing, J. Gong, X. Lin, N. Yao, W. Shen, A. Rahimi-Iman, W. Fang, L. Tong, A simple approach to fiber-based tunable microcavity with high coupling efficiency. Appl. Phys. Lett. **114**, 021106 (2019). https://doi.org/10.1063/1.5083011
26. J.A. Smyder, T.D. Krauss, Coming attractions for semiconductor quantum dots. Mater. Today **14**, 382–387 (2011)
27. X. Dai, Y. Deng, X. Peng, Y. Jin, Quantum-dot light-emitting diodes for large-area displays: towards the dawn of commercialization. Adv. Mater. **29**, 1607022 (2017)
28. X. Dai, Z. Zhang, Y. Jin, Y. Niu, H. Cao, X. Liang, L. Chen, J. Wang, X. Peng, Solution-processed, high-performance light-emitting diodes based on quantum dots. Nature **515**, 96–99 (2014)
29. J. Li, L. Gan, Z. Fang, H. He, Z. Ye, Bright tail states in blue-emitting ultrasmall perovskite quantum dots. J. Phys. Chem. Lett. **8**, 6002–6008 (2017)
30. Z. Shi, Y. Li, Y. Zhang, Y. Chen, X. Li, D. Wu, T. Xu, C. Shan, G. Du, High-efficiency and air-stable perovskite quantum dots light-emitting diodes with an all-inorganic heterostructure. Nano Lett. **17**, 313–321 (2017)
31. T. Yu, F. Wang, Y. Xu, L. Ma, X. Pi, D. Yang, Graphene coupled with silicon quantum dots for high-performance bulk-silicon-based Schottky-junction photodetectors. Adv. Mater. **28**, 4912–4919 (2016)
32. W. Gu, X. Liu, X. Pi, X. Dai, S. Zhao, L. Yao, D. Li, Y. Jin, M. Xu, D. Yang, G. Qin, Silicon-quantum-dot light-emitting diodes with interlayer-enhanced hole transport. IEEE Photonics J. **9**, 1–10 (2017)
33. S. Du, Z. Ni, X. Liu, H. Guo, A. Ali, Y. Xu, X. Pi, Graphene/silicon-quantum-dots/Si Schottky-PN cascade heterojunction for short-wavelength infrared photodetection, in *2017 IEEE International Electron Devices Meet.* (IEEE, 2017), , pp. 8.7.1–8.7.4
34. Z. Ni, L. Ma, S. Du, Y. Xu, M. Yuan, H. Fang, Z. Wang, M. Xu, D. Li, J. Yang, W. Hu, X. Pi, D. Yang, Plasmonic silicon quantum dots enabled high-sensitivity ultrabroadband photodetection of graphene-based hybrid phototransistors. ACS Nano **11**, 9854–9862 (2017)
35. X. Liu, S. Zhao, W. Gu, Y. Zhang, X. Qiao, Z. Ni, X. Pi, D. Yang, Light-emitting diodes based on colloidal silicon quantum dots with octyl and phenylpropyl ligands. ACS Appl. Mater. Interfaces **10**, 5959–5966 (2018)

36. S. Zhao, X. Liu, W. Gu, X. Liang, Z. Ni, H. Tan, K. Huang, Y. Yan, X. Yu, M. Xu, X. Pi, D. Yang, Al_2O_3 -interlayer-enhanced performance of all-inorganic silicon-quantum-dot near-infrared light-emitting diodes. IEEE Trans. Electron Devices **65**, 577–583 (2018)
37. G. Konstantatos, M. Badioli, L. Gaudreau, J. Osmond, M. Bernechea, F.P.Gd Arquer, F. Gatti, F.H.L. Koppens, Hybrid graphene–quantum dot phototransistors with ultrahigh gain. Nat. Nanotechnol. **7**, 363–368 (2012)
38. D. Zhang, L. Gan, Y. Cao, Q. Wang, L. Qi, X. Guo, Understanding charge transfer at PbS-decorated graphene surfaces toward a tunable photosensor. Adv. Mater. **24**(20), 2715–2720 (2012)
39. Y.Q. Huang, R.J. Zhu, N. Kang, J. Du, H.Q. Xu, Photoelectrical response of hybrid graphene-PbS quantum dot devices. Appl. Phys. Lett. **103**, 143119 (2013)
40. F.H.L. Koppens, T. Mueller, P. Avouris, A.C. Ferrari, M.S. Vitiello, M. Polini, Photodetectors based on graphene, other two-dimensional materials and hybrid systems. Nat. Nanotechnol. **9**, 780–793 (2014)
41. K.J. Tielrooij, L. Orona, A. Ferrier, M. Badioli, G. Navickaite, S. Coop, S. Nanot, B. Kalinic, T. Cesca, L. Gaudreau, Q. Ma, A. Centeno, A. Pesquera, A. Zurutuza, H.d. Riedmatten, P. Goldner, F.J.G.d. Abajo, P. Jarillo-Herrero, F.H.L. Koppens, Electrical control of optical emitter relaxation pathways enabled by graphene. Nat. Phys. **11**, 281–287 (2015)
42. P. Lu, J. Lang, Z. Weng, A. Rahimi-Iman, H. Wu, Hybrid structure of 2D layered GaTe with Au nanoparticles for ultrasensitive detection of aromatic molecules. ACS Appl. Mater. Interfaces **10**, 1356–1362 (2018). https://doi.org/10.1021/acsami.7b14121
43. O. Abdulmunem, K. Hassoon, M. Gaafar, A. Rahimi-Iman, J.C. Balzer, TiN nanoparticles for enhanced THz generation in TDS systems. J. Infrared Milli. Terahz Waves **38**, 1206–1214 (2017). https://doi.org/10.1007/s10762-017-0412-z
44. S. Lepeshov, A. Gorodetsky, A. Krasnok, N. Toropov, T.A. Vartanyan, P. Belov, A. AlúA, E.U. Rafailov, Boosting terahertz photoconductive antenna performance with optimised plasmonic nanostructures. Sci. Rep. **8**, 6624 (2018)
45. K.S. Novoselov, Graphene: materials in the Flatland (Nobel lecture). Angew. Chem. Int. Ed. Engl. **50**, 6986–7002 (2011)
46. A.K. Geim, I.V. Grigorieva, Van der Waals heterostructures. Nature **499**, 419 (2013)
47. Z. Lin, A. McCreary, N. Briggs, S. Subramanian, K. Zhang, Y. Sun, X. Li, N.J. Borys, H. Yuan, S.K. Fullerton-Shirey, A. Chernikov, H. Zhao, S. McDonnell, A.M. Lindenberg, K. Xiao, B.J. LeRoy, M. Drndić, J.C.M. Hwang, J. Park, M. Chhowalla, R.E. Schaak, A. Javey, M.C. Hersam, J. Robinson, M. Terrones, 2D materials advances: from large scale synthesis and controlled heterostructures to improved characterization techniques, defects and applications. 2D Mater. **3**, 042001 (2016)
48. K.S. Novoselov, A. Mishchenko, A. Carvalho, A.H. Castro Neto, 2D materials and van der Waals heterostructures. Science **353**, aac9439 (2016)
49. E. Gibney, 2D or not 2D. Nature **522**, 274 (2015). news feature
50. O. Mey, F. Wall, L.M. Schneider, D. Günder, F. Walla, A. Soltani, H. Roskos, N. Yao, P. Qing, W. Fang, A. Rahimi-Iman, Enhancement of the monolayer tungsten disulfide exciton photoluminescence with a two-dimensional material/air/gallium phosphide in-plane microcavity. ACS Nano **13**, 5259–5267 (2019). https://doi.org/10.1021/acsnano.8b09659
51. O.A. Ajayi, J.V. Ardelean, G.D. Shepard, J. Wang, A. Antony, T. Taniguchi, K. Watanabe, T.F. Heinz, S. Strauf, X.Y. Zhu, J.C. Hone, Approaching the intrinsic photoluminescence linewidth in transition metal dichalcogenide monolayers. 2D Mater. **4**, 031011 (2017)
52. F. Cadiz, E. Courtade, C. Robert, G. Wang, Y. Shen, H. Cai, T. Taniguchi, K. Watanabe, H. Carrere, D. Lagarde, M. Manca, T. Amand, P. Renucci, S. Tongay, X. Marie, B. Urbaszek, Excitonic linewidth approaching the homogeneous limit in MoS_2-based van der Waals heterostructures. Phys. Rev. X **7**, 021026 (2017)
53. L.M. Schneider, S. Lippert, J. Kuhnert, O. Ajayi, D. Renaud, S. Firoozabadi, Q. Ngo, R. Guo, Y.D. Kim, W. Heimbrodt, J.C. Hone, A. Rahimi-Iman, The influence of the environment on monolayer tungsten diselenide photoluminescence. Nano-Struct. Nano-Objects **15**, 84–97 (2018), online 2017. https://doi.org/10.1016/j.nanoso.2017.08.009

54. L. Meckbach, T. Stroucken, S.W. Koch, Influence of the effective layer thickness on the ground-state and excitonic properties of transition-metal dichalcogenide systems. Phys. Rev. B **97**, 035425 (2018)
55. L.M. Schneider, S.S. Esdaille, D.A. Rhodes, K. Barmak, J.C. Hone, A. Rahimi-Iman, Shedding light on exciton's nature in monolayer quantum material by optical dispersion measurements. Opt. Express **27**, 37131–37149 (2019). https://doi.org/10.1364/OE.27.037131
56. L.M. Schneider, S.S. Esdaille, D.A. Rhodes, K. Barmak, J.C. Hone, A. Rahimi-Iman, Direct measurement of the radiative pattern of bright and dark excitons and exciton complexes in encapsulated tungsten diselenide. Sci. Rep. **10**, 8091 (2020). https://doi.org/10.1038/s41598-020-64838-z
57. L.M. Schneider, S.S. Esdaille, D.A. Rhodes, K. Barmak, J.C. Hone, A. Rahimi-Iman, Optical dispersion of valley-hybridised coherent excitons with momentum-dependent valley polarisation in monolayer semiconductor. 2D Materials **8**, 015009 (2020)
58. Y. Cao, V. Fatemi, S. Fang, K. Watanabe, T. Taniguchi, E. Kaxiras, P. Jarillo-Herrero, Unconventional superconductivity in magic-angle graphene superlattices. Nature **556**, 43–50 (2018)
59. M. Yankowitz, S. Chen, H. Polshyn, Y. Zhang, K. Watanabe, T. Taniguchi, D. Graf, A.F. Young, C.R. Dean, Tuning superconductivity in twisted bilayer graphene. Science **363**, 1059–1064 (2019)
60. Y. Cao, V. Fatemi, A. Demir, S. Fang, S.L. Tomarken, J.Y. Luo, J.D. Sanchez-Yamagishi, K. Watanabe, T. Taniguchi, E. Kaxiras, R.C. Ashoori, P. Jarillo-Herrero, Correlated insulator behaviour at half-filling in magic-angle graphene superlattices. Nature **556**, 80 (2018)
61. A.L. Sharpe, E.J. Fox, A.W. Barnard, J. Finney, K. Watanabe, T. Taniguchi, M.A. Kastner, D. Goldhaber-Gordon, Emergent ferromagnetism near three-quarters filling in twisted bilayer graphene. Science **365**, 605–608 (2019)
62. N. Born, R. Gente, I. Al-Naib, M. Koch, Laser beam machined free-standing terahertz metamaterials. Electron. Lett. **51**, 1012–1014 (2015)

Chapter 8
Conclusion and Outlook

Abstract Semiconductor photonics and quantum technologies are strongly in the focus of current research endeavours to guarantee steady progress on the scientific and engineering frontier. The study as well as fabrication of quantum structures, the discovery as well as utilisation of novel materials, and the understanding as well as harnessing of quantum phenomena are an essential part of research in these two key scientific domains. Thus, interdisciplinary work at the crossroads of quantum physics, nanotechnology, materials science and photonics has been inevitable with regard to the development of optoelectronic and quantum technologies, as well as the research performed within the author's Habilitation project settled in this topical landscape. This final chapter of the book provides, both, general concluding remarks by the author and an overview on ongoing research projects carried out by the author with the help of his team and cooperation partners. Hereby, endeavours concerning investigations on self-mode-locked VECSELs, manipulation and control of cavity–polaritons, as well as explorations in the field of optoelectronics and light–matter interactions with 2D semiconductors are wrapped up with brief outlook parts.

8.1 Summary

Semiconductor photonics remains an exciting subject with yet increasing importance for future technologies ranging from classical to quantum optics. Various types of low-dimensional structures for quantum confinement and the engineering of opto-electronic properties are employed. Novel device concepts for applications, such as communications, computing or metrology, are delivered. In this domain, the field of semiconductor optics covers the fundamental aspects, whereas the field of semiconductor devices makes use of them. These subjects are strongly linked to advances in nanotechnology and quantum physics. Their utilisation opened up the field of quantum technologies, promising unmatched computational powers, new sensing capabilities or secure communications.

A. Rahimi-Iman, *Semiconductor Photonics of Nanomaterials and Quantum Structures*, Springer Series in Solid-State Sciences 196,
https://doi.org/10.1007/978-3-030-69352-7_8

Semiconductor nanomaterials research has entered the stage of industrial applications with quantum dots being used as high-brightness, high colour-purity luminescent nanostructures for display technologies, or even acting as single-photon emitters for quantum information systems, or as light-absorbing structures for highly-sensitive photodetectors. Synthesis of quantum materials has evolved to that extent that colloidal quantum dots and other nanoparticles are synthesised in solution with prospects of mass producibility and device integration. Remarkably, institutes worldwide are pioneering many topics in this field and numerous companies use highly innovative concepts to exploit the benefits of such quantum structures. For instance, prominent display manufacturers are currently developing new display technologies based on colloidal quantum dots light sources or graphene-based lighting and flexible electrodes.

In addition, natural quantum materials, such as layered two-dimensional (2D) systems, have been in the focus for more than a decade, with graphene as the prominent conductive example, and transition-metal dichalcogenides (TMDCs) and hexagonal boron nitride (hBN) as the semiconducting and insulating counterparts, respectively. 2D-materials research is a hot topic of functional nanomaterials science and has received unrivalled attention in the international science community. Recently, 2D-materials research was considered among the top 10 topics of interest by key scientific journals. Their properties in the monolayer and few-layer regimes give access to intriguing physics, while the possibility to stack 2D materials vertically enables enormous variability in terms of van-der-Waals epitaxy, which is layer-by-layer arrangement or growth of 2D materials that are linked to their neighbouring material sheets by van-der-Waals forces. The high surface-to-bulk ratio, quantum confinement and the broken symmetry of 2D semiconductors, such as TMDCs or the post-TMDC materials (e.g. group-III monochalcogenides), are responsible for considerable changes in optoelectronic properties when the layer number approaches the monolayer limit, or the environment of ultrathin flakes is altered, or the stacking of layered structures is varied. Those changes can be of great importance to device design principles.

From this, innovative potential applications for such materials can be deduced, covering the fields of electronics, optoelectronics, energy, or sensing. 2D materials with their mechanical strength and flexibility also allow one to envisage flexible devices and to target applications, which require bendable substrates or materials. Most strikingly, 2D semiconductors in the monolayer and few-layer regime offer a wide range of band gaps, huge exciton binding energies, strong light–matter interaction and even polarisation dichroism. Band gaps from a few milli-electron-volts to several electron-volts were reported for 2D materials, and peculiarities, such as the valley Hall effect, enable the design of future valleytronic devices that exploit the spin-selectivity of the time-reversal-symmetry-broken electronic band structure. In recent years, different applications were addressed or suggested, depending on the material properties, ranging from electrodes in batteries and channel-materials in transistors to ultrasensitive photodetectors or molecular sensors. For instance, 2D materials such as the conducting graphene can be used for the generation of broadband tunable THz sources or sensors, TMDCs employed as active medium and the semiconducting post-TMDC material gallium telluride utilised for sensing applications. The broad variety of material properties and applications is currently the main driving force for extensive research worldwide in this domain.

In this work, the endeavours to characterise novel nanomaterial systems as well as quantum structures with the help of optical techniques and the achievements are highlighted within the context of international and domestic ongoing research efforts.

In recent years, the influences of the substrate material, dielectric environment and stacking configuration or order have been discussed for these fascinating monolayer semiconductors by various works. A number of them conducted by the author and co-workers in optical studies at cryogenic temperatures and room temperature have addressed energetics, dynamics and vibronics of the monolayers investigated with spectroscopic methods. Based on excellent samples from partner groups at the Columbia University and the Stevens Institute of Technology from the U.S., various interesting effects were reported and important material comparisons enabled. Among others, the impact of the substrate choice on biexciton observation in WSe_2, the difference in polarisation anisotropy for AA' and AB stacked WS_2 bilayers, and the occurrence of meV energy–momentum dispersion for excitons within the light cone—understood as exciton–polaritons—for high-quality monolayer samples have been demonstrated. In addition, heterostructuring capabilities with regard to a modification of electronic properties and charge-transfer states were explored, making primarily use of homemade van-der-Waals stacks.

For efficient and sensitive photodetection or sensing, new concepts employing 2D materials, such as graphene and related materials, are being intensively investigated in the whole research community. One approach to combine colloidal quantum dots and a graphene-based field-effect transistor led to a hybrid photodetector with gate-tunable Förster energy transfer between dots and graphene channel, as demonstrated in a joint study of the collaborating physicists from Marburg, Germany, and Hangzhou, China. In another study, it was demonstrated that gold-nanoparticle-decorated GaTe can act as a very sensitive probe for certain aromatic molecules and could become a promising candidate for surface-enhanced Raman spectroscopy.

Also, other materials such as the material class of the perovskites have shown very interesting properties that render them attractive candidates for optoelectronic devices, such as solar cells or lasers. Typically, lead–halide perovskites provide a fruitful testbed for optical studies, which were performed collaboratively by the author with domestic and international partners. In a first step, a very promising material of this class was analysed on a wide relevant temperature range to probe optoelectronic properties, enabled by the efforts of the partner group in Hangzhou. Following that, further characterisation of the charge-carrier dynamics and photoluminescence properties was achieved. Recently, nonlinear optics in single-crystalline perovskites provided by a group from Tübingen, Germany, were probed with a Z-scan technique, which had been previously employed by the author for the systematic investigations on nonlinear lensing (and in the same breath on nonlinear absorption) in semiconductor disk laser chips that are also highlighted in this work.

In addition to the conventional optoelectronic microdevices, future on-chip optoelectronic circuitry for computing or possibly biomedical applications will highly benefit from advances in the field of functionalised nanomaterials, and light–matter coupling phenomena. It is also worth noting that, among many pathways to the use of novel materials for photonics, the employment of optical microcavities in combination with quantum materials has become very attractive for cavity quantum-electrodynamics studies and nanophotonics, which nowadays go beyond the use of common semiconductor materials and epitaxially-grown quantum structures. Such concepts are for instance subject of the research activities jointly performed by the author with partners from the Zhejiang University, Hangzhou, with eyes towards efficient quantum emitters for future quantum communication purposes. The principal investigators performed, both, jointly and individually scientific activities to identify new application potentials of 2D materials, as well as zero-dimensional structures, and to investigate light–matter interactions as well as a variety of novel materials in optical cavities. For instance, the author explored concepts to smoothly tune the coupling regime from weak to strong across the exceptional point. Also, a new approach was targeted to deliver good enough optical microresonators based on fibre technologies and high-reflectivity interfaces for efficient in- and out-coupling of signal from flexible and tunable light–matter-coupled systems, as produced and studied in the Hangzhou nanophotonics laboratories and further investigated in Marburg. In addition, to bring 2D semiconductors closer to applications and to improve the emission properties, photoluminescence (PL) enhancement was recently studied using WS_2 on nano-fabricated optical microresonators, which are achieved from ring patterns (in-plane Bragg gratings) in a dielectric substrate. This approach pursued for controlled emission rates and profiles by the author, and jointly published with co-workers from Frankfurt, Hangzhou, and Marburg, allows combining both vertical and horizontal interference effects to improve both in-coupling into and out-coupling of light from 2D materials. Such structures could be at the core of future on-chip optical interconnects for light-based information processing purposes. Photodetectors, photovoltaics, nonlinear optics and light-emitting diodes involving 2D materials could strongly benefit from these advances for the concept introduced.

In the context of strong light–matter interactions, polariton physics offers an exciting playground for the study of Bose–Einstein-like condensation phenomena, as well as the related effects, such as superfluidity. Employing 2D materials in this domain has further enriched the field due to spin- and valley-sensitivity of TMDCs. The manipulation and control of hybrid quantum states in matter with external photons is an interesting direction to continue polariton research, after having studied various examples of polariton condensates, their generation and their correlations. Studies by the author on polaritons and excitons in external fields were deepened in the past years as summarised in this work.

Other international and domestic collaborations have so far enabled effective studies on semiconductor disk lasers, for instance with regard to the self-mode-locking effect and performance boosts for single-frequency or THz-generating devices, as well as on material spectroscopy and quantum optics. In one case, a record flux from a single-photon source was achieved by combining self-built pulsed lasers with quantum emitters from Berlin, Germany.

The highlighted example of past interactions between the partners from Hangzhou and Marburg can reveal the fruitful nature of such collaborations. In this context, the first Sino–German Symposium on Functional Nano-Materials Science (FNMS2018), which was jointly held at the Zhejiang University in 2018, paved the way for further research cooperation within a binational group (FNMS-COOP) including the author's team—briefly addressed in this work's outlook part.

From the discussed subjects, it can be easily evidenced that nanostructuring capabilities are the key to the success in the modern world of micro- and nanotechnological devices, whereas optical characterisation remains an indispensable tool for the study of all the discussed semiconductor quantum structures and nanomaterials. Numerous useful methods are summarised within this work in the context of the pursued research. The role of quantisation effects is also briefly highlighted for pedagogical purposes. Without proper spectroscopic techniques one would literally be in the dark, unable to tap many of the possibilities. This would be a great deficit given the fact that one would miss the chance to explore a wonderful world of many-particle effects, excitations and their correlations, as well as complex electronic compounds in solids. All these give rise to a plethora of sophisticated applications for semiconductor structures and devices that have recently entered the second phase of a quantum technological revolution.

8.2 Concluding Remarks

Semiconductor photonics and quantum technologies have been major contributors to modern scientific and industrial achievements and have enabled optical, electronical and optoelectronic tools that have a strong impact on our daily lives ranging from information processing, communications and industrial production to security and medical applications. Consequently, they are strongly in the focus of current research endeavours to guarantee steady progress on the scientific and engineering frontier. The study of quantum structures and the understanding of quantum phenomena are an essential part of research in these two key scientific domains. Similarly crucial for the fast development of the aforementioned technologies is the discovery of novel (semiconductor) materials and their proper utilisation. Thus, interdisciplinary work at the crossroads of quantum physics, nanotechnology, materials science and photonics has been inevitable with regard to, both, the development of these technologies and the research performed within this Habilitation project settled in this topical landscape.

In the past six years[1], a plethora of relevant investigations had been begun as well as successfully concluded in Marburg, Germany, and the way to many more explorations has been paved by the project works summarised within this book. While the focus and emphasis was shaped by the previously existing as well as ongoing research activities and the rich measurement capabilities of the author's host group, a big flexibility, diversity and independence in the study of interesting nanomaterials,

[1]Expressing the author's view in early summer 2020.

laser systems and physics phenomena evolved throughout this Habilitation period, fortunately. Indeed, the author's personal research interests have always guided him back towards nanophotonics and quantum optics subjects, and in the wider sense light–matter interactions, although the interlude of semiconductor/functional (nano-) materials studies with various optical spectroscopy techniques can only be regarded as an invaluable advantage and enriching experience.[2]

In summary, the author's recent endeavours included, on the one hand, investigations on various semiconductor disk laser systems towards self-mode-locking as well as nonlinear lensing, and even their employment for the high-frequent excitation of quantum-dot single-photon sources, the achievement of high-power single-frequency lasing and of tunable THz generation. On the other hand, they included studies on nanoparticles, van-der-Waals (vdW) heterostructures, and monolayer quantum materials, such as transition-metal dichalcogenides (TMDCs), with a focus on light–matter interactions and optical properties, as well as the continuation of polariton-physics research towards ultrafast control and manipulation of coherent quantum states and Bose–Einstein-like condensates in optical microcavities. Motivated by a long-term experience in these fields, several preliminary works have been pursued by the author with the help of his research team in preparation of major research projects and possible future activities. The privilege, that no lack of fundamental science, engineering-related and device-oriented work was given, can be evidenced throughout the previous chapters of this work and in the following overview of ongoing research within the scope of the author's projects.

8.3 Exploring the Mechanism Behind Self-Mode-Locking in VECSELs

For many years, semiconductor disk lasers (SDLs), also often referred to as vertical-external-cavity surface-emitting lasers (VECSELs) [1–3], have been seen as an ideal platform for the realisation of compact, robust and cost-efficient fs-pulsed lasers [4, 5]. Naturally, saturable-absorber mirrors, which are embedded in the external-resonator device, have been widely employed to achieve mode-locking with SDLs due to their maturity and the well-established understanding of key design parameters for ultrashort pulsing [6, 7]. Nonetheless, in many regards the development of saturable-absorber-free devices has been envisaged as a desirable engineering goal. This is because the recently obtained and already much discussed self-mode-locking effect [2, 5, 8–17] has promised the design of such less complex and more flexible mode-locked VECSELs, provided that the technology is matured and reliable.

[2]Amendment: Simultaneously, a unique opportunity was given the author with respect to independent student teaching and supervision, as evidenced by numerous courses on "Quantum Technology" (full Master level course introduced in 2014 and since then more than five times lectured), "Laser Spectroscopy" (2019), "Semiconductor Quantum Structures for Photonic Devices" (2017 and 2018, Zhejiang University), "Semiconductor Physics and Devices" (2014) and many more delivered in Marburg, as well as in Hangzhou.

The aim to harness self-mode-locking is understandable, as saturable-absorber mirrors[3] have to be individually designed and produced for the targeted operation wavelength, even though for growth experts it appears to be a reasonably little additional effort to deliver such extra chip once high-quality gain mirror growth is accomplished in a certain spectral range. However, self-mode-locked SDLs are expected to circumvent other restrictions—i.e. other than an additional design, optimisation and growth process—naturally set by saturable-absorber based devices, such as thermal management of SESAMs, degradation at high powers and thereby peak-power limitations, and additional costs and space for the implementation of SESAM chips in VECSELs.

In fact, self-mode-locking is an appealing pathway towards ultrashort pulses, but the mechanisms behind this phenomenon in VECSELs still remain unclear and are not understood. Nevertheless, self-mode-locking has been successfully obtained for different devices, which are based on quantum-well [13] or quantum-dot [12] gain media, with advanced characterisation in support of a mode-locked operation claim (for an overview on self-mode-locked VECSELs, see [18]). Owing to the fact that SDL research in Marburg contributed to these achievements with key demonstrations in 2014, the aim of the author's first major VECSEL project has been to perform important investigations regarding self-mode-locking within the scope of the first—and in the meanwhile also second—funding period. Several experiments have been envisioned and partly realised in order to uncover the mechanisms and nonlinear effects responsible for mode-locking, and to boost the development of fs-pulsed SDLs towards powerful, cost-efficient devices for a variety of applications.

In summary, within the author's VECSEL project, pump–probe experiments on a running SDL's active region are to be performed in order to acquire a time-resolved picture of the gain dynamics. This may reveal the influence of gain saturation on an intensity-dependent refractive index, which causes self-phase modulation and self-focusing. Further experimental studies on the laser chip aim at providing a direct measurement of a possible Kerr-lensing effect, which is assumed to play a major role. Given the fact that the chip exhibits such an intensity-dependent lensing, and shows little influence of the optical pumping situation, as demonstrated so far by the author's Z-scan experiments, one can already draw a conclusion on the effective nonlinear refractive index of a VECSEL chip (see preliminary measurements [19–21] and advanced studies [22, 23]). However, only time-resolved measurements of the nonlinearity may provide important insights into the origin of nonlinear lensing that can be attributed to either an ultrafast bound-electronic Kerr effect (BEKE)—if taking place on the fs scale—or to free-carrier nonlinearities (FCN)—on the ps scale. In fact, an interplay of both effects, which needs to be unravelled still, may be the reason for stable self-mode-locking, acting as both an ultrafast and slow artificial saturable absorber, respectively (see discussion in [22]). Additional measurements regarding the phase information of the optical pulses may finally allow for a clear identification of the main mechanism behind mode-locking and a better understanding of the properties of the pulsed light source.

---

[3]SAMs, usually semiconductor SAMs, known as SESAMs.

For the successful employment of self-mode-locked SDLs in applications, such as spectroscopy, multi-photon microscopy, material processing and more, it is essential to reveal the nature of the nonlinearity which causes pulse formation in the self-mode-locked device. Thereby, significant improvements can be achieved with respect to the pulse duration, spectral versatility and performance. The recently understood role of the microcavity on the nonlinearities in an SDL chip in combination with a better understanding of the group-delay dispersion in VECSELs may support the sophisticated design of gain chips for self-mode-locking VECSELs, as well as dispersion compensation efforts. The author expects even peak powers in the order of several kilo-Watts and pulse durations as short as 100 fs to come into reach as a consequence of the optimisation of such saturable-absorber-free mode-locked VECSELs.

8.4 Manipulating and Controlling Cavity–Polaritons with Terahertz Waves

The rich world of light–matter interactions has enabled numerous technological advances, some widely known as the products of semiconductor photonics achievements, such as LEDs and lasers, others anticipated to bring the next generation of quantum technologies to commercially-available products, such as quantum communication and information processing devices. Among the interaction effects between matter and light states, the strong quantum coupling of excitons and photons in semiconductors remains fascinating, that results in coherent superpositions of the two components, or quasi-particles called exciton–polaritons [24–26]. The appearance of such mixed modes is interesting from, both, the point of view of fundamental and applied research. The demonstration of these hybrid quasi-particles in optical microcavities [27–30] in the semiclassical ensemble regime based on quantum wells (QWs) [31] and in the quantum limit for quantum dots (QDs) [32–34] opened up new possibilities with regards to cavity-QED studies [35, 36] and coherent light generation [37, 38]. Strong inter-particle interactions result from the exciton component, whereas extremely low effective mass and high mobility are due to the photonic component.

The field of condensation of exciton–polaritons [39–41] emerged in the previous decade shortly after the demonstration of true Bose–Einstein condensates of dilute atomic gases [42–44] which opened a new chapter in the field of quantum optics. This major breakthrough on quasi-particles in solids led to a vast exploration of condensation-related phenomena and phase transitions [45–48], such as a spontaneous coherence build-up, superfluidity and superconductivity. Even the use of qubits based on polariton Rabi oscillations [49] and novel light sources emitting in the visible or THz range, such as polariton lasers [37, 50–54], as well as THz bosonic lasers and the like [55–61], have been proposed.

Within the last few years, a few independent efforts have been made to combine THz radiation and polariton systems in experiments, with the aim to study the interactions of THz radiation and QW–microcavity systems [62–65]—yet, the envisioned THz generation has remained unachieved. Among others, THz probe experiments addressed the dark side of exciton–polariton condensates, i.e. the uncondensed optically-dark exciton fraction in reservoir states outside the light cone [63]. Also, THz "reset" pulses were used to manipulate the quantum state of the strongly-coupled exciton–photon system in microcavities by disturbing the coherent energy exchange by depopulation of 1s-excitonic polarisation [64].

On the one hand, interactions of THz waves with excitonic systems have been widely studied and utilised in the literature [66–69]. On the other hand, control of the polariton's quantum state is not only achieved with THz pulses, but has been demonstrated with optical pulses [70, 71]. However, in the author's current polariton research project, a better understanding of how THz radiation and polaritonic gases interact with each other, both in the linear and nonlinear regime, shall be gained. Thereby, the way will be paved for ultrafast manipulation and control of light–matter coupling, as well as the development of future practical THz-generation schemes involving polariton systems. This has motivated a systematic investigation of THz-induced effects in various configurations involving ultrafast spectroscopy experiments and microcavity polaritons, which are envisioned in the author's project.

The author's aims are to directly measure Rabi oscillations using a digital holography technique for ultrafast time-resolved (TR) luminescence recordings used by his cooperation partners [70, 71] from Lecce, Italy, and to investigate the effects of transient THz pulses on the coherent state of the light–matter coupled system. TR-PL studies will further give access to the dynamics of polariton condensates influenced by the presence of pulsed THz radiation. Thereby, the effects of transient electric fields on condensates of polaritons, which are a unique testbed for condensation studies in solids, will be probed.

In this context, light is going to be shed on the manipulation and control of coherent states, and the ultrafast switching between a condensed and uncondensed polariton gas or a polaritonic and photonic regime will be explored. Consecutive investigations of the second-order temporal autocorrelation function of THz-disturbed polariton gases and condensates/superfluids will give important insights into the photon statistics and, thereby, reveal the impact of transient external (THz) fields on the correlations within the polariton cloud and the degree of coherence. The author is convinced that this project will deepen the understanding of the phenomena related to THz–exciton–polariton coupling and enhance the developments at the crossroads of two disciplines with novel and sophisticated experiments, with the ultimate goal of enabling further utilisation of light–matter interactions for novel optical quantum technologies and light-source concepts.

8.5 Towards Optoelectronic Devices and Microcavity Experiments with 2D Materials

As light-based technologies have become indispensable in our modern world, a demand for high-performing novel optical materials is growing and optimisation of optical devices with respect to energy efficiency and cost effectiveness receives much attention. Low-dimensional materials feature unique advantages over their bulk counterparts and are expected to enable next-generation innovative technologies. Particularly, the opportunity to vertically stack 2D materials offers vast possibilities in terms of 'heterostructuring' and "band-gap engineering" [72, 73]. Nowadays, 2D materials are popular not only for conventional optoelectronics [74], but also for valleytronics due to the optical and electronic properties related to the valley-pseudo-spin [75, 76].

Naturally, for the successful use in devices, the optoelectronic properties of, both, individual monolayers (MLs) and stacked ML-based heterosystems (HSs) need to be thoroughly explored. Previous works on monolayer materials have clearly shown how the substrate [77–80] and the environment [81–85] alter the optical properties of 2D semiconductors such as WS_2, WSe_2 and so forth of the family of TMDCs. With regard to high-quality monolayer-system studies, it has been shown that hBN encapsulation has become an indispensable tool [83, 86–89]. In addition to the use of the surrounding materials' dielectric properties (e.g. electronic screening) or interface-related effects (e.g. electronic doping), optical properties can be further tailored through light–matter interactions with the help of patterned substrate structures resembling in-plane optical microcavities [80], or vertically constructed (monolithic or open, tunable) Fabry-Pérot (FP) optical microcavities [90–95]. Last, but not least, homobilayers of different symmetry or heterostructures comprising different monolayers with certain layer-to-layer orientations can open up new possibilities in the modification of material properties [97–100]. This can for instance happen via band structure modifications [101–104], or the formation of moiré patterns and thereby in-plane superlattices in the 2D potential landscape that can have a strong influence on the spectral features [105–110], to name but a few.

In a current research project, a joint theoretical and experimental study of optical properties of the (yet little explored) $MoSe_2/MoTe_2$ and $WSe_2/MoTe_2$ ML–ML heterostructures is conducted. Primarily, the understanding of this material class and their possible applications is going to be demonstrated by investigating whether type-II heterostructures of $MoSe_2/MoTe_2$ retain the direct band gap of their corresponding monolayers and are promising candidates for photovoltaic devices and detectors due to the presence of interlayer (i.e. spatially indirect) excitons—also known as charge-transfer excitons. Similarly, it can be clarified whether type-I heterostructures of $WSe_2/MoTe_2$ are ideal structures for utilisation in low-threshold 2D-materials diode lasers due to the presence of intralayer (i.e. spatially direct) excitons. Moreover, it is important to understand whether heterostructuring deteriorates the optical properties of monolayers, and whether lattice mismatch and the substrate choice significantly affect the optical properties of these heterostructures.

In addition, alternative heterostructures such as the ML–ML WSe_2/WS_2 type-II structure are currently under investigation [110], and the use of such heterojunctions in 2D photovoltaic devices will be explored. Also, in order to see if intersubband transitions for few-layer TMDC [111] can be used for device concepts employing out-of-plane fields for possible tunable resonance-tunnelling effects, which requires that the Bloch wave nature perseveres under bias, cooperative work on quantum wells made from van-der-Waals materials has been targeted together with the research group of H. G. Roskos in Frankfurt.

Furthermore, given the strong interest in light–matter coupling experiments, the overarching working title "2D-materials optical RESOnator NAnoTEchnologies (2RESONATE)" was defined by the author in August 2017, with the (long-term) aim to explore open tunable resonator concepts with 2D materials as active medium (Fig. 8.1).

An experimental testbed for strong light–matter interactions has been in development over the last years, also in collaboration with the research group of W. Fang in Hangzhou [112] (see Fig. 7.7), and complementary theoretical studies by the author's team have proposed continuously tunable light–matter coupling in microcavities with 2D materials [95] (see Figs. 3.2 and 4.8). With such, a transition from the strong to the weak coupling regime can be investigated and the properties of such system around the exceptional point of the light–matter coupling explored.

Moreover, it is the author's aim to utilise piezo-driven tunable optical microcavities with low mode volume (operational at very-low and room temperature), which incorporate suitable 2D-material systems (currently used also in the context of Purcell effect studies in the author's team), and (ideally) with high-quality polarisation-selective optical mirrors. These systems are targeted for polariton condensation studies at elevated temperatures up to room temperature and (ideally) for helicity-dependent coupling effects (Fig. 8.2). Future work also aims at the exploration of 2D–nanophotonic structures with nontrivial optical topological properties.

To obtain helicity-dependent reflectivities, first efforts by the author's team involved nanopatterning highly-reflecting mirrors for valley-selective nanophotonics (cf. Fig. 8.3 for an early concept, with spectra calculated using a free electromag-

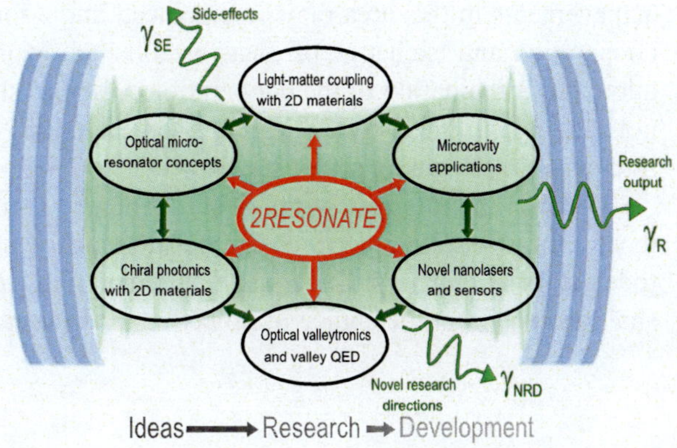

Fig. 8.1 Overview of the goals within the frame of "2RESONATE"

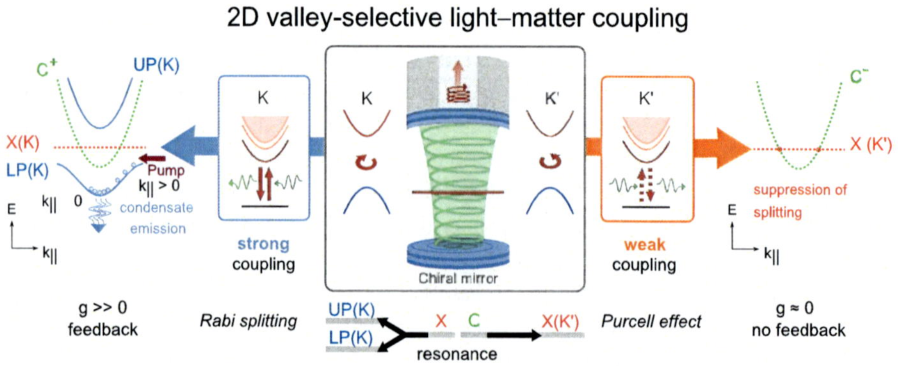

Fig. 8.2 Schematic diagram of valley-selective light–matter coupling in chiral microcavities. Helicity is indicated by circled arrows and +/− superscript of the cavity mode C. Individual coupling to distinct valleys (K/K') gives rise to different spectral features: polariton energy–momentum dispersion branches (left), or Purcell regime (right). The centre box shows the simplified band structure of a 2D membrane in an open cavity. The boxes at the side represent the excitonic picture in which polarised cavity light interacts with matter (coupling strength indicated by arrow style and label)

netic solver [96]) and the design of the required chiral photonic structures with the help of an evolutionary neural learning algorithm. The obtained preliminary results with reflectance differences in the few-percent regime promise a considerable impact on coupling situations in future light–matter interaction experiments involving optimised chiral reflectors.

8.6 Functional Nanomaterials Sciences Cooperation Group

2D-materials research is a hot topic of functional nanomaterials science. The combination of 2D and 0D systems, as well as 3D materials, delivers a measureless scope to explore materials science and promotes the performance of electronic, optoelectronic and thermoelectric devices. With a collaboration project dedicated to the utilisation and understanding of various nanostructured heterojunctions or material systems, the achievements in this area may be enhanced and a long-term network for scientific cooperation and exchange of ideas established. Moreover, the joint research may allow the combination of the cooperation partners' skills to obtain new capabilities not available in the individual groups' environment, and train a new generation of interdisciplinary young scientists.

Based on the interests of the cooperation group members and their key competences, the scientific objectives can be summarised as follows: (1) 2D-system photodetectors and sensors; (2) 2D- and nanomaterials light–matter coupled systems. It shall be noted that the choice of objectives and the range of related scientific topics

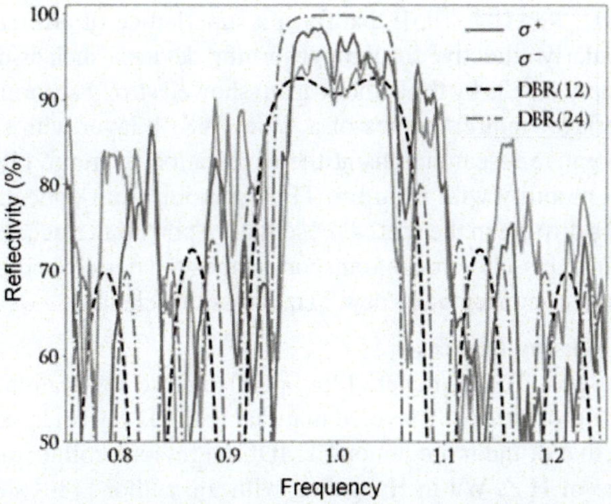

Fig. 8.3 Exemplary reflectivity modelling of a preliminary chiral mirror design, performed within research work in the team of the author in 2017. Simulation executed by O. Mey using a Fourier modal method (S4), preliminary data for a design suggestion by the author based on patterned DBRs for exploratory purposes. Note that the quality of simulated spectra strongly depends on the modelling resolution/Fourier coefficients. Nonetheless, the coarse spectra for opposite circular polarisation $\sigma^{\pm}$ indicate sufficient reflectivity differences. For comparison, the expected stop band for DBRs with 12 and 24 mirror pairs are displayed in dashed and dash-dotted lines, respectively

crystallised out of the wish to enable two things: (1) a broader interaction among the various participants of the 1st Sino–German Symposium on Functional Nano-Materials Sciences in Hangzhou (FNMS2018) [113, 114] from which the equally-labelled cooperation group (FNMS-COOP) has been forged; (2) the involvement of groups with different expertise and background (with, both, various synthesis and characterisation capabilities), as well as project members with mutual interest in the performed research work. Naturally, existing collaborations and constructive interactions have been at the core of forming the group. The scope of the project including the organisation of multiple symposia may serve as a great platform for enhanced collaboration providing many interaction and connection opportunities for the involved parties and their co-workers.

TMDC-Based Optoelectronic Devices

For effficient and sensitive photodetection, new concepts employing 2D materials, such as graphene and related materials, are being intensively investigated in the whole research community (as discussed in the literature, such as [115–118], to name but a few). Based on the experience obtained from previous work on a hybrid graphene-based photodetector [119] and polarisation-sensitive absorption from high-symmetry TMDC bilayers [104], and within the scope of this cooperation group, photodetector concepts using monolayer or few-layer 2D semiconductors are going to be explored further, particularly with a focus on heterostructures [72, 73] and twisted homo-/hetero-bilayers. Twisted bilayers can give rise to moiré patterns (see for instance [97, 98, 105, 106]) that form a superlattice of artificial quantum dots and are particularly attractive for light absorbing devices, such as detectors or solar cells. Preliminary results by the author's team showed strong absorption in the visible range for interlayer moiré features of a WSe_2/WS_2 bilayer with small twist angle [110]. Collaborative research aims at the exploration of moiré bilayers with good charge-carrier mobility with regard to THz emission from optically or electrically excited carriers drifting in the spatially oscillating potential. Such work may provide fundamental insights into charge transport in these materials and pave the way for the achievement of conceptually new THz emission schemes.

GaTe-Based Sensing Devices

One promising post-TMDC material for optoelectronic applications is GaTe, which has remarkable properties as a layered material with 2D crystal lattice that is slightly different than that of the more famous TMDCs. Previous collaborative work by the research group of H.Z. Wu in Hangzhou with the author [120] showed that GaTe multilayers decorated by gold nanoparticles act as very sensitive probe systems for certain aromatic molecules. This renders them attractive for surface-enhanced Raman spectroscopy, which is explored by partners from Hangzhou. With this "Golden-GaTe" towards ultra-sensitive sensing in sight, currently, further research is going on cooperatively to understand charge-carrier dynamics and temperature dependencies of Au-GaTe in comparison to pristine and annealed flakes of GaTe. Moreover, the influence of gold decoration on the optical properties is studied.

Mid-Infrared and THz Detectors

High-speed mid-infrared photodetectors with a two-dimensional electron gas (2DEG) in CdTe/PbTe heterojunctions are a topic of further investigation within the cooperation group FNMS-COOP. Such 2DEG is formed at the heterojunction's interface. With this platform of a novel 2DEG established in the group of H.Z. Wu in Hangzhou, the cooperation partners aim at the development of a high-speed and room-temperature operating mid-IR photodetector. Making use of the conduction-band and valence-band offsets of the heterojunction, the photoresponse of the detectors originating from the intrinsic response of PbTe shall open the path to broadband detection schemes spanning the IR and THz range (studied jointly via ultrafast spectroscopy). Moreover, the development of suitable optoelectronic structures is pursued and other concepts involving the thermoelectric effect in graphene may be explored with partners. The underlying work by the group of H. G. Roskos in Frankfurt builds on a careful evaluation of the performance of THz detectors based on graphene field-effect transistors [121], which has led to the development of a circuit simulator, which takes rectification by both resistive mixing and the hot-carrier thermoelectric effect in graphene into account [122].

2D–3D Interface and Nanoparticle-Enhanced Perovskite-Based Structures

The optical characterisation of perovskite solar cell materials was targeted with the work carried out by Kong et al. involving the current cooperation-group coordinators from Hangzhou and Marburg and continued by W. Kong and co-workers in a new research environment at the Southern University of Science and Technology, Shenzhen, and at Hebei University afterwards [123–125]. The cooperation partners' next efforts will be directed to a better understanding of the charge-carrier dynamics in perovskites combined with 2D materials for improved charge-carrier extraction, and the optical (optoelectronic) characterisation of ultrathin 2D-perovskite layers with eyes towards future energy harvesting applications. This research will benefit from the core competences of several FNMS-COOP project members, the exchange of sample material and the use of different characterisation tools. Exploiting 2D and 3D hybrid perovskite materials with their high mobilities as well as good stability and a high absorption coefficient, respectively, the combination of 2D and 3D hybrids can be of benefit to the production of solar cells. Such combination will be investigated jointly. In addition, the employment of 0D perovskite nanoparticles in such schemes, which have to undergo systematic characterisation, may add further benefits to photodetection/photovoltaic schemes.

2D-Materials Nanoscopy

To bring 2D semiconductors closer to applications and to improve the emission properties, PL enhancement has been studied since 2017 by the author's team with the help of co-workers from Frankfurt and Hangzhou using WS_2 on nanofabricated optical microresonators, which are achieved from ring patterns in a dielectric substrate [80]. Previous works of the cooperation partners allowed combining both vertical and horizontal interference effects to improve both incoupling into and outcoupling of light from 2D materials. Photodetectors, 2D-material photovoltaics, nonlinear optics and light-emitting diodes could strongly benefit from advances based on these "bull's-eye" ring patterns (or in certain applications such as integrated photonic circuits also from linear grating patterns), and this is currently under investigation. Such structures could be at the core of future on-chip optical interconnects for light-based information processing purposes. The cooperation partners plan to further investigate the nanoscale structure with regard to confinement effects using nanoscopy, nanopatterning techniques and nanomaterials in order to obtain new insights into the applicability of such in-plane microcavities (employing linear or circular Bragg grating patterns). Here, nanoscopy (e.g. SNOM) on common 2D materials can particularly reveal the effects of patterned substrates on the spatially-modified nanoscale emission patterns of the covering 2D-material flake.

Electrical Contacts, Microoptics and Plasmonics

Nanostructuring capabilities are the key to the success in the modern world of micro- and nanotechnological devices. In this context, one aim pursued within the cooperation group is to fabricate, characterise and develop sub-micrometer and micrometer systems involving 2D materials. In fact, micro-/nanofabrication techniques are essential for realising designs of optoelectronic devices, particularly when it comes to the achievement of electrical interconnects and integrated optics for sub-micrometer (nano-sized) systems. Jointly, the production of nano-sized electrical contacts and nanofabricated polymer optics shall be examined for use in connection with 2D materials, in order to improve the design of photosensitive field-effect transistors used in electrostatically-gated photodetectors under investigation by the project partners. Furthermore, controlled positioning of nanoparticles on 2D materials by combing top-down electron-beam lithographic techniques with the meniscus-force-deposition technique by the group of P. J. Klar in Giessen promises benefits to such developments [126–129]. Using metallic or other-type nanoparticles, and depositing them on the surface of the 2D material, may allow one to tune the optical response of 2D materials, with eyes towards improving the optoelectronic properties of 2D devices such as detectors or solar cells.

2D-Materials Synthesis and Stacking

The development as well as improvement of 2D materials syntheses remains a key goal of FNMS-COOP project partners working together with J. He at the National Center for Nanoscience and Technology (short: Nanocenter) in Beijing. The demonstration of high-quality large-area synthesised monolayer and multilayer crystals and their comparison with their mechanically exfoliated counterparts will improve the capabilities in terms of nanodevice fabrication. Direct synthesis of vdW heterostructures remains one goal on that path. Also, the development of improved electrical contacting schemes is currently under investigation. Future endeavours will address van-der-Waals epitaxy, which is necessary for more complex vertical and horizontal alignment of different 2D materials, and which requires joint characterisation efforts. Such research is beneficial to activities in both host countries of the cooperation group regarding 2D devices.

High-Quality Tunable Open Optical Microcavities

For practical photonic devices, concepts of optics and materials sciences need to be combined. This typically leads to fascinating light–matter interaction studies, where quantum emitters are placed inside optical microcavities or waveguide structures. Microcavities confine the light field in a small mode volume (with high free spectral range (FSR)) and bring cavity photons into coupling with excitons in the respective quantum material by tuning individual modes into resonance with each other. Indeed, if the resonator quality (Q-factor) and the FSR are too small, the cavity mode is not sharp enough (and, thus, photon lifetimes not long enough) and the optical resonances are not separated enough, respectively, to enable the clear observation of the desired light–matter coupling effects. Recently, collaborative work has demonstrated a new approach to produce optical microresonators based on fibre technologies for nanophotonics [112], offering the team different variations of open tunable microcavities with Q-factors between few hundreds up to few thousands. As part of ongoing project activities, high-quality 2D materials and different colloidal quantum dots are going to be employed inside these fibre-based microcavities and optical properties of such structures systematically investigated by the author and co-workers (group of W. Fang in Hangzhou) jointly with eyes towards weak and strong coupling, and single-photon generation.

Strong Light–Matter Coupling with 2D Materials

Quasi-particle formation in light–matter-coupled microresonators gives access to the intriguing physics of Bose–Einstein condensates at elevated temperatures up to room temperature, provided that the underlying matter excitations for the so-called cavity–polaritons, the excitons, prevail at the respective temperature and at the needed particle densities for stimulated scattering effects [45, 48]. 2D materials have promised a

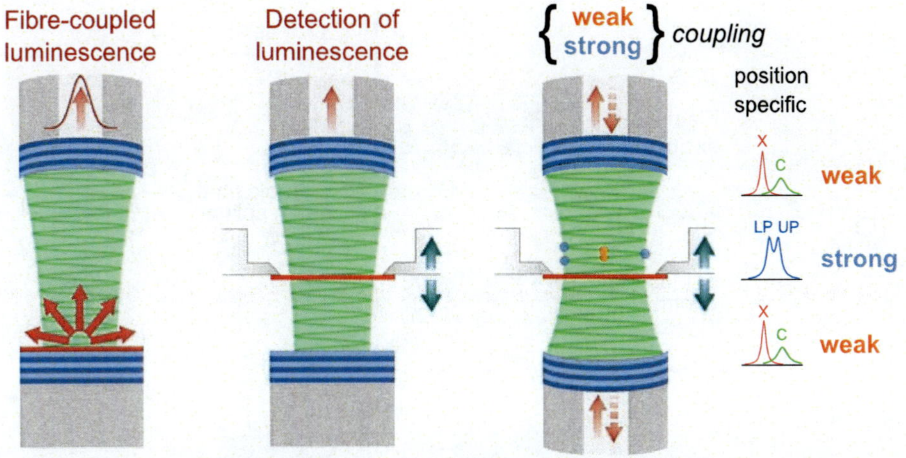

Fig. 8.4 Tunable open-cavity microresonator designs for variable light–matter coupling with 2D membranes, employing dielectric mirror coatings on planar transparent substrates or fibre tips. Different monolayer positions with respect to the cavity mode field result in different coupling strengths, indicated spectrally to the right. Corresponding experimental studies were defined within the frame of "2RESONATE" and supported by recent theoretical modelling [95] for two WS_2–microcavity configurations

route towards condensation studies at room temperature owing to their extraordinarily large exciton binding energies for a quantum-well-like semiconducting material [74, 130]. Based on recently reported microcavities [112], the author plans to observe polaritons by coupling excitons of suitable 2D semiconductors with reasonably-well-confined light fields in tunable and open (fully or partially fibre-based) microcavities (Fig. 8.4). Based on calculations for such 2D–cavity systems [95], even a smooth transition between the weak and the strong coupling of the cavity mode with the 2D exciton resonance (across the exceptional point) could be in principle probed with the two proposed design concepts—by smoothly modifying the effective coupling constant $g = \hbar\Omega_0(\theta, \mathbf{r}, L_{\text{cav}}) = \hbar\Omega_{\text{Rabi}}$, with θ the adjustable incidence angle, $\mathbf{r}$ the variable emitter location in the field, and L_{cav} the tunable cavity length.

It is envisioned that joint development and experimentation on these 2D–cavity systems—with internal feedback loops on production and characterisation results, as well as with advanced in/out-coupling and contacting schemes (Fig. 8.5)—will enable advanced achievements in light–matter coupling experiments[4]. These could one day lead to the observation of (valley-polarised) Bose–Einstein-like polariton condensation with 2D materials (and ideally with helicity-selective cavities), at room temperature. These goals could be seen in the frame of the overarching "2RESONATE" subject of the author's research, promising further interesting developments (Fig. 8.6).

[4]In general terms light–matter interfaces.

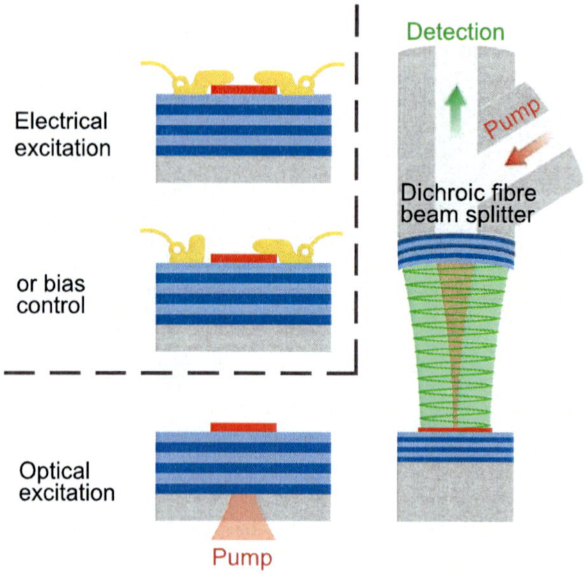

Fig. 8.5 Sketches of electrical contacting schemes (e.g. involving TMDCs and graphene contacts), optical excitation and optical read-out

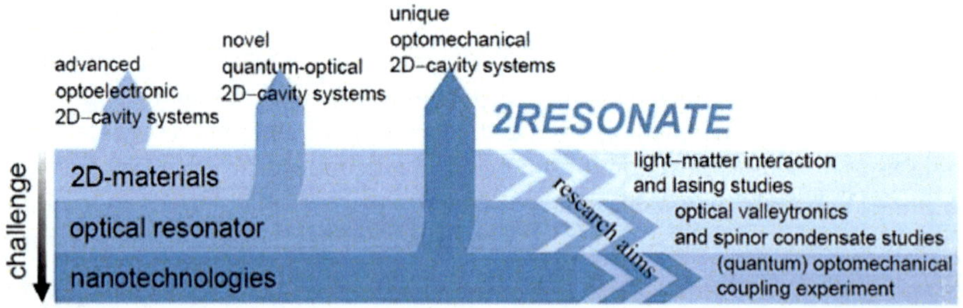

Fig. 8.6 Diagram of the research aims and the envisioned outcomes within the "2RESONATE" frame

References

1. M. Kuznetsov, F. Hakimi, R. Sprague, A. Mooradian, High-power (>0.5-W CW) diode-pumped vertical-external-cavity surface-emitting semiconductor lasers with circular TEM_{00} beams. IEEE Photon. Technol. Lett. **9**(8), 1063–1065 (2014)
2. A. Rahimi-Iman, Recent advances in VECSELs. J. Opt. **18**, 093003 (2016)
3. M. Guina, A. Rantamäki, A. Härkönen, Optically pumped VECSELs: review of technology and progress. J. Phys. D: Appl. Phys. **50**, 383001 (2017)
4. U. Keller, A.C. Tropper, Passively mode locked surface-emitting semiconductor lasers. Phys. Rep. **429**, 67–120 (2006)
5. M.A. Gaafar, A. Rahimi-Iman, K.A. Fedorova, W. Stolz, E.U. Rafailov, M. Koch, Mode-locked semiconductor disk lasers. Adv. Opt. Photonics **8**, 370–400 (2016)

6. P. Klopp, U. Griebner, M. Zorn, M. Weyers, Pulse repetition rate up to 92 GHz or pulse duration shorter than 110 fs from a mode-locked semiconductor disk laser. Appl. Phys. Lett. **98**, 071103 (2011)
7. D. Waldburger, S.M. Link, M. Mangold, C.G.E. Alfieri, E. Gini, M. Golling, B.W. Tilma, U. Keller, High-power 100 fs semiconductor disk lasers. Optica **3**, 844–852 (2016)
8. Y.F. Chen, Y.C. Lee, H.C. Liang, K.Y. Lin, K.W. Su, K.F. Huang, Femtosecond high-power spontaneous mode-locked operation in vertical-external cavity surface-emitting laser with gigahertz oscillation. Opt. Lett. **36**, 4581–4583 (2011)
9. L. Kornaszewski, G. Maker, G.P.A. Malcolm, M. Butkus, E.U. Rafailov, C.J. Hamilton, SESAM-free mode-locked semiconductor disk laser. Laser Photon. Rev. **6**, L20–L23 (2012)
10. A.R. Albrecht, Y. Wang, M. Ghasemkhani, D.V. Seletskiy, J.G. Cederberg, M. Sheik-Bahae, Exploring ultrafast negative Kerr effect for mode-locking vertical external-cavity surface-emitting lasers. Opt. Express **21**, 28801–28808 (2013)
11. M. Gaafar, C. Möller, M. Wichmann, B. Heinen, B. Kunert, A. Rahimi-Iman, W. Stolz, M. Koch, Harmonic self-mode-locking of optically pumped semiconductor disc laser. Electron. Lett. **50**, 542–543 (2014)
12. M. Gaafar, D.A. Nakdali, C. Möller, K.A. Fedorova, M. Wichmann, M.K. Shakfa, F. Zhang, A. Rahimi-Iman, E.U. Rafailov, M. Koch, Self-mode-locked quantum-dot vertical-external-cavity surface-emitting laser. Opt. Lett. **39**, 4623–4626 (2014)
13. M. Gaafar, P. Richter, H. Keskin, C. Möller, M. Wichmann, W. Stolz, A. Rahimi-Iman, M. Koch, Self-mode-locking semiconductor disk laser. Opt. Express **22**, 28390–28399 (2014)
14. A. Rahimi-Iman, M. Gaafar, W. Stolz, E.U. Rafailov, M. Koch, Self-mode-locked semiconductor disk lasers. Opt. Photon. News **25**, 28–28 (2014)
15. A. Rahimi-Iman, M. Gaafar, D.A. Nakdali, C. Möller, F. Zhang, M. Wichmann, M.K. Shakfa, K.A. Fedorova, W. Stolz, E.U. Rafailov, M. Koch, Recent advances in the field of vertical-external-cavity surface-emitting lasers, in *Vertical External Cavity Surface Emitting Lasers (VECSELs) V*, vol. 9349 (International Society for Optics and Photonics, 2015), p. 934906
16. A. Rahimi-Iman, M. Gaafar, C. Möller, M. Vaupel, F. Zhang, D. Al-Nakdali, K.A. Fedorova, W. Stolz, E.U. Rafailov, M. Koch, Self-mode-locked vertical-external-cavity surface-emitting laser, in *Vertical External Cavity Surface Emitting Lasers (VECSELs) VI*, vol. 9734 (International Society for Optics and Photonics, 2016), p. 97340M
17. R. Bek, M. Grossmann, H. Kahle, M. Koch, A. Rahimi-Iman, M. Jetter, P. Michler, Self-mode-locked AlGaInP-VECSEL. Appl. Phys. Lett. **111**, 182105 (2017)
18. A. Rahimi-Iman, Self-mode-locked vertical-external-cavity surface-emitting laser, in *Vertical External Cavity Surface Emitting Lasers: VECSEL Technology and Applications*, 2nd edn., ed. by M. Jetter, P. Michler (Wiley-VCH, 2020). In press. ISBN 978-3527413621
19. A.H. Quarterman, M.A. Tyrk, K.G. Wilcox, Z-scan measurements of the nonlinear refractive index of a pumped semiconductor disk laser gain medium. Appl. Phys. Lett. **106**, 011105 (2015)
20. E.A. Shaw, A.H. Quarterman, A.P. Turnbull, T.C. Sverre, C.R. Head, A.C. Tropper, K.G. Wilcox, Nonlinear lensing in an unpumped antiresonant semiconductor disk laser gain structure. IEEE Photon. Technol. Lett. **28**(13), 1395–1398 (2016)
21. A.H. Quarterman, S. Mirkhanov, C.J.C. Smyth, K.G. Wilcox, Measurements of nonlinear lensing in a semiconductor disk laser gain sample under optical pumping and using a resonant femtosecond probe laser. Appl. Phys. Lett. **109**, 121113 (2016)
22. C. Kriso, S. Kress, T. Munshi, M. Grossmann, R. Bek, M. Jetter, P. Michler, W. Stolz, M. Koch, A. Rahimi-Iman, Microcavity-enhanced Kerr nonlinearity in a vertical-external-cavity surface-emitting laser. Opt. Express **27**, 11914–11929 (2019). https://doi.org/10.1364/OE.27.011914
23. C. Kriso, S. Kress, T. Munshi, M. Grossmann, R. Bek, M. Jetter, P. Michler, W. Stolz, M. Koch, A. Rahimi-Iman, Wavelength and pump-power dependent nonlinear refraction and absorption in a semiconductor disk laser. IEEE Photon. Technol. Lett. **32**, 85–88 (2020). Published online, 2019

24. S.I. Pekar, The theory of electromagnetic waves in a crystal in which excitons are produced. Sov. Phys. JETP **6**, 785 (1958)
25. J.J. Hopfield, Theory of the contribution of excitons to the complex dielectric constant of crystals. Phys. Rev. **112**, 1555 (1958)
26. C.F. Klingshirn, *Semiconductor Optics* (Springer, Berlin, 2012)
27. Y. Yamamoto, A. Imamoglu, *Mesoscopic Quantum Optics* (Wiley, New Jersey, 1999)
28. K. Vahala, *Optical Microcavities* (World Scientific, Singapore, 2004)
29. A. Kavokin, G. Malpuech, *Cavity Polaritons* (Academic, New York, 2003)
30. B. Deveaud, *The Physics of Semiconductor Microcavities* (WILEY-VCH, New Jersey, 2007)
31. C. Weisbuch, M. Nishioka, A. Ishikawa, Y. Arakawa, Observation of the coupled exciton-photon mode splitting in a semiconductor quantum microcavity. Phys. Rev. Lett. **69**, 3314 (1992)
32. J.P. Reithmaier, G. Sek, A. Löffler, C. Hofmann, S. Kuhn, S. Reitzenstein, L.V. Keldysh, V.D. Kulakovskii, T.L. Reinecke, A. Forchel, Strong coupling in a single quantum dot-semiconductor microcavity system. Nature **432**, 197–200 (2004)
33. T. Yoshie, A. Scherer, J. Hendrickson, G. Khitrova, H.M. Gibbs, G. Rupper, C. Ell, O.B. Shchekin, D.G. Deppe, Vacuum Rabi splitting with a single quantum dot in a photonic crystal nanocavity. Nature **432**, 200–203 (2004)
34. E. Peter, P. Senellart, D. Martrou, A. Lemaître, J. Hours, J.M. Gérard, J. Bloch, Exciton-photon strong-coupling regime for a single quantum dot embedded in a microcavity. Phys. Rev. Lett. **95**, 067401 (2005)
35. Y. Yamamoto, F. Tassone, H. Cao, *Semiconductor Cavity Quantum Electrodynamics* (Springer, Berlin, 2000)
36. A.V. Kavokin, J.J. Baumberg, G. Malpuech, F.P. Laussy, *Microcavities* (Oxford University Press, Oxford, 2007)
37. A. Imamoglu, R.J. Ram, S. Pau, Y. Yamamoto, Nonequilibrium condensates and lasers without inversion: exciton-polariton lasers. Phys. Rev. A **53**, 4250 (1996)
38. A. Kavokin, G. Malpuech, F.P. Laussy, Polariton laser and polariton superfluidity in microcavities. Phys. Lett. A **306**, 187–199 (2003)
39. H. Deng, G. Weihs, C. Santori, J. Bloch, Y. Yamamoto, Condensation of semiconductor microcavity exciton polaritons. Science **298**(5591), 199–202 (2002)
40. J. Kasprzak, M. Richard, S. Kundermann, A. Baas, P. Jeambrun, J.M.J. Keeling, F.M. Marchetti, M.H. Szymańska, R. André, J.L. Staehli, V. Savona, P.B. Littlewood, B. Deveaud, L.S. Dang, Bose-Einstein condensation of exciton polaritons. Nature **443**, 409–414 (2006)
41. R. Balili, V. Hartwell, D. Snoke, L. Pfeiffer, K. West, Bose-Einstein condensation of microcavity polaritons in a trap. Science **316**, 1007–1010 (2007)
42. M.H. Anderson, J.R. Ensher, M.R. Matthews, C.E. Wieman, E.A. Cornell, Observation of Bose-Einstein condensation in a dilute atomic vapor. Science **269**, 198–201 (1995)
43. K.B. Davis, M.O. Mewes, M.R. Andrews, N.J. van Druten, D.S. Durfee, D.M. Kurn, W. Ketterle, Bose-Einstein condensation in a gas of sodium atoms. Phys. Rev. Lett. **75**, 3969 (1995)
44. C.C. Bradley, C.A. Sackett, J.J. Tollett, R.G. Hulet, Evidence of Bose-Einstein condensation in an atomic gas with attractive interactions. Phys. Rev. Lett. **75**, 1687 (1995)
45. H. Deng, H. Haug, Y. Yamamoto, Exciton-polariton Bose-Einstein condensation. Rev. Mod. Phys. **82**, 1489 (2010)
46. D. Sanvitto, V. Timofeev, *Exciton Polaritons in Microcavities: New Frontiers* (Springer, Berlin, 2012)
47. A.V. Kavokin, J.J. Baumberg, G. Malpuech, F.P. Laussy, *Microcavities*, vol. 1 (Oxford University Press, Oxford, 2017)
48. A. Rahimi-Iman, *Polariton Physics: From Dynamic Bose–Einstein Condensates in Strongly-Coupled Light–Matter Systems to Polariton Lasers*, Springer Series in Optical Sciences, vol. 229 (Springer International Publishing, Cham, 2020)
49. S.S. Demirchyan, I.Y. Chestnov, A.P. Alodjants, M.M. Glazov, A.V. Kavokin, Qubits based on polariton rabi oscillators. Phys. Rev. Lett. **112**, 196403 (2014)

50. C. Schneider, A. Rahimi-Iman, N.Y. Kim, J. Fischer, I.G. Savenko, M. Amthor, M. Lermer, A. Wolf, L. Worschech, V.D. Kulakovskii, I.A. Shelykh, M. Kamp, S. Reitzenstein, A. Forchel, Y. Yamamoto, S. Höfling, An electrically pumped polariton laser. Nature **497**, 348–352 (2013)
51. P. Bhattacharya, B. Xiao, A. Das, S. Bhowmick, J. Heo, Solid state electrically injected exciton-polariton laser. Phys. Rev. Lett. **110**, 206403 (2013)
52. P. Bhattacharya, T. Frost, S. Deshpande, M.Z. Baten, A. Hazari, A. Das, Room temperature electrically injected polariton laser. Phys. Rev. Lett. **112**, 236802 (2014)
53. S. Kim, B. Zhang, Z. Wang, J. Fischer, S. Brodbeck, M. Kamp, C. Schneider, S. Höfling, H. Deng, Coherent polariton laser. Phys. Rev. X **6**, 011026 (2016)
54. M. Ramezani, A. Halpin, A.I. Fernández-Domínguez, J. Feist, S.R.-K. Rodriguez, F.J. Garcia-Vidal, J. Gómez Rivas, Plasmon-exciton-polariton lasing. Optica **4**, 31–37 (2017)
55. K.V. Kavokin, M.A. Kaliteevski, R.A. Abram, A.V. Kavokin, S. Sharkova, I.A. Shelykh, Stimulated emission of terahertz radiation by exciton-polariton lasers. Appl. Phys. Lett. **97**, 201111–3 (2010)
56. I.G. Savenko, I.A. Shelykh, M.A. Kaliteevski, Nonlinear terahertz emission in semiconductor microcavities. Phys. Rev. Lett. **107**, 027401 (2011)
57. A.V. Kavokin, I.A. Shelykh, T. Taylor, M.M. Glazov, Vertical cavity surface emitting terahertz laser. Phys. Rev. Lett. **108**, 197401 (2012)
58. T.C.H. Liew, M.M. Glazov, K.V. Kavokin, I.A. Shelykh, M.A. Kaliteevski, A.V. Kavokin, Proposal for a Bosonic cascade laser. Phys. Rev. Lett. **110**, 047402 (2013)
59. S. Huppert, O. Lafont, E. Baudin, J. Tignon, R. Ferreira, Terahertz emission from multiple-microcavity exciton-polariton lasers. Phys. Rev. B **90**, 241302(R) (2014)
60. K. Kristinsson, O. Kyriienko, I.A. Shelykh, Terahertz laser based on dipolaritons. Phys. Rev. A **89**, 023836 (2014)
61. K. Rojan, Y. Léger, G. Morigi, M. Richard, A. Minguzzi, Enhanced second-order nonlinearity for THz generation by resonant interaction of exciton-polariton rabi oscillations with optical phonons. Phys. Rev. Lett. **119**, 127401 (2017)
62. J.L. Tomaino, A.D. Jameson, Y.-S. Lee, G. Khitrova, H.M. Gibbs, A.C. Klettke, M. Kira, S.W. Koch, Terahertz excitation of a coherent λ-type three-level system of exciton-polariton modes in a quantum-well microcavity. Phys. Rev. Lett. **108**, 267402 (2012)
63. J.M. Ménard, C. Poellmann, M. Porer, U. Leierseder, E. Galopin, A. Lemaître, A. Amo, J. Bloch, R. Huber, Revealing the dark side of a bright exciton-polariton condensate. Nat. Commun. **5**, 4648 (2014)
64. A.D. Jameson, J.L. Tomaino, Y.-S. Lee, G. Khitrova, H.M. Gibbs, C.N. Böttge, A.C. Klettke, M. Kira, S.W. Koch, Direct measurement of light-matter energy exchange inside a microcavity. Optica **1**, 276 (2014)
65. B. Piętka, N. Bobrovska, D. Stephan, M. Teich, M. Król, S. Winnerl, A. Pashkin, R. Mirek, K. Lekenta, F. Morier-Genoud, H. Schneider, B. Deveaud, M. Helm, M. Matuszewski, J. Szczytko, Doubly dressed bosons: exciton polaritons in a strong terahertz field. Phys. Rev. Lett. **119**, 077403 (2017)
66. R.A. Kaindl, M.A. Carnahan, D. Hägele, R. Lövenich, D.S. Chemla, Ultrafast terahertz probes of transient conducting and insulating phases in an electron-hole gas. Nature **423**, 734–738 (2003)
67. M. Wagner, H. Schneider, D. Stehr, S. Winnerl, A.M. Andrews, S. Schartner, G. Strasser, M. Helm, Observation of the intraexciton autler-townes effect in GaAs/AlGaAs semiconductor quantum wells. Phys. Rev. Lett. **105**, 167401 (2010)
68. B. Ewers, N.S. Köster, R. Woscholski, M. Koch, S. Chatterjee, G. Khitrova, H.M. Gibbs, A.C. Klettke, M. Kira, S.W. Koch, Ionization of coherent excitons by strong terahertz fields. Phys. Rev. B **85**, 075307 (2012)
69. M.J. Drexler, R. Woscholski, S. Lippert, W. Stolz, A. Rahimi-Iman, M. Koch, Disturbing the coherent dynamics of an excitonic polarization with strong terahertz fields. Phys. Rev. B **90**, 195304 (2014)
70. L. Dominici, D. Colas, S. Donati, J.P. Restrepo Cuartas, M. De Giorgi, D. Ballarini, G. Guirales, J.C. López Carreño, A. Bramati, G. Gigli, E. Del Valle, F.P. Laussy, D. Sanvitto, Ultrafast control and Rabi oscillations of polaritons. Phys. Rev. Lett. **113**, 226401 (2014)

71. D. Colas, L. Dominici, S. Donati, A.A. Pervishko, T.C. Liew, I.A. Shelykh, D. Ballarini, M. De Giorgi, A. Bramati, G. Gigli, E. Del Valle, F.P. Laussy, A.V. Kavokin, D. Sanvitto, Polarization shaping of Poincaré beams by polariton oscillations. Light Sci. Appl. **4**, e350 (2015)
72. A.K. Geim, I.V. Grigorieva, Van der Waals heterostructures. Nature **499**, 419 (2013)
73. K.S. Novoselov, A. Mishchenko, A. Carvalho, A.H. Castro Neto, 2D materials and van der Waals heterostructures. Science **353**, aac9439 (2016)
74. K.F. Mak, J. Shan, Photonics and optoelectronics of 2D semiconductor transition metal dichalcogenides. Nat. Photon. **10**, 216–226 (2016)
75. K.F. Mak, D. Xiao, J. Shan, Light-valley interactions in 2D semiconductors. Nat. Photon. **12**, 451–460 (2018)
76. J.R. Schaibley, H. Yu, G. Clark, P. Rivera, J.S. Ross, K.L. Seyler, W. Yao, X. Xu, Valleytronics in 2D materials. Nat. Rev. Mater. **1**, 16055 (2016)
77. Y. Yu, Y. Yu, C. Xu, Y.Q. Cai, L. Su, Y. Zhang, Y.W. Zhang, K. Gundogdu, L. Cao, Engineering substrate interactions for high luminescence efficiency of transition-metal dichalcogenide monolayers. Adv. Funct. Mater. **26**, 4733–4739 (2016)
78. S. Lippert, L.M. Schneider, D. Renaud, K.N. Kang, O. Ajayi, J. Kuhnert, M.-U. Halbich, O.M. Abdulmunem, X. Lin, K. Hassoon, S. Edalati-Boostan, Y.D. Kim, W. Heimbrodt, E.-H. Yang, J.C. Hone, A. Rahimi-Iman, Influence of the substrate material on the optical properties of tungsten diselenide monolayers. 2D Mater. **4**, 025045 (2017). https://doi.org/10.1088/2053-1583/aa5b21
79. L.M. Schneider, S. Lippert, J. Kuhnert, D. Renaud, K.N. Kang, O. Ajayi, M.-U. Halbich, O.M. Abdulmunem, X. Lin, K. Hassoon, S. Edalati-Boostan, Y.D. Kim, W. Heimbrodt, E.H. Yang, J.C. Hone, A. Rahimi-Iman, The impact of the substrate material on the optical properties of 2D WSe_2 monolayers. Semiconductors **52**, 565–571 (2018). https://doi.org/10.1134/S1063782618050275
80. O. Mey, F. Wall, L.M. Schneider, D. Günder, F. Walla, A. Soltani, H. Roskos, N. Yao, P. Qing, W. Fang, A. Rahimi-Iman, Enhancement of the monolayer tungsten disulfide exciton photoluminescence with a two-dimensional material/air/gallium phosphide in-plane microcavity. ACS Nano **13**, 5259–5267 (2019). https://doi.org/10.1021/acsnano.8b09659
81. Y. Lin, X. Ling, L. Yu, S. Huang, A.L. Hsu, Y.-H. Lee, J. Kong, M.S. Dresselhaus, T. Palacios, Dielectric screening of excitons and trions in single-layer MoS_2. Nano Lett. **14**, 5569–5576 (2014)
82. A.V. Stier, N.P. Wilson, G. Clark, X. Xu, S.A. Crooker, Probing the influence of dielectric environment on excitons in monolayer WSe_2: insight from high magnetic fields. Nano Lett. **16**, 7054–7060 (2016)
83. O.A. Ajayi, J.V. Ardelean, G.D. Shepard, J. Wang, A. Antony, T. Taniguchi, K. Watanabe, T.F. Heinz, S. Strauf, X.Y. Zhu, J.C. Hone, Approaching the intrinsic photoluminescence linewidth in transition metal dichalcogenide monolayers. 2D Mater. **4**, 031011 (2017)
84. L.M. Schneider, S. Lippert, J. Kuhnert, O. Ajayi, D. Renaud, S. Firoozabadi, Q. Ngo, R. Guo, Y.D. Kim, W. Heimbrodt, J.C. Hone, A. Rahimi-Iman, The influence of the environment on monolayer tungsten diselenide photoluminescence. Nano-Struct. Nano-Objects **15**, 84–97 (2018), online 2017. https://doi.org/10.1016/j.nanoso.2017.08.009
85. L. Meckbach, T. Stroucken, S.W. Koch, Influence of the effective layer thickness on the ground-state and excitonic properties of transition-metal dichalcogenide systems. Phys. Rev. B **97**, 035425 (2018)
86. F. Cadiz, E. Courtade, C. Robert, G. Wang, Y. Shen, H. Cai, T. Taniguchi, K. Watanabe, H. Carrere, D. Lagarde, M. Manca, T. Amand, P. Renucci, S. Tongay, X. Marie, B. Urbaszek, Excitonic linewidth approaching the homogeneous limit in MoS_2-based van der Waals heterostructures. Phys. Rev. X **7**, 021026 (2017)
87. J. Wierzbowski, J. Klein, F. Sigger, C. Straubinger, M. Kremser, T. Taniguchi, K. Watanabe, U. Wurstbauer, A.W. Holleitner, M. Kaniber, K. Müller, J.J. Finley, Direct exciton emission from atomically thin transition metal dichalcogenide heterostructures near the lifetime limit. Sci. Rep. **7**, 12383 (2017)

88. D. Rhodes, S.H. Chae, R. Ribeiro-Palau, J. Hone, Disorder in van der Waals heterostructures of 2D materials. Nat. Mater. **18**, 541–549 (2019)
89. L.M. Schneider, S.S. Esdaille, D.A. Rhodes, K. Barmak, J.C. Hone, A. Rahimi-Iman, Shedding light on exciton's nature in monolayer quantum material by optical dispersion measurements. Opt. Express **27**, 37131–37149 (2019). https://doi.org/10.1364/OE.27.037131
90. X. Liu, T. Galfsky, Z. Sun, F. Xia, E.C. Lin, Y.H. Lee, S. Kéna-Cohen, V.M. Menon, Strong light-matter coupling in two-dimensional atomic crystals. Nat. Photon. **9**, 30–34 (2014). https://doi.org/10.1038/nphoton.2014.304
91. S. Dufferwiel, S. Schwarz, F. Withers, A.A. Trichet, F. Li, M. Sich, O. Del Pozo-Zamudio, C. Clark, A. Nalitov, D.D. Solnyshkov, G. Malpuech, K.S. Novoselov, J.M. Smith, M.S. Skolnick, D.N. Krizhanovskii, A.I. Tartakovskii, Exciton-polaritons in van der Waals heterostructures embedded in tunable microcavities. Nat. Commun. **6**, 8579 (2015)
92. L.C. Flatten, Z. He, D.M. Coles, A.A. Trichet, A.W. Powell, R.A. Taylor, J.H. Warner, J.M. Smith, Room-temperature exciton-polaritons with two-dimensional WS_2. Sci. Rep. **6**, 33134 (2016)
93. N. Lundt, S. Klembt, E. Cherotchenko, S. Betzold, O. Iff, A.V. Nalitov, M. Klaas, C.P. Dietrich, A.V. Kavokin, S. Höfling, C. Schneider, Room-temperature Tamm-plasmon exciton-polaritons with a WSe_2 monolayer. Nat. Commun. **7**, 13328 (2016)
94. S. Wang, S. Li, T. Chervy, A. Shalabney, S. Azzini, E. Orgiu, J.A. Hutchison, C. Genet, P. Samorì, T.W. Ebbesen, Coherent coupling of WS_2 monolayers with metallic photonic nanostructures at room temperature. Nano Lett. **16**(7), 4368–74 (2016)
95. F. Wall, O. Mey, L.M. Schneider, A. Rahimi-Iman, Continuously-tunable light-matter coupling in optical microcavities with 2D semiconductors. Sci. Rep. **10**, 8303 (2020). https://doi.org/10.1038/s41598-020-64909-1
96. V. Liu, S. Fan, S4: A free electromagnetic solver for layered periodic structures. Comput. Phys. Commun. **183**(10), 2233–2244 (2012)
97. B. Hunt, J.D. Sanchez-Yamagishi, A.F. Young, M. Yankowitz, B.J. LeRoy, K. Watanabe, T. Taniguchi, P. Moon, M. Koshino, P. Jarillo-Herrero, R.C. Ashoori, Massive dirac fermions and hofstadter butterfly in a van der Waals heterostructure. Science **340**(6139), 1427–1430 (2013)
98. C.R. Dean, L. Wang, P. Maher, C. Forsythe, F. Ghahari, Y. Gao, J. Katoch, M. Ishigami, P. Moon, M. Koshino, T. Taniguchi, K. Watanabe, K.L. Shepard, J. Hone, P. Kim, Hofstadter's butterfly and the fractal quantum Hall effect in moiré superlattices. Nature **497**, 598–602 (2013)
99. Y. Cao, V. Fatemi, A. Demir, S. Fang, S.L. Tomarken, J.Y. Luo, J.D. Sanchez-Yamagishi, K. Watanabe, T. Taniguchi, E. Kaxiras, R.C. Ashoori, P. Jarillo-Herrero, Correlated insulator behaviour at half-filling in magic-angle graphene superlattices. Nature **556**, 80 (2018)
100. Z. Wang, D.A. Rhodes, K. Watanabe, T. Taniguchi, J.C. Hone, J. Shan, K.F. Mak, Evidence of high-temperature exciton condensation in two-dimensional atomic double layers. Nature **574**, 76–80 (2019)
101. T. Jiang, H. Liu, D. Huang, S. Zhang, Y. Li, X. Gong, Y.-R. Shen, W.-T. Liu, S. Wu, Valley and band structure engineering of folded MoS_2 bilayers. Nat. Nanotechnol. **9**, 825–829 (2014)
102. K. Liu, L. Zhang, T. Cao, C. Jin, D. Qiu, Q. Zhou, A. Zettl, P. Yang, S.G. Louie, F. Wang, Evolution of interlayer coupling in twisted molybdenum disulfide bilayers. Nat. Commun. **5**, 4966 (2014)
103. M. Xia, B. Li, K. Yin, G. Capellini, G. Niu, Y. Gong, W. Zhou, P.M. Ajayan, Y.H. Xie, Spectroscopic signatures of AA' and AB stacking of chemical vapor deposited bilayer MoS_2. ACS Nano **9**(12), 12246–12254 (2015)
104. L.M. Schneider, J. Kuhnert, S. Schmitt, W. Heimbrodt, U. Huttner, L. Meckbach, T. Stroucken, S.W. Koch, S. Fu, X. Wang, K. Kang, E.-H. Yang, A. Rahimi-Iman, Spin-layer and spin-valley locking in CVD-grown AA'- and AB-stacked tungsten-disulfide bilayers. J. Phys. Chem. C **123**, 21813–21821 (2019). https://doi.org/10.1021/acs.jpcc.9b07213
105. H. Yu, G.-B. Liu, J. Tang, X. Xu, W. Yao, Moiré excitons: from programmable quantum emitter arrays to spin-orbit-coupled artificial lattices. Sci. Adv. **3**, e1701696 (2017)

106. K.L. Seyler, P. Rivera, H. Yu, N.P. Wilson, E.L. Ray, D.G. Mandrus, J. Yan, W. Yao, X. Xu, Signatures of moiré-trapped valley excitons in $MoSe_2/WSe_2$ heterobilayers. Nature **567**, 66–70 (2019)
107. K. Tran, G. Moody, F. Wu, X. Lu, J. Choi, K. Kim, A. Rai, D.A. Sanchez, J. Quan, A. Singh, J. Embley, A. Zepeda, M. Campbell, T. Autry, T. Taniguchi, K. Watanabe, N. Lu, S.K. Banerjee, K.L. Silverman, S. Kim, E. Tutuc, L. Yang, A.H. MacDonald, X. Li, Evidence for moiré excitons in van der Waals heterostructures. Nature **567**, 71–75 (2019)
108. E.M. Alexeev, D.A. Ruiz-Tijerina, M. Danovich, M.J. Hamer, D.J. Terry, P.K. Nayak, S. Ahn, S. Pak, J. Lee, J.I. Sohn, M.R. Molas, M. Koperski, K. Watanabe, T. Taniguchi, K.S. Novoselov, R.V. Gorbachev, H.S. Shin, V.I. Fal'ko, A.I. Tartakovskii, Resonantly hybridized excitons in moiré superlattices in van der Waals heterostructures. Nature **567**, 81–86 (2019)
109. X. Lu, P. Stepanov, W. Yang, M. Xie, M.A. Aamir, I. Das, C. Urgell, K. Watanabe, T. Taniguchi, G. Zhang, A. Bachtold, A.H. MacDonald, D.K. Efetov, Superconductors, orbital magnets and correlated states in magic-angle bilayer graphene. Nature **574**, 653–657 (2019)
110. M. Shah, L.M. Schneider, A. Rahimi-Iman, Observation of intralayer and interlayer excitons in monolayered WSe_2/WS_2 heterostructure. Semiconductors **53**(16), 2140–2146 (2019). https://doi.org/10.1134/S1063782619120273
111. P. Schmidt, F. Vialla, S. Latini, M. Massicotte, K.-J. Tielrooij, S. Mastel, G. Navickaite, M. Danovich, D.A. Ruiz-Tijerina, C. Yelgel, V. Fal'ko, K.S. Thygesen, R. Hillenbrand, F.H.L. Koppens, Nano-imaging of intersubband transitions in van der Waals quantum wells. Nat. Nanotechnol. **13**, 1035–1041 (2018). https://doi.org/10.1038/s41565-018-0233-9
112. P. Qing, J. Gong, X. Lin, N. Yao, W. Shen, A. Rahimi-Iman, W. Fang, L. Tong, A simple approach to fiber-based tunable microcavity with high coupling efficiency. Appl. Phys. Lett. **114**, 021106 (2019). https://doi.org/10.1063/1.5083011
113. Sino-German Center for Research Promotion, First Sino-German Symposium on Functional Nano-Materials Sciences, http://www.sinogermanscience.org.cn/de/aktuelles/de_2018/201806/t20180626_489284.html. Online 2018, http://www.sinogermanscience.org.cn/de/aktuelles/de_2018/201806/t20180626_489284.html. Online 2020
114. A. Rahimi-Iman, Advances in functional nano-materials science. Ann. Phys. **532**(9), 2000015 (2020). https://doi.org/10.1002/andp.202000015
115. F.H.L. Koppens, T. Mueller, P. Avouris, A.C. Ferrari, M.S. Vitiello, M. Polini, Photodetectors based on graphene, other two-dimensional materials and hybrid systems. Nat. Nanotechnol. **9**, 780–793 (2014)
116. Y. Yu, F. Miao, J. He, Z. Ni, Photodetecting and light-emitting devices based on two-dimensional materials. Chin. Phys. B **26**, 036801 (2017)
117. D. Unuchek, A. Ciarrocchi, A. Avsar, K. Watanabe, T. Taniguchi, A. Kis, Room-temperature electrical control of exciton flux in a van der waals heterostructure. Nature **560**, 340–344 (2018)
118. R. Ribeiro-Palau, C. Zhang, K. Watanabe, T. Taniguchi, J. Hone, C.R. Dean, Twistable electronics with dynamically rotatable heterostructures. Science **361**, 690–693 (2018)
119. R. Li, L.M. Schneider, W. Heimbrodt, H. Wu, M. Koch, A. Rahimi-Iman, Gate tuning of Förster resonance energy transfer in a graphene - quantum dot FET photo-detector. Sci. Rep. **6**, 28224 (2016). https://doi.org/10.1038/srep28224
120. P. Lu, J. Lang, Z. Weng, A. Rahimi-Iman, H. Wu, Hybrid structure of 2D layered GaTe with Au nanoparticles for ultrasensitive detection of aromatic molecules. ACS Appl. Mater. Interfaces **10**, 1356–1362 (2018). https://doi.org/10.1021/acsami.7b14121
121. A. Zak, M.A. Andersson, M. Bauer, J. Matukas, A. Lisauskas, H.G. Roskos, J. Stake, Antenna-integrated 0.6 THz FET direct detectors based on CVD graphene. Nano Lett. **14**, 5834–5838 (2014)
122. M. Bauer, M. Andersson, A. Zak, P. Sakalas, D. Cibiraite, A. Lisauskas, M. Schröter, J. Stake, H.G. Roskos, The potential for sensitivity enhancement by the thermoelectric effect in carbon-nanotube and graphene Tera-FETs. J. Phys. Conf. Ser. **647**, 012004 (2015)
123. W. Kong, Z. Ye, Z. Qi, B. Zhang, M. Wang, A. Rahimi-Iman, H. Wu, Characterization of an abnormal photoluminescence behavior upon crystal-phase transition of perovskite $CH_3NH_3PbI_3$. Phys. Chem. Chem. Phys. **17**, 16405–16411 (2015)

124. W. Kong, A. Rahimi-Iman, G. Bi, X. Dai, H. Wu, Oxygen intercalation induced by photocatalysis on the surface of hybrid lead halide perovskites. J. Phys. Chem. C **120**, 7606–7611 (2016)
125. W. Kong, T. Ding, G. Bi, H. Wu, Optical characterizations of the surface states in hybrid lead-halide perovskites. Phys. Chem. Chem. Phys. **18**(18), 12626–12632 (2016)
126. M. Weinhold, S. Chatterjee, P.J. Klar, Modifying graphene's lattice dynamics by hot-electron injection from single gold nanoparticles. Commun. Phys. **2**, 18 (2019)
127. A. Fabian, M. Czerner, C. Heiliger, M.T. Elm, D.M. Hofmann, P.J. Klar, Domain formation in rectangular magnetic nanoparticle assemblies. Phys. Rev. B **98**, 054401 (2018)
128. A. Fabian, M.T. Elm, D.M. Hofmann, P.J. Klar, Hierarchical structures of magnetic nanoparticles for controlling magnetic interactions on three different length scales. J. Appl. Phys. **121**, 224303 (2017)
129. Z. Xie, S. Feng, P. Wang, L. Zhang, X. Ren, L. Cui, T. Zhai, J. Chen, Y. Wang, X. Wang, W. Sun, J. Ye, P. Han, P.J. Klar, Y. Zhang, Demonstration of a 3D radar-like SERS sensor micro- and nanofabricated on an optical fiber. Adv. Opt. Mater. **3**, 1232–1239 (2015)
130. C. Schneider, M.M. Glazov, T. Korn, S. Höfling, B. Urbaszek, Two-dimensional semiconductors in the regime of strong light-matter coupling. Nat. Commun. **9**, 2695 (2018)

Index

A. Rahimi-Iman, *Semiconductor Photonics of Nanomaterials and Quantum Structures*, Springer Series in Solid-State Sciences 196,
https://doi.org/10.1007/978-3-030-69352-7

L

M